# Soil Chemical Pollution, Risk Assessment, Remediation and Security

# NATO Science for Peace and Security Series

This Series presents the results of scientific meetings supported under the NATO Programme: Science for Peace and Security (SPS).

The NATO SPS Programme supports meetings in the following Key Priority areas: (1) Defence Against Terrorism; (2) Countering other Threats to Security and (3) NATO, Partner and Mediterranean Dialogue Country Priorities. The types of meeting supported are generally "Advanced Study Institutes" and "Advanced Research Workshops". The NATO SPS Series collects together the results of these meetings. The meetings are co-organized by scientists from NATO countries and scientists from NATO's "Partner" or "Mediterranean Dialogue" countries. The observations and recommendations made at the meetings, as well as the contents of the volumes in the Series, reflect those of participants and contributors only; they should not necessarily be regarded as reflecting NATO views or policy.

**Advanced Study Institutes (ASI)** are high-level tutorial courses intended to convey the latest developments in a subject to an advanced-level audience

**Advanced Research Workshops (ARW)** are expert meetings where an intense but informal exchange of views at the frontiers of a subject aims at identifying directions for future action

Following a transformation of the programme in 2006 the Series has been re-named and re-organised. Recent volumes on topics not related to security, which result from meetings supported under the programme earlier, may be found in the NATO Science Series.

The Series is published by IOS Press, Amsterdam, and Springer, Dordrecht, in conjunction with the NATO Public Diplomacy Division.

### Sub-Series

| | | |
|---|---|---|
| A. | Chemistry and Biology | Springer |
| B. | Physics and Biophysics | Springer |
| C. | Environmental Security | Springer |
| D. | Information and Communication Security | IOS Press |
| E. | Human and Societal Dynamics | IOS Press |

http://www.nato.int/science
http://www.springer.com
http://www.iospress.nl

**Series C: Environmental Security**

# Soil Chemical Pollution, Risk Assessment, Remediation and Security

edited by

## Lubomir Simeonov

Solar-Terrestrial Influences Laboratory,
Bulgarian Academy of Sciences,
Sofia, Bulgaria

and

## Vardan Sargsyan

Yerevan State Institute of Economy,
Yerevan, Armenia

 Springer

Published in cooperation with NATO Public Diplomacy Division

Proceedings of the NATO Advanced Research Workshop on
Soil Chemical Pollution, Risk Assessment, Remediation and Security
Sofia, Bulgaria
23–26 May 2007

A C.I.P. Catalogue record for this book is available from the Library of Congress.



ISBN 978-1-4020-8256-6 (PB)
ISBN 978-1-4020-8255-9 (HB)
ISBN 978-1-4020-8257-3 (e-book)

---

Published by Springer,
P.O. Box 17, 3300 AA Dordrecht, The Netherlands.

*www.springer.com*



*Printed on acid-free paper*

---

# CONTENTS

# PREFACE

The book contains the contributions at the NATO Advanced Research Workshop on Soil Chemical Pollution, Risk Assessment, Remediation and Security, which took place in Sofia, Bulgaria, May 23–26, 2007. Scientists representing fields of chemistry, biology, toxicology, physics, risk assessment, ecology, environmental protection, remediation and public health, from Armenia, Azerbaijan, Belarus, Belgium, Bulgaria, Germany, France, Hungary, Italy, Kazakhstan, Moldova, Netherlands, Poland, Romania, Russian Federation, Spain and Ukraine participated in the work of the ARW.

The main objective of the ARW was to contribute to the existing knowledge on soil pollution and remediation. Stress was given to: critical assessment of the used analyses and methods for study effects in combined chemical pollution (organic pollutants and pesticides, metals) on soil biota and fertility; to evaluate specific aspects of the risk assessment; to assess the most advanced technologies for soil remediation used for different purposes; to evaluated the economical involvement; to evaluate national policies for soil security in different NATO and partner countries.

Conclusions are based on the discussion during the ARW and related to the proposed objectives and expected outcomes:

1.  Laboratory and field methods for soil pollutants analyses: Laboratory analytical methods and techniques are mostly studied, compared with the sampling and sample preparation techniques; reference materials, standards, quality control procedures. Reference laboratories and accreditation still are not well harmonized among the different countries. A lot of methods are not standardised. The quality control procedures are not completely set-up. In different countries the ISO EN and CEN analytical standards are not available in native language or even not introduced.
2.  Strategy of sampling and monitoring of polluted sites: Development in the sampling and analytical physical, chemical and biological methods to characterise the quality of soils is already done, but more efforts should be put on this domain. Sampling and measurement uncertainty data processing contribute to the analytical results improvement. The sampling methods for soil monitoring are properly used.
3.  Sources, movement, interaction and transformation of soil pollutants: Information about the sources of soil pollution is available. Soil background data

not enough available, in terms of: compounds, soil types, and regions. Knowledge about pollutants pathways of migration in/from soils is not satisfactory. The monitoring valid data for soil quality control are not enough covering the needs in respect to modeling and remediation. Information about interactions and transformation of the single pollutant in soils are available, but not enough centralized and systematized (books, monographs with theory and exercises). Soil-biological methods will help the elucidation of the combined effects of combined chemical pollution.

4. Risk assessment of combined and complex exposure and effects of soil chemical pollution: The use of risk assessment methods to start a remediation process was emphasized, but they are not commonly used. Risk assessment methods, independent on contaminants and the pathways, are available. Methods for determinate the ecological risk for special situations, such as multimedia pollution are needed.

5. Management of soil pollution. Prevention strategies and remediation: A lot of soil remediation methods and technologies are available (physical, chemical and biological) and among them, phytoremediation have been found to be an efficient one and less expensive. Phytoremediation contributes to natural attenuation, and regenerate the soil quality. Bioremediation (microorganisms) is also a promising technique for soil remediation of different contaminants. Most of the remediation methods and techniques are costly, therefore are avoided or improperly used. The validation of using these methods for different soils is insufficient.

Recommendation were formulated as concerning the following fields.

*Research needs*
Continued research is needed in order to develop new or improved methods for sampling strategies to examine the spatial and temporal variability of the chemical pollutants in contaminated sites. Research priorities should focus on: (i) development and improvement of sampling strategies to examine the spatial and temporal variability of the chemical pollutants in contaminated sites, (ii) the harmonization of the use of bioindicators for soil pollution monitoring especially for unknown and combined chemicals, (iii) development of non-invasive (geophysical) tools to characterize the spatial temporal distribution of the soil physical and chemical properties related to pollution at the laboratory and field scales and the establishment and (iv) validation of new analytical methods for determination of specific soil emerging pollutants.

Research projects for management of chemical pollution to the soil should be multidisciplinary, combining the expertise from the various specialties related to understanding and management of these chemical impacts. A specific need has been identified for research funding to support (i) improvement of the knowledge of processes involved in pollutants fate with especial focus on bioavailability, (ii) improvement of modeling approaches for prediction of pollutants' fate and concentrations in the soil, (iii) the improvement of risk assessment methodologies for combined and complex exposure and (iv) effects of soil chemical pollution harmonization of approaches and methods for establishment of soil pollution quality standards on the base of eco toxicity and health data related to different soil use types, (v) development of new ecologically safe methods for effective remediation of polluted soils to restore original multifunctionality of the soils.

*Management of chemical polluted soils*
The design of field experiments concerning soil remediation should be based on previous laboratory tests. A key requirement for the efficient management of chemical polluted soils is the development of financial mechanisms for optimal control of pollution and pricing of land and soil resources and the availability of information concerning the distribution of chemicals in the environment and the associated health impacts. To facilitate the wide-spread availability of information, databases of environmental monitoring data and human health impacts (i.e., geomedical databases) should be developed and shared.

*Environmental and environmental health policies*
Coordination of environmental policies among and between nations is essential for the management of chemical polluted soils. Requirements for the control and management of chemical concentration limits of pollutants in soils should be harmonized across nations. In addition, policies should be implemented risk assessment in order to protect the public health and to prevent the transport of pollutants from soils to water and food.

*Public communication and education*
The general public plays a critical role in the protection of the environment through direct actions and support of public policies for environmental protection. In addition, environmental scientists and researchers have an obligation to facilitate public education and to ensure that the public is provided with accurate information on chemical threats to the environment. International collaboration exchange of experience in the field of soil management should be encouraged by Science and Peace NATO Program.

Lubomir Simeonov
Solar-Terrestrial Influences Laboratory
Bulgarian Academy of Sciences
Sofia, Bulgaria

Vardan Sargsyan
Yerevan State Institute of Economy
Yerevan, Armenia

# ACKNOWLEDGEMENTS

The organizers wish to express their gratitude to the NATO Science Committee for the approval of the ARW and the Financial Award provided, which made the event possible as special thanks go to Dr. Deniz Beten, Programme Director of Panel Environmental Security of the NATO Public Diplomacy Division.

The editors are especially grateful also to the Publishing Editor of the book Mrs. Annelies Kersbergen and to the personal of Springer/NATO Publishing Unit for the excellent collaboration for the publication of the results from the NATO Advanced Research Workshop on Soil Chemical Pollution, Risk Assessment, Remediation and Security, held in Sofia, Bulgaria, May 23–26, 2007.

# LIST OF CONTRIBUTORS

Nurlan Kazkenovic Almaganbetov
  Land Relation Department
  Kazakh Scientific Institute of Economy
  Agro-Industrial Complex and Rural Area Development
  Flat 14, House 51, District Koktem-1, 480090 Almaty
  Kazakhstan

Ivan Atanassov
  Institute of Sustainable Development-Foundation
  Jaroslav Vechin Str., Block 10 A, 13 Apart.; 1408 Sofia
  Bulgaria

Elisabeta Chirila
  Chemistry Department
  Ovidius University of Constanta
  124 Mamaia Blvd, 900527 RO, Constanta
  Romania

Ala Cojocaru
  National Institute of Ecology and Geography
  Moldavian Academy of Sciences
  Ministry of Ecology and Natural Resources
  5 Gh. Tudor Str., 2028-MD Chisinau
  Moldova

Anna Dimitrova
  Chemical Analyses in the Environment
  Laboratory of Hygiene Soil and Waste
  National Center of Public Health Protection
  15 Academic Gechov Bul., 1431 Sofia
  Bulgaria

Radka Donkova
  Department of Soil Microbiology
  Soil Science Institute Puchkarov
  7 Shossee Bankja, 1080 Sofia
  Bulgaria

Camelia Draghici
    Chemistry Department
    Transilvania University of Brasov
    50 Iuliu Maniu Str., Brasov
    Romania

Gyula Dura
    National Institute of Environmental Health
    Gyali ut. 2–6, H-1097 Budapest
    Hungary

Ayaz Efendiev
    Azerbaijan National Academy of Sciences
    10 Istiglaliyyet Str., 1001 Baku AZ
    Azerbaijan

Stefano Girotti
    Dipartimento di Scienta dei Metalli
    Electrochimica e Tecniche Chimice
    University of Bologna
    15 Via San Donato, Bologna
    Italy

Beata Janecka
    Institute of Environmental Engineering
    Czestochowa University of Technology
    Breznicka 60 A, 42–200 Czestochowa
    Poland

Mathieu Javaux
    Department of Environmental Sciences and Land Use Planning
    Universite Catholique de Louvain
    2 Croix du Sud, Box 2; 1348 Louvain-la-Neuve
    Belgium

Monika Kedziorek-Dupuy
    Environmental HydroGeochemistry Laboratory: Soil and Groundwater
    LHGE-JE 2397, University of Pau, Bordeaux
    France

Mykola Kharytonov
    Dnipropetrovsk State Agrarian University
    25 Voroshilov Str., 49600 Dnipropetrovsk
    Ukraine

Valery Khomich
Institute for Problems of Natural Resources Use and Ecology
National Academy of Sciences of Belarus
10 Skoriny, 220114 Minsk
Belarus

Rafal Kucharski
Land Management Department
Institute for Ecology of Industrial Areas
6 Kossutha Str., Katowice 40–844
Poland

Sebastien Lambot
Department of Environmental Sciences and Land Use Planning
Universite Catholique de Louvain
2 Croix du Sud, Box 2; 1348 Louvain-la-Neuve
Belgium

Victoria Lokhanska
Ukrainian Laboratory of Quality and Safety of Agricultural Products
National Agricultural University
15 Geroiv Oborony, Str., 03041 Kiev
Ukraine

Fliur Macaev
Laboratory of Organic Synthesis
Institute of Chemistry
Academy of Sciences of Moldova
3 Academy Str., MD-2028 Chisinau
Moldova

Roser Rubio
Departament de Quimica Analitica
Facultat de Quimica
Universitat de Barcelona
1–11 Marti i Franques, 3a Planta, 08028 Barcelona
Spain

Vardan Sargsyan
Armenian State University of Economics
128 Nalbandyan, Yerevan 375025
Armenia

Aleksandra Sas-Nowosielska
    Land Management Department
    Institute for Ecology of Industrial Areas
    6 Kossutha Str., Katowice 40-844
    Poland

Lubomir Simeonov
    Solar Terrestrial Influences Laboratory
    Bulgarian Academy of Sciences
    G.Bonchev Str. Block 29, 1113 Sofia
    Bulgaria

Dina Solodoukhina
    Department of Public Health
    Kursk State Medical University
    2–17 Yamskaya Str., 305004 Kursk
    Russian Federation

Hachadur Tchuldjian
    Soil Pollution Department
    Soil Science Institute Puchkarov
    7 Shossee Bankja, 1080 Sofia
    Bulgaria

Konstantin Terytze
    Federal Environment Agency
    General Affairs of Soil Protection
    Wörlitzer Platz 1, 06844 Dessau
    Germany

Miquel Vidal
    Departament de Quimica Analitica
    Facultat de Quimica
    Universitat de Barcelona
    1–11 Marti i Franques, 3a Planta, 08028 Barcelona
    Spain

# DETOXIFICATION OF SOILS, POLLUTED JOINTLY BY HEAVY METALS, ACID WASTES AND ACID PRECIPITATIONS

HACHADUR TCHULDJIAN[*]
*Institute of Soil Science "N. Poushkarov"*
*7, Shosse Bankya Str., 1080 Sofia, Bulgaria*

**Abstract.** Some theoretical premises on detoxification of soils, acidified and contaminated by heavy metals from mining activities are subject of general discussion on the results obtained in Bulgaria. Presented results are obtained, both, by model pot experiments and field trials in some mining areas, contaminated mainly with copper and others including radioactive elements, cadmium and lead. The methods described are given briefly in their chemistry of treatments. They are restricted to: (1) Suppression the availability of conta-minants to plants, based to the neutralization of exchangeable acidity of soil (liming ); (2) Capsulation of a polyvalent metal between the structural composites in an organic – mineral complex, i.e. liming and addition of an organic or/and mineral waste (peat, coal dust, wasted ferric hydroxide); (3) Microbial in situ detoxification based on a biochemical acidicy or slightly alkaline dissolution of the ore waste on the surface layer of soil and biochemical reductive precipi-tation of metal ions, transferred and located in a deeper soil horizon ($B_2$). Each one of methods discussed, reveals its effectiveness, depending on the material used for detoxification, its dose, level of contamination, and the biology of tested plant. Liming as an independent action versus the toxic acidity of soil, promises to become enough for restoration of soils at moderate levels of heavy metal pollution. However, at higher pollution levels, liming must be done versus the total acidity of soil and by addition to soil other amendments also. The microbial treatment, compare to the traditional ones, outlines as a new highly effective approach, because of its opportunity to remove although a part of heavy metals from ploughed soil surfaces. It is a convenient method, parti-cularly for soils, strongly polluted with ore wastes.

**Keywords:** toxicity, amendments, soil acidity, heavy metals, detoxification

---

[*] To whom correspondence should be addressed: hachko_ch@mail.bg

L. Simeonov and V. Sargsyan (eds.),
*Soil Chemical Pollution, Risk Assessment, Remediation and Security.*
© Springer Science+Business Media B.V. 2008

## 1. Introduction

Soil pollution adopts like an entrance of substances from outside in soil and whose concentrations and behaviours in it do not respond to the local background of soil. Introduced or alien to soil, these substances bring to its damage at a definite accumulation rate.

It is known, that the soil is remarkable with its high receptiveness, high of decay potential and functions distributing the substances entering in its environment. Accordingly them, the wastes, entering from outside in soil, are subject to transformations and remove as harmless for environment. However, between the wastes there are some, who contain heavy metals and who in biological and in geochemical context are determined like micronutrients. They are indecomposable, and at a definite rate of accumulation in soil above their background, turn toxic up a nutritional chain: soil – plant (animal) – man.

## 2. Heavy Metals in Bulgarian Soils

The term "heavy metals" in biological context means iron and all metals denser than it. The likes of iron are copper, zinc, lead, cadmium, etc. For some of them, the natural quantities in this country are usually higher than in the world. Copper, lead and nickel, for instance are rarely above 40 ppm, whereas for cadmium and mercury, they are one tenth and one hundredth parts of ppm only. For most of them, it is essential to exercise properties of micronutrients upon the different forms of life.

It is known also, that the necessity for plants of copper, zinc or manganese fulfils in strictly prescribed doses only. And, what is more in their function of micronutrients, particularly for plants, is their chemical reactivity within the soil.

The background levels in ppm, for some heavy metals in Bulgarian soils (Tchuldjian, 1989), vary mostly in following limits:

| Cu | Zn | Pb | Cd | As | Co | Ni | Hg | Cr | Mn |
|----|----|----|----|----|----|----|----|----|----|
| $30 \pm 25$ | $75 \pm 20$ | $25 \pm 15$ | $0.30 \pm 0.25$ | $8 \pm 7$ | $12 \pm 10$ | $25 \pm 20$ | $0.07 \pm 0.05$ | $60 \pm 30$ | $850 \pm 400$ |

In the event of heavy metal accumulation in soil, i.e. in excess toward other soil properties, there appear signs of phytotoxicity or plants themselves, eventually adapted to such an environment, become toxic to animals and man. That mean, that plant has a "tolerance" to the accumulated excess.

## 3.  Plant Tolerance to the Excess of Heavy Metals

The tolerance is not bound, by all means, with the metal quantity in organism which tolerates harmless. Some selected genotypes of grasses (Antonovics, 1971) which are widespread upon old mines, distinguish with high tolerance (*agrostis tenuis, agrostis stolonifera, Anthoxanum odoratum*). Some other species like *Becium Hombley or Thlaspi Caerulescense* accumulate very high amounts of metals in grams per kilogram dry matter. Similarly to them, genetically modified forms called "phyto-extractors of heavy metals" are created (Baker et al., 1994). In any case, if the specified grasses turn the land use to reclamation of mine wastes or grassing against erosion, the phyto-extractors perform an accelerated heavy metal export from soil, i.e. their removing. This approach of plant adaptation to soil is convenient at a specific tolerance to a single metal or group of chemical analogs like zinc, lead and cadmium. However, in most cases, pollution has a polymetalic character and includes arsenic, selenium, sulfide ore wastes or such from pyrometallurgy emitting acid gases and toxic dusts.

Such a reality demands other approach for soil detoxification and do not associates solely by the biology of plant. That calls for outside interventions to change some soil properties in direction of suppressing heavy metal availability to plants or other means like direct chemical extraction of metals, transference in deeper soil horizons, etc. An approach like this, attain an opposite sense – accommodation of soil to plant. To work out, such an approach requires a good knowledge on the soil-chemical behaviour of metal and acid contaminants in subject, the interactions between them and soil composites.

## 4.  Heavy Metals and Soil Acidity

It is known, that the toxic properties of the metal excess in soil belongs only to some single part of its total amount, which is available to plants in abnormal concentrations. The availability has a causal relationship with different processes, who involve pollutants – dissolution, ion exchange, sedimentation, complex formation, migration in depth, oxidation, reduction, etc. The most substantial are outlined the heavy metal interactions with the surface structures of mineral and organic colloids and their distribution as different chemical species between the solid and liquid phases of soil.

Among the factors controlling that distribution, respectively the different rates of availability to plants, the most important are the soil acidity, cation exchange capacity and soil organic matter. The acidity of soil solution (pH) determine the possibility for an equilibrium transition of not readily soluble hydroxide and steadily bounded forms of heavy metals with humic substances to water soluble ion forms and slightly bounded exchangeable forms upon the clay colloids. An exampled simple scheme, controlling this equilibrium could be presented by copper:

exch. Cu on soil adsorbent (solid phase)
$$\updownarrow$$
$$Cu\,(OH)_2 \; + \; 2H^+ \quad \longleftrightarrow \quad Cu^{2+} \; + \; 2H_2O$$
slightly available                        in soil
$$\downarrow$$                               solution
$$CuO \; + \; H_2O$$                    (toxic Cu)
unavailable

The shown example suggests that the acid pollutants (pH < 6) will shift the ion – hydroxide equilibrium to right by formation of "mobile" species meaning increasing toxicity. That means formation of readily available to plants varieties. Such are the ion, ion exchangeable and soluble complexes of heavy metals with the soil organic fulvic acids. Conversely, if hydrogen ions be removed (pH > 6), the equilibrium will be shift in left and not available copper hydroxide and copper oxide to plants will form. It was found, that there was not observed any signs of copper toxicity to alfalfa up to 400 mg/kg in soil (Tchuldjian, 1978).

On such a basis, norms of permissible concentration levels for some heavy metals, depending on pH in agricultural soils were worked out in this country (Regulation 3, 1996). For slightly acid soils, i.e. with optimal acidity pH ~ 6 the following values are in ppm:

| Cu | Zn | Pb | Cd | Ni | Cr | Hg |
|-----|-----|-----|-----|-----|-----|-----|
| 120 | 200 | 70 | 1.5 | 60 | 190 | 1 |

## 5.  Detoxification of Heavy Metals in Soil Through Liming

The removal of acidic pollution of soil appears to be the basic way for detoxification of heavy metal excess in soils. This achieves by liming of soil. The main theoretical premise for that is the complete removal of its exchangeable acidity (Ganev et al., 1985; Ganev, 1990; Tchuldjian, 1978). It is composed mainly by $Al^{3+}$ and partly by heavy metal ions with acidic properties adsorbed on the strong acid surfaces of clay colloids ($T_{ca}$). So, their removal should mean, their precipitation as unavailable forms in soil. It was shown (Ganev,

1990) that heavy metal precipitation in soil as hydroxides, brings to an end at pH $\sim$ 6. The following equations exemplify neutralization of exchangeable acidity trough liming of soil:

1. $2CuS + 3O_2 + 2H_2O = 2Cu^{2+} + 2H_2SO_4$
   pollutant
2. $(T_{ca})Al_2 + 3Cu^{2+} + 3 H_2O \leftrightarrow 2 Al(OH)_3 + (T_{ca})Cu_3$
   exch. acidity                                          exch. acidity
3. $(Tca)Cu + CaCO_3 + H_2O = (Tca)Ca + Cu(OH)_2 + CO_2$

As to the free copper ions in soil solution, which are in equilibrium with exch. Cu, their precipitation will end as copper oxide via copper hydroxide:

4. $Cu^{2+} + H_2O + CaCO_3 \leftrightarrow Cu(OH)_2 + Ca^{2+} + CO_2$
5. $Cu(OH)_2 \rightarrow CuO + H_2O$

## 6. Restrictions

Obviously, this approach confines to conversion of excess heavy metal's actual toxicity in potential one. The higher pollution levels, requires liming to a slightly alkaline pH $\sim$ 7.5–8. Then, the permissible pollution levels of heavy metals accept higher values (Regulation 3, 1996).

So, permissible concentration values, the so called "trigger values", are changed to (ppm):

| Cu | Zn | Pb | Cd | Ni | Cr | Hg |
|----|-----|----|----|----|-----|----|
| 80 | 370 | 80 | 3 | 70 | 200 | 1 |

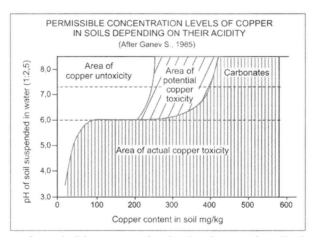

Figure 1. A curve of permissible concentration levels of copper in soils depending on their acidity (After Ganev et al., 1985).

At very high pollution rates, more than five permissible pollution units, it is recommended to change the land use for other purposes than agriculture (Instruction, 1994). As an example in pot experiment with alfalfa, Figure 1 shows the areas of copper non-toxicity, potential toxicity and actual toxicity in a soil polluted with copper.

Figures, such like the shown one, was worked out also for zinc and lead. It was observed, that the presence of copper excess in low available forms could stay in soil up to 400 ppm at pH 7.4. Yet, in presence of free carbonates (pH 8.2) it would be possible a little above 400 ppm.

## 7. Results Obtained in Field Trials

Two field trials on a deluvial meadow fluvisols, contaminated with copper (~250 ppm) and acid precipitations from a copper smelter, were carried out producing potatoes and green corn (Ganev et al., 1985). Different treatments and norms of lime and manure were applied to soil plots. Data obtained was shown on Table 1.

The yield obtained after liming of soil versus its exchangeable acidity, gives some priority compare to other treatments as the cheapest way for remediation of acid soils polluted by heavy metals at the existing level of contamination (~250 ppm). The highest yield effect obtained from addition of farmyard manure (field I) has no relation to soil acidity. Its pH value was not altered! Probably,

TABLE 1. Data obtained from field trials with potato and silage maize on a copper contaminated soil (after Ganev et al., 1985)

| Treatments | Field I (potato tubers – t/ha) | | | | | Field II (silaged maize – t/ha) | | | | |
|---|---|---|---|---|---|---|---|---|---|---|
| | pH H$_2$O | exch. acidity | Yield t/ha | % | Cu tubers ppm | pH H$_2$O | exch. acidity | Yield t/ha | % | Cu plants ppm |
| control | 4.0 | 2.0 | 5.1 | 100 | 18 | 4.1 | 1.9 | 1.1 | 100 | 77 |
| control-fertilized | 4.0 | 2.9 | 8.7 | 170 | 17 | 4.1 | 2.2 | 2.6 | 250 | 67 |
| CaO vs. exch. acidity | 5.5 | 0.4 | 15.0 | 293 | 15 | 5.7 | 0.1 | 17.5 | 1650 | 107 |
| CaO vs. total acidity | 6.2 | 0.0 | 13 2 | 256 | 14 | 6.4 | 0.0 | 14.9 | 1410 | – |
| CaO in slight excess | 6.6 | 0.0 | 12.3 | 240 | 15 | 6.8 | 0.0 | 12.7 | 1200 | 95 |
| farm. Manure 200 t/ha | 4.3 | 1.9 | 19.6 | 381 | 12 | 4.7 | 1.2 | 12.4 | 1127 | 91 |
| | | | | | | 6.3 | 0.0 | 21.3 | 2010 | 94 |
| CaO + manure 200 t/ha | 6.0 | 0.0 | 17.7 | 344 | 14 | | | | | |
| | GD 5% - 2.1 | | | | | GD 5% - 2.9 | | | | |
| | GD 1% - 2.8 | | | | | GD 1% - 3.9 | | | | |
| | GD 0.1% - 3.6 | | | | | GD 0.1% - 5.1 | | | | |

this effect belongs to the high nutritional status of treated soil plot only. Similar results were obtained in field II with green maize. There, the treatment with manure, after removing of the total acidity of soil, the yield effect was the best. However, all treatment show high concentrations of copper in plants.

## 8.   Detoxification of Heavy Metals by Mixed Organic – Mineral Additives

Addition of peat, organic wastes (manure, coal dust), flotation ferric hydroxide or combined with liming to heavy metal polluted soils, are accepted as effective means of detoxification. They are based on the modern ideas suggesting formation of three-layer charged colloidal structures (Raychev, 1996). The latter are considered as a final result from organic-mineral interactions neutralizing the toxic acidity of soil and blocking heavy metals between the clay lattice and organic structure like a "sandwich". This assumption shown on Figure 2 was an object of verification trough bringing into proper correlation the above indicated additives in pot experiments.

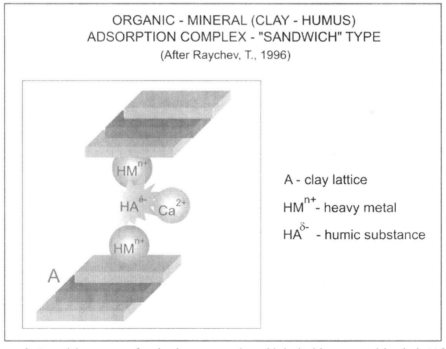

*Figure 2.* A model structure of a clay-humus complex with locked heavy metal ion in it (After Raychev, 1996).

In addition, the elaborated model structure contributes to single out the appropriate amendments and especially the calculation of their proportions and treatment doses (Raychev, 2004). It was based on the well known Ganev equation (Ganev, 1990) for liming of acid soils. The latter was modified so that the liming to be combined with various organic and/or mineral additives responding to above mentioned concepts upon the soil adsorbent. The aim was, to optimize the norm of lime, at considerably higher pollution levels. Some results from pot experiments based on that model are given below (Benkova, 2005).

## 9.  Results Obtained by Pot Experiments (Benkova, 2005)

Pot experiments with alfalfa were carried out on strongly polluted loamy sand fluvisol, (Cu – 845 mg/kg; pH 4.0; exch. acidity – 3.7 mev/100 g and total acidity – 11.2 mev/100 g) from the copper refinery, near to Pirdop (Bulgaria). Soil was treated with CaO, wasted ferric hydroxide, peat and coal dust calculated in different combinations based on the modified Ganev equation for liming of acid soils. Some results, extracted from pot experiments on the ways discussed up here, are presented below on Table 2. The shown values, obtained on heavy copper accumulations in soil, suggest the applied lime dose, versus exchangeable acidity of soil, was not so effective compare to the yields obtained from the other treatments. Despite the high yield effect (+ 64%), copper concentration in plants stays as before much the same.

TABLE 2. Plant yield and Cu content in its dry matter (3)

| Treatments | Yield g/pot | % | Cu in plant ppm |
|---|---|---|---|
| CaO vs. exch. acidity | 11 | 100 | 59 |
| CaO vs. total acidity | 13 | 118 | 53 |
| CaO + peat | 17 | 155 | 47 |
| CaO + peat + Fe(OH)$_3$ | 18 | 164 | 57 |
| CaO + coal | 16 | 145 | 45 |
| CaO + coal + Fe(OH)$_3$ | 18 | 164 | 47 |

## 10. Microbial Detoxification of Heavy Metals in Polluted Soils

Exertions to overcome the toxicity of heavy metals to plants have found new decisions. One of them appears the possibility to transfer metal contaminants downwards the soil profile, out of the nutritional zone of plant roots. The method could summarize in several steps of biochemical impacts beginning through microbial oxidation of waste and putting it to solution. After, move dissolved metals in dept by continuous flushing with suitable solvent and, at last,

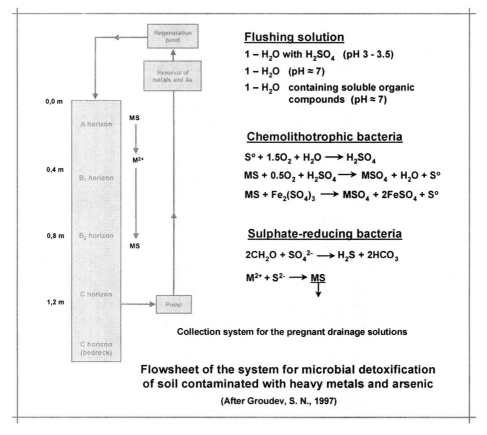

**Flushing solution**

$1 - H_2O$ with $H_2SO_4$  (pH 3 - 3.5)

$1 - H_2O$  (pH ≈ 7)

$1 - H_2O$  containing soluble organic
  compounds  (pH ≈ 7)

**Chemolithotrophic bacteria**

$S^o + 1.5O_2 + H_2O \longrightarrow H_2SO_4$

$MS + 0.5O_2 + H_2SO_4 \longrightarrow MSO_4 + H_2O + S^o$

$MS + Fe_2(SO_4)_3 \longrightarrow MSO_4 + 2FeSO_4 + S^o$

**Sulphate-reducing bacteria**

$2CH_2O + SO_4^{2-} \longrightarrow H_2S + 2HCO_3$

$M^{2+} + S^{2-} \longrightarrow \underline{MS}$

Collection system for the pregnant drainage solutions

**Flowsheet of the system for microbial detoxification
of soil contaminated with heavy metals and arsenic**

**(After Groudev, S. N., 1997)**

*Figure 3.* Flowsheet of the system for microbial detoxification of soil contaminated with heavy metals and arsenic (After Groudev, 1997).

subjection the metal ions back to microbial precipitation turning them in solid phase as an insoluble, mineral compound at a suitable dept in soil.

As an example, oxidation of *covelline* (CuS) in aerobic environment by chemolithotrophic sulfur – oxidizing bacteria, genus *Acidithiobacillus* follows:

1. $CuS + 0.5O_2 + H_2SO_4 = Cu^{2+} + SO_4^{2-} + H_2O + S^0$
2. $S^0 + 1.5O_2 + H_2O = H_2SO_4$

After that, follows, a microbial reduction of sulphates in deeper horizon (B_2), i.e. in anaerobic environment, and formation of insoluble copper sulphide (Ls = $6.10^{-36}$):

3. $SO_4^{2-} + HCHO = H_2S + 2HCO3^-$
4. $Cu^{2+} + S^{2-} = CuS$

This approach has got a methodological realization since 1993 by Groudev, S. and associates (Groudev, 1997; Groudev et al., 2001) mainly for treatment of mine wastes. A block scheme, shown below, presents the biochemical oxidation of the non-soluble metal sulfides in horizon $A$ of soil and turning them in soluble metal sulfates. And contrary, to turn back the soluble sulfates in $B_2$ horizon in non-soluble sulfides. A collector system to catch the dissolved heavy metals was designed to pump them up with aim to prevent from pollution the ground waters (Figure 3).

## 11. Field Trials on Heavily Polluted Soils from Mining Activities (Groudev, 1996)

### 11.1.  MICROBIAL TREATMENT OF ACID SOIL, POLLUTED WITH COPPER AND CADMIUM

A ploughed soil surface (pH 4.8), polluted with acid depositions, Cu and Cd, was subjected to inoculation, mainly by acidophilic ($Fe^{2+}$ and $S^{2-}$)-oxidizing chemolithotrophic bacteria and anaerobic sulfate-reducing bacteria, respectively. The soil processing was carried out in situ with acidified flushing solution ($H_2SO_4$, pH 3–3.5), washing by periodical irrigation to transfer the dissolved metals into the deeply located soil horizons. A part of results obtained were shown on Table 3 (Groudev, 1997). The values obtained after treatment, compare to those before treatment show, without any suspicious, a good effectiveness of the examined in situ method of detoxification. The copper content in ploughed soil horizon has decreased almost five times. However, cadmium content decrecases not so sharply and needs a repeated processing.

TABLE 3. Data obtained after in situ bacterial treatment of heavy metal contaminated soil near a copper smelter (after Groudev, 1997)

| Pollutant | Before treatment | | After treatment | |
|---|---|---|---|---|
| ppm | Horizon A | Horizon $B_2$ | Horizon A | Horizon $B_2$ |
| Copper | 820 | 240 | 170 | 880 |
| Cadmium | 28 | 12 | 12 | 51 |

Formation of copper and cadmium immobilized phases in $B_2$ horizon takes about three months. Soil grassing, mineral fertilization and farmyard manure, liming, periodical soil loosen and irrigation were applied to promote the examined "in situ" method.

The values obtained after treatment, compare to those before treatment show, without any suspicious, a good effectiveness of the examined in situ method of detoxification. The copper content in ploughed soil horizon has decreased almost five times. However, cadmium content decreases not so sharply and needs a repeated processing.

Formation of copper and cadmium immobilized solid phases in $B_2$ horizon takes about three months. Soil treatment was promoted by grassing, mineral fertilization, addition of farmyard manure, liming, periodical soil loosen and irrigation.

## 11.2. MICROBIAL TREATMENT OF SLIGHTLY POLLUTED WITH URANIUM

A similar field experiment was carried out with such a technique, yet on a slightly alkaline soil (pH ~ 7.7) contaminated with radioactive and other heavy metals (Groudev, 2001). Polluted ploughed layer (hor.A) of a soil profile was subjected to continuous 8 month microbial oxidation and leaching with carbonate ions and dissolved organic compounds:

1. $UO_2 + H_2O_2 \rightarrow (UO_2)^{2+} + 2OH^-$
2. $UO_2 + 2O_2 + 3HCO_3^- + H_2O \rightarrow [UO_2(CO_3)_3]^{4-} + 5OH^-$
3. $(UO2)^{2+} + CH_3\text{-}CH(OH)\text{-}COO^- \rightarrow [CH_3\text{-}CH(OH)\text{-}COO(UO_2)]^+$

The toxic heavy metals were solubilized mainly as organic acid microbial secretion complexes. The dissolved uranyl species and other dissolved contaminants, removed from surface layer, were transferred mainly to horizon $B_2$. They were immobilized as final insoluble compounds by anaerobic sulphate-reduction, enhanced by injected soluble organics and ammonium phosphate. Uranium precipitation was brought to the mineral uraninite (Lovely and Phillips, 1992). The most of radium was adsorbed on clay minerals in horizon $B_2$. The other ions of toxic metals (Cu, Cd, Pb) were converted mainly in non-soluble sulphides. Table 4 show, with close similarity to Table 3, the extent to which were removed metals from horizon A and accumulated in horizon $B_2$.

TABLE 4. Total amounts of metals in soil, before and after treatments

| Parameter | Before treatment | | After treatment | |
|-----------|------------------|--|-----------------|--|
| ppm | Horizon A | Horizon B | Horizon A | Horizon B |
| Uranium | 35 | 17 | 9 | 27 |
| Radium* | 540 | 280 | 90 | 460 |
| Copper | 648 | 230 | 240 | 378 |
| Cadmium | 7 | 3 | 3 | 4.6 |
| Lead | 275 | 190 | 95 | 247 |

* Ra in Bq/kg dry soil

The values on the table make clear, that part of metal ions have left in the effluents. They were efficiently treated by a passive system located near the experimental plot.

All these methods under discussion bring to the conclusion, that the heavy metal toxicity to plants could be removed. The methods have considerable opportunities for elaboration of adequate technology. However, their effectiveness will requires a choice, according to the rate of accumulation the excess of heavy metals, the plant tolerance and mostly, the purpose of plant production.

## References

Antonovics J., A.D. Bradshaw, R.G. Turner, Heavy metal tolerance in plants, Adv. Ecol. Res., 7: 1–85, 1971

Baker A.J.M., S.P. McGrath, C.M.D. Sidoly, R.D. Reeves, The possibility of in situ heavy metal decontamination of polluted soils using crops of metal-accumulating plants, Resources, Conserv. Recycl., 11, 41–49, 1994, Elsevier Sc. B.V

Benkova M., Colloid-chemical interactions at organo-mineral amelioration of heavy metal polluted soils, Ph.D. Thesis, pp 140, Soil Sc. Inst., Sofia, 2005 (bg)

Ganev, St, H. Chouldjiyan, P. Petrov, G. Ilkov, Chemical amelioration of soils polluted with copper and acid sediments, Soil Sc., Agrochem, and Pl. Protect., v. XX, 6, 3–17, 1985 (bg)

Ganev, St., Modern Soil Chemistry, Science Art Publ., 1990 (bg)

Groudev, St., Microbial detoxification of heavy metals and arsnic in a contaminated soil, in "Proceed. XX Int. Min. Process. Congr.", Aachen, Ger, Sept. 21–26, 1997, v. 5, 729–736, 1997

Groudev, S., P. Georgiev, I. Spasova, K. Komnitsas, Bioremediation of a soil contaminated with radioactive elements, Hydrometallurgy, 59, 311–318, 2001

Instruction on the determination of pollutants and pollution rate of agricultural lands and the regime of their use, Dep. of Agricult., Inf. Bulet. 27, Sofia 1994 (bg)

Lovely, D.R., E.P. Phillips, Reduction of uranium by Desulfovibrio desulfuricans, Appl. Environ. Microbiol., 58: 850–856, 1992

Regulation 3: Norms related to permissible contents of harmful substances in soil, State J. 36, 1979 and 5, 1996 (bg)

Raychev, T., Influence of the organo-mineral interaction on the colloid-chemical state of soil adsorbent, Doct. Thesis, pp 161, Soil Sc. Inst., Sofia, 1996 (bg)

Raychev, T., A. Arsova, M. Benkova, Estimation of quantitative rates for organo-mineral liming of heavy metal polluted acid soils, Bulg. J. Agric. Sci., 8: 539–550, 2004

Tchuldjian H., Chemical states of copper in soil and their toxicity to plants in the industrial pollution processes, Ph.D. Thesis, Soil Sc. Inst, Sofia, 1978 (bg)

Tchuldjian H., Soil pollution, In "Lectures on Soil Science", Project FAO/Bulgaria, TCP/BUL4502 (T), 1989, 307–324, Sofia

# CONTAMINATION OF SOILS BY WASTE DEPOSITS

ELISABETA CHIRILA[1*] AND CAMELIA DRĂGHICI[2]
*[1]Chemistry Department, Ovidiu University*
*124 Mamaia Blvd., 900527 Constanta, Romania*
*[2]Transilvania University of Brasov*
*29 Eroilor Bldv., 500036 Brasov, Romania*

**Abstract.** The purpose of this paper is to present original results concerning concentrations of eight metals in soils adjacent to municipal waste deposits from Constanta County, Romania, during April-October 2006. The surface and depth soil solutions were obtained using aqua regia and the applied analytical technique for metal determination was flame atomic absorption spectrometry (FAAS). The mean measured values ranged as follows: Cd: 0.09–0.15 mg/kg d.w.; Co: 7.92–9.27 mg/kg d.w.; Cr: 11.37–13.86 mg/kg d.w.; Cu: 16.91–20.92 mg/kg d.w.; Mn: 379–441 mg/kg d.w.; Ni: 20.58–28.95 mg/kg d.w.; Pb: 7.24–9.08 mg/kg d.w. and Zn: 44.28–49.93 mg/kg d.w. Except nickel all other metals concentrations have been founded below the normal limits accepted by the Romanian regulations. As a general observation, in depth soil concentrations are higher (Cr, Cu, Ni, Pb) or similar (Mn, Zn) than in surface samples (except cadmium and cobalt).

**Keywords:** heavy metals, soils, FAAS, waste deposits

## 1. Introduction

Since the start of industrialization in the nineteenth century, the material cycle amounts of heavy metals, have increased considerable and influence the quality of soil, plants, animals and people.

Industry and trade represent significant sources of heavy metal emissions, particularly metal processing operations and refineries. Special sources of emissions include cement works (thallium), battery manufacturers (lead) and electroplating operations (copper, nickel, chrome, zinc, cadmium, etc.). Through agrochemicals, like phosphate-containing fertilizers are, metals such cadmium

---

[*] To whom correspondence should be addressed: echirila@univ-ovidius.ro

L. Simeonov and V. Sargsyan (eds.),
*Soil Chemical Pollution, Risk Assessment, Remediation and Security.*
© Springer Science+Business Media B.V. 2008

are accumulated in the soil. Moreover, heavy metal emissions occur due to targeted production for particular commercial purposes from coal, cement or glass production.

Through refuse, wastewater or sludge and exhausted air (dust particles), they also enter into or on soil and, thereby, into the food chain (from plants to animals and finally to humans).

The degree of antropogenous effect of metal cycles can be represented as a global interference factor: it indicates the ratio of the antropogenously – induced amount of material to that of the natural (geochemical) material cycle.

The geochemical cycle of metals begins with the plutonic rock, which then enters into the water and the atmosphere via the Earth's surface or the ocean floor due to volcanic activity. The geochemical cycle of heavy metals consists in common processes of dissolution-precipitation equilibria, the transition of metal compounds in aerosols and the return to the soil and water via precipitations (Schwedt, 2001).

Soil pollution is defined as the build-up in soils of persistent toxic compounds, chemicals, salts, radioactive materials, or disease-causing agents, which have adverse effects on plant growth and animal health.

Soil contamination is the presence of man-made chemicals or other alteration of the natural soil environment. This type of contamination typically arises from the rupture of underground storage tanks, application of pesticides, percolation of contaminated surface water to subsurface layer, leaching of wastes from landfills or direct discharge of industrial wastes to the soil. The most common chemicals involved are petroleum hydrocarbons, solvents, pesticides, and heavy metals. The occurrence of this phenomenon is correlated with the degree of industrialization and intensity of chemical usage.

Soils and sediments are the solid components of terrestrial and aquatic ecosystems which serve as sources and sinks for nutrients and solid chemicals. The use of soils for industrial, agriculture and urban activities always involves a drastic modification of their composition and can eventually create enormous problems for its future use, involving high capital investments and health risks (Manea, 2004).

The term municipal solid waste (MSW) is used to describe most non-hazardous solid waste from a city, town or village that requires routine collection and transport to a processing or disposal site. MSW is not generally considered hazardous. But certain types of commercial and industrial wastes like those poisonous, explosive or dangerous can cause immediate and direct harm to people and the environment if they are not disposed of properly. The data about chemical composition of MSW are useful in waste management planning (Burnley, 2007).

Soil contamination as well as surface water and ground water pollution can be caused by the disposal of solid waste in improperly built landfills. This kind of pollution has important public health consequences (Manea, 2003).

The vulnerability of soils and groundwater to pollutants depends on soil pollutants and, consequently, the vulnerabilities are pH, redox potential, cation exchange capacity and humus content of soil (Briggs et al., 1997).

The soil is characterized by the natural concentration of heavy metals (that depend on the soil type and its composition) and by soil contamination with heavy metals, provided by human activity. Soil pollution with heavy metals can also be determined by infiltration of highly contaminated storm water.

Heavy metals or trace metals is the term applied to a large group of trace metals which are both industrially biologically important. Agricultural productivity can be limited by deficiencies of essential trace elements such as Cu, Mn and Zn in crops and Co, Cu, Mn and Zn in livestock. However, when certain heavy metals are presented in excessive concentrations, they give rise to concern with regard to human health and agriculture (Dobra and Viman, 2006; Statescu and Cotiusco-Zauca, 2006) and their accurate analytical determination remains a challenge for chemists (Baiulescu et al., 1990; Anderson, 1999; Crompton, 2001; Draghici et al., 2003; Simeonov et al., 1998; Simeonov and Managadze, 2006). The purpose of this paper is to present original results concerning concentrations of eight heavy metals in soils adjacent to municipal waste deposits from Constanta County, Romania, during April-October 2006.

## 2.  Experimental

Total Cd, Co, Cr, Cu, Mn, Ni, Pb and Zn concentrations in soils using flame atomic absorption spectrometry (FAAS) have been determined.

The studied sites were adjacent soils of two improperly built landfills located in Constanta County, Romania: Eforie Sud waste landfill (corresponding to 8,650 people) and Techirghiol waste landfill (corresponding to 7,150 people).

In order to determine metals concentration from soils, five samples were collected using a special device from the surface and depth of 20–40 cm from each location at 1–2.5 m distance of the landfill boundary (Chirila, 2004) and sampling campaign was carried out during April and October 2006. Samples from surface and depth have been obtained each month by the appropriate homogenization of collected samples, previously dried for 16 hours at room temperature.

In order to obtain the soil solutions, 3 grams of soil sample have been extracted with aqua regia in 250 mL volumetric flask (ISO 11466). The supernatant, the filtrate and the washing solution have been collected in 100 mL calibrated flask (Chirila and Draghici, 2003).

The spectrometric measurements have been done using a flame atomic absorption spectrometer (FAAS) Spectr AA220, provided by Varian Company. The measurements have been done in triplicate and the mean values are reported.

For the background correction, the zero calibration solution was done using aqua regia and deionised water (for Cd, Co, Cu, Ni, Pb and Zn); for Cr and Mn, the zero calibration solution was prepared by adding of 3.7 mg/L La, using a Lanthanum chloride solution. All used reagents were of spectral purity grade. Table 1 presents the concentration ranges and characteristics of the obtained calibration curves.

TABLE 1. Calibration characteristics for metal determination with FAAS technique

| Metal | Concentration range (mg/L) | Correlation coefficient |
|-------|----------------------------|-------------------------|
| Cd    | 0.3–1.0                    | 0.9991                  |
| Co    | 0.4–1.2                    | 0.9998                  |
| Cr    | 0.4–1.2                    | 0.9992                  |
| Cu    | 0.5–1.5                    | 0.9994                  |
| Mn    | 0.6–6.0                    | 0.9999                  |
| Ni    | 0.4–1.2                    | 0.9993                  |
| Pb    | 0.4–3.6                    | 0.9999                  |
| Zn    | 0.5–2.0                    | 0.9995                  |

## 3. Results and Discussion

The studies were performed in order to observe the heavy metal concentration evolution in adjacent soils to solid waste deposits. Once metals are introduced and contaminate the environment, they will remain. Metals neither are nor degraded like carbon-based (organic) molecules.

The measured concentrations have been compared with the Romanian regulations (Table 2).

TABLE 2. Regulatory limits for heavy metals in soils (after 756/1997 – Romanian regulation of environment pollution evaluation)

| Metal | Concentration, mg/kg dry weight | | | | |
|-------|--------------|-------|-------|-------|-------|
|       | Normal value | Alert limit | | Intervention limit | |
|       |              | S     | NS    | S     | NS    |
| Cd    | 1            | 3     | 5     | 5     | 10    |
| Co    | 15           | 30    | 100   | 50    | 250   |
| Cr    | 30           | 100   | 300   | 300   | 600   |
| Cu    | 20           | 100   | 250   | 200   | 500   |
| Mn    | 900          | 1,500 | 2,000 | 2,500 | 4,000 |
| Ni    | 20           | 75    | 200   | 150   | 500   |
| Pb    | 20           | 50    | 250   | 100   | 1,000 |
| Zn    | 100          | 300   | 700   | 600   | 1,500 |

S – sensible utilization, NS – non-sensible

Cadmium concentration in soil depends on the geological origin of the parent material, texture, intensity of weathering processes, organic matter and other factors. Cadmium enters the soil in smaller quantities than lead and it reach the soil through air. It is derived from incinerator exhaust gases and from phosphate fertilizers.

Generally, in acidic soils, cadmium is very mobile and does not accumulate. Compounds with humic acids are less stable. Under reductive conditions and in presence of sulfate ion, CdS is formed. The mean reported Cd concentration in lithosphere is 0.1 mg/kg d.w.

Figure 1 presents the mean cadmium concentration in studied soil samples. In Eforie Sud soil samples Cd concentration ranged between 0–0.9 mg/kg d.w. in surface samples and 0–0.46 mg/kg d.w. in depth samples. Soil samples from Techirghiol registered lower Cd concentrations (0–0.54 mg/kg d.w. in surface samples and 0–0.37 mg/kg d.w. in depth samples). All determined Cd concentrations were below the normal limits for soil.

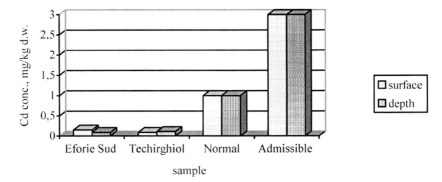

*Figure 1.* Cadmium concentration evolution in soil adjacent to waste deposits during April-October 2006 (mean values, mg/kg dry weight).

Cobalt is a trace component in the earth's crust (0.003%), an element that occurs naturally in many different chemical forms throughout our environment. Small amounts of cobalt are essential for good health. Cobalt usually occurs in association with other metals such as copper, nickel, manganese and arsenic. Small amounts are found in most rocks, soil, surface and underground water, plants and animals. Natural sources of cobalt in the environment are soil, dust, seawater, volcanic eruptions and forest fires. It is also release to the environment from burning coal and oil, from car, truck and airplane exhausts, and from industrial processes that use the metal or its compounds. All types of soils contain some amount of cobalt. The average concentration of cobalt in soils around the world is 8 mg/kg d.w. Toxic effects on plants are unlikely to occur below soil cobalt concentrations of 40 ppm. Plant species vary in their sensitivity to cobalt, and soil type while soil chemistry greatly influences cobalt toxicity. The more acidic the soil, the greater are the potential for cobalt toxicity, at any concentration. Where cobalt concentrations in soil are greater than 40 ppm, and cobalt toxicity to vegetation is suspected; cobalt uptake into plants can be reduced by liming the soil and by incorporating uncontaminated soil, peat moss, compost or manure into the soil.

The mean cobalt concentration in studied soil samples are presented in Figure 2. All the determined values are lower than the normal Co concentrations in soil.

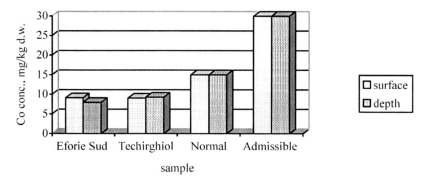

*Figure 2.* Cobalt concentration evolution in soil adjacent to waste deposits during April-October 2006 (mean values, mg/kg dry weight).

Chromium is a trace component in the earth's crust (0.02%), a unique element in soil, because of essentiality to human and animal life and non-essentiality for the vegetable kingdom and its possible presence in two main oxidation forms, trivalent and hexavalent which show opposite properties. The reported mean total chromium concentration in lithosphere is 69 mg/kg d.w. The two forms have completely different effects on living organisms: Cr(III) is apparently useful or harmless at reasonable concentrations, while Cr(VI) is extremely toxic. In addition, Cr(III) is not mobile in soil, therefore the risks of leaching are negligible, while Cr(VI), mainly present in the forms of chromates ($CrO_4^{2-}$) and dichromates ($Cr_2O_7^{2-}$), is generally mobile and often is part of crystalline minerals.

Conversion of Cr(III) to Cr(VI) has been shown in some particular soils, that are rich in manganese oxides, poor in organic matter and have high redox potential. On the contrary, the reverse transformation of Cr(VI) to Cr(III) is very common and easier, so that it is difficult to find hexavalent chromium forms in soil solution or in leaching waters. The problem of Cr enrichment in soil has been often discussed not only in relation to the discharge of tannery wastes, but also to the possibility of Cr presence in soil amendments, mainly organics, and to the existence of excellent organic fertilizers produced from leather residues or wastes.

Figure 3 presents the mean total chromium concentration in the studied soil samples. In Eforie Sud soil samples Cr concentration ranged between 3.24–28.72 mg/kg d.w. in surface samples and 7.25–30.06 mg/kg d.w. in depth samples. Soil samples from Techirghiol registered similar Cr concentrations (7.50–28.55 mg/kg d.w. in surface samples and 8.40–19.32 mg/kg d.w. in depth samples). All the determined Cr concentrations were below the normal limit in soil.

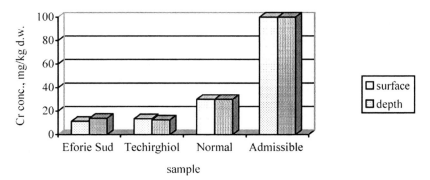

*Figure 3.* Chromium concentration evolution in soil adjacent to waste deposits in April-October 2006 (mean values, mg/kg dry weight).

Copper is also a trace element in the earth's crust (0.007%); is more mobile than Cd and its solubility increases at pH < 5. Although Cu is among the trace elements essential for life, in the case of plants toxic effects occur at 20 or more mg/kg d.w. In the past, the major source of Cu pollution was smelters that contributed vast quantities of Cu–S particulates to the atmosphere. Presently, the burning of fossil fuels and waste incineration are the major sources of Cu to the atmosphere and the application of sewage sludge, municipal composts, pig and poultry wastes are the primary sources of anthropogenic Cu contributed to the land surface. The amount of Cu available to plants varies widely by soils.

Available Cu can vary from 1 to 200 ppm (parts per million) in both mineral and organic soils depending on the soil pH and texture. The finer-textured mineral soils generally contain the highest amounts of Cu. The lowest concentrations are associated with the organic or peat soils. Availability of Cu is related to soil pH. As soil pH increases, the availability of this nutrient decreases. It is attracted to soil organic matter and clay minerals. Toxic at high doses; excess Cu can lead to Zn deficiencies and vice-versa.

Figure 4 presents the mean copper concentration in the studied soil samples. In Eforie Sud soil samples Cu concentration ranged between 15.80–21.75

mg/kg d.w. in surface samples and 13.37–47.60 mg/kg d.w. in depth samples. Soil samples from Techirghiol registered similar Cu concentrations (15.80–19.25 mg/kg d.w. in surface samples and 16.25–22.11 mg/kg d.w. in depth samples). The determined Cu concentrations in Eforie Sud soils sometimes slowly exceeded the normal limit in soil and copper concentration is higher in depths samples than in the surface samples.

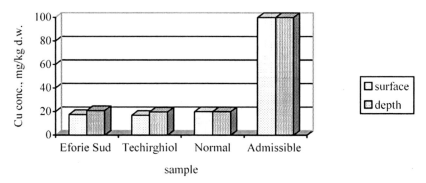

*Figure 4.* Copper concentration evolution in soil adjacent to waste deposits during April-October 2006 (mean values, mg/kg dry weight).

Manganese is a less abundant major component (0.1%) in the earth's crust. The major anthropogenic sources of environmental manganese include municipal wastewater discharges, sewage sludge, mining and mineral processing (particularly nickel), emissions from alloy, steel, and iron production, combustion of fossil fuels, and, to a much lesser extent, emissions from the combustion of fuel additives. Mean reported Mn concentration in soils is 300–600 mg/kg d.w. Availability of Mn increases as soil pH decreases. As soil pH dips below 5.5, Mn toxicity maybe evident as pH increases above 6.5 deficiencies are more likely. At lower pH the Mn(II) is dominant and is more readily plant available. At higher pH the Mn(III), Mn(IV) and Mn(VII) species dominate and are less plant available. Liming acid soils changes the availability of Mn by changing soil solution pH and the species of manganese. Organic matter plays a role in determining the fate of Mn. Soils with a high organic matter and neutral pH will present low concentrations of Mn. As the organic matter increases the complexing of Mn with organic matter also increases. Combine this with high soil pH and the Mn availability decreases further. Soils high in organic matter will usually be low in available Mn. Higher organic matter also encourages more microbiological activity which can further decrease the availability of Mn. These conversions are also influenced by soil aeration and moisture levels. Poor soil oxygen level caused by high moisture conditions coupled with a high rate of microbe activity fuelled by organic matter, consume oxygen and convert less available forms of Mn to more reduced or available forms. Under prolonged wet conditions available Mn can leach out of the root zone. Under continued waterlogged conditions these forms of Mn can plug the tile when exposed to the air in the tile runs. The role of Mn in plants was discovered in 1922. It is essential for photosynthesis, production of chlorophyll and nitrate reduction. Plants which are deficient in Mn exhibit a slower rate of photosynthesis by as much as half of a normal plant. Plants which are low in Mn causes other metals such as iron to exist in an oxidized and unavailable form the reduced form of metals are available for metabolism (WHO, 2004).

The mean total manganese concentrations in the studied soil samples are presented in Figure 5.

In Eforie Sud soil samples Mn concentration ranged between 235–665 mg/kg d.w. in surface samples and 247–628 mg/kg d.w. in depth samples.

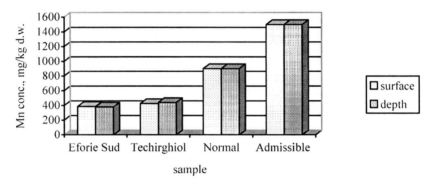

*Figure 5.* Manganese concentration evolution in soil adjacent to waste deposits during April-October 2006 (mean values, mg/kg dry weight).

Soil samples from Techirghiol registered higher Mn concentrations (343–684 mg/kg d.w. in surface samples and 382–616 mg/kg d.w. in depth samples). All founded values are lower than the normal Mn concentrations in soil.

Trace component on the earth's crust (0.008%), nickel combined with other elements occurs naturally in the earth's crust, is found in all soils, and is also emitted from volcanoes. Nickel compounds are used for nickel plating, to color ceramics, to batteries production and as catalyst to increase the rate of chemical reactions. Nickel may be released to the environment from the stacks of large furnaces used to make alloys or from power plants and trash incinerators.

Soil generally contains between 4 and 80 mg/kg d.w. nickel. The highest soil concentrations (up to 9.000 ppm) are found near industrial areas where nickel is extracted from ore. High concentrations of nickel occur because dust released from stacks during processing settles out of the air.

Nickel is essential to maintain health in animals. Although a lack of nickel has not been found to affect the health of humans, a small amount of nickel is probably also essential for humans.

Figure 6 presents the mean nickel concentration in the studied soil samples. In Eforie Sud soil samples Ni concentration ranged between 16.62–26.55 mg/kg d.w. in surface samples and 15.68–30.95 mg/kg d.w. in depth samples. Soil samples from Techirghiol registered higher Ni concentrations (11.30–28.55 mg/kg d.w. in surface samples and 15.44–49.47 mg/kg d.w. in depth samples). The determined Ni concentrations in both analyzed soils sometimes exceeded the normal limit in soil and nickel concentration is higher in depths samples than in the surface samples.

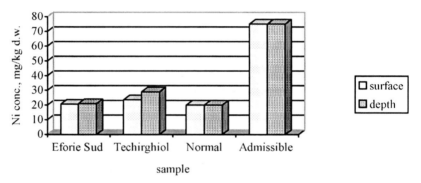

*Figure 6.* Nickel concentration evolution in soil adjacent to waste deposits during April-October 2006 (mean values, mg/kg dry weight).

Lead is a trace component in the earth's crust; the average reported lead concentration in the lithosphere is 14 mg/kg d.w.

The most important environmental sources for Pb are gasoline combustion (presently a minor source, but in the past 40 years a major contributor to Pb pollution), Cu–Zn–Pb smelting, battery factories, sewage sludge, coal combustion, and waste incineration.

Lead exhibits a pronounced tendency for accumulation in the soil, because it is minimally mobile even at low pH value. In soil with phosphate contents lead forms deposits of sparingly soluble lead phosphates ($Pb_3(PO_4)_2$, $Pb_4O(PbO_4)_2$, $Pb_5(PO_4)_3(OH)$. In soils with carbonate content lead is present as $PbCO_3$ and under reductive conditions as $PbS$. High levels of lead pollution still occur in the vicinity of industrial facilities and waste incinerators that have insufficient elimination of suspended dust.

The mean lead concentrations in studied soil samples are presented in the Figure 7.

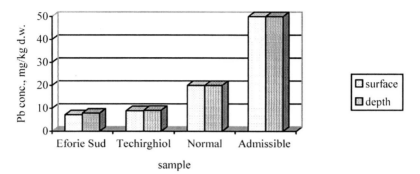

*Figure 7.* Lead concentration evolution in soil adjacent to waste deposits during April-October 2006 (mean values, mg/kg dry weight).

In Eforie Sud soil samples Pb concentration ranged between 4.69–12.40 mg/kg d.w. in surface samples and 5.98–11.42 mg/kg d.w. in depth samples. Soil samples from Techirghiol registered higher Pb concentrations (1.02–16.27 mg/kg d.w. in surface samples and 5.05–17.93 mg/kg d.w. in depth samples). Lead concentration is higher in depths samples than in the surface samples and all determined values are lower than the normal Pb concentrations in soil.

Zinc is an essential element, a trace component in the earth's crust (0.013%); the average reported Zn concentration in lithosphere is 80 mg/kg d.w. Zinc occurs naturally in air, water and soil, but zinc concentrations are rising unnaturally, due to addition of zinc through human activities. Most zinc is added during industrial activities, such as mining, coal and waste combustion and steel processing. Some soils are heavily contaminated with zinc, and these are to be found in areas where zinc has to be mined or refined, or were sewage sludge from industrial areas has been used as fertilizer. When the soils of farmland are polluted with zinc, animals will absorb concentrations that are damaging to their health. Water-soluble zinc that is located in soils can contaminate the groundwater.

Zinc cannot only be a threat to cattle, but also to plant species. Plants often have a zinc uptake that their systems cannot handle, due to the accumulation of zinc in soils. On zinc-rich soils only a limited number of plants have a chance of survival. That is why there is not much plant diversity near zinc-disposing factories. Due to the effects upon plants, zinc is a serious threat to the productions of farmlands. Despite of this, zinc-containing manures are still applied. Finally, zinc can interrupt the activity in soils, as it negatively influences the activity of microorganisms and earthworms. The breakdown of organic matter may seriously slow down because of this.

Zinc is one of the most mobile metals in the soil. The solubility of zinc in soil increases especially at pH < 6. At higher pH and in the presence of phosphates, the zinc appropriated by plants can be significantly reduced. The pH-dependent process of adsorption in clay minerals and in various oxides constitutes the most significant regulatory process for the availability of Zn in soils. It can result from the smelter industry. Environmental and food chain risk caused by zinc seems to play an important role in controlling the cadmium up-take in cadmium contaminated soils, whenever the ratio Zn:Cd is higher than 100.

Figure 8 presents the mean zinc concentration in the studied soil samples.

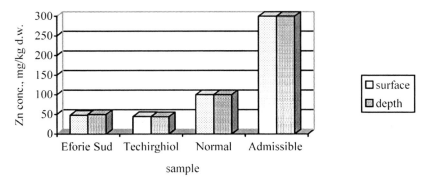

*Figure 8.* Zinc concentration evolution in soil adjacent to waste deposits during April-October 2006 (mean values, mg/kg dry weight).

In Eforie Sud soil samples Zn concentration ranged between 28.27–96.14 mg/kg d.w. in surface samples and 23.16–88.80 mg/kg d.w. in depth samples. Soil samples from Techirghiol registered lower Zn concentrations (29.70–79.08 mg/kg d.w. in surface samples and 31.64–66.18 mg/kg d.w. in depth samples). All determined Zn concentrations were below the normal limit in soil.

**Conclusions**

Cd, Co, Cr, Cu, Mn, Ni, Pb and Zn determination in soils adjacent to municipal solid waste deposits from Constanta County has been done using FAAS technique.

The mean measured values ranged as follows: for cadmium between 0.09–0.15 mg/kg d.w. in surface samples and 0.09–0.11 mg/kg d.w. in depth samples; for cobalt between 9.02–9.11 mg/kg d.w. in surface samples and 7.92–9.27 mg/kg d.w. in depth samples; for chromium between 11.37–13.46 mg/kg d.w. in surface samples and 12.45–13.86 mg/kg d.w. in depth samples; for copper between 16.91–17.55 mg/kg d.w. in surface samples and 19.65–20.92 mg/kg d.w. in depth samples; for manganese between 385–425 mg/kg d.w. in surface samples and 379–441 mg/kg d.w. in depth samples; for nickel between 20.58–23.52 mg/kg d.w. in surface samples and 20.90–28.95 mg/kg d.w. in depth samples; for lead between 7.24–8.94 mg/kg d.w. in surface samples and 7.90–9.08 mg/kg d.w. in depth samples and for zinc between 44.55–47.71 mg/kg d.w. in surface samples and 44.28–49.93 mg/kg d.w. in depth samples.

Except nickel all other metals concentrations have been founded below the normal limits from Romanian regulations. As a general observation, in depth soil concentrations are higher (Cr, Cu, Ni, Pb) or similar (Mn, Zn) than in surface samples (except cadmium and cobalt).

# References

Anderson K.A., 1999, Analytical Techniques for Inorganic Contaminants, AOAC International.

Baiulescu G.E., Dumitrescu P. and Zugrăvescu Gh., 1990, Sampling, Ellis Horwood, London.

Briggs D., Smithson P., Addison K., Atkinson K., 1997, Fundamentals of the physical environment, 2nd Edition, Butler & Tauner, London, 322–336.

Burnley S.J., 2007, The use of chemical composition data in waste management planning – A case study, Waste Management, 27, 327–336.

Chirila E. and Draghici C., 2003, Pollutants analysis. I. Water Quality Control, Transilvania University Press, Brasov (Romanian).

Chirila E., 2004, Sampling, in Colbeck I, Drăghici C. and Perniu D. (eds), Polution and Enviromental Monitoring, The Publishing House of the Romanian Academy, Bucharest, 109–128.

Crompton T.R., 2001, Determination of Metals and Anions in Soils, Sediments and Sludges, Spon Press, Taylor & Francis.

Dobra M. and Viman V., 2006, Determination of the concentration of heavy metals in soils and plants by ICP-MS, EEMJ, 5, 1197–1203.

Draghici C., Coman Gh., Sica M., Perniu D., Tica R. and Badea M., 2004, Capilary electrophoresis for soil analysis, Proceedings Bramat 2003, 494–501.

Manea F., 2003, Solid waste management, in Waste management, Pretty J., Oros V., Draghici C. (Eds), The Publishing House of the Romanian Academy, Bucharest, 87–93.

Manea F., 2004, Soil monitoring, in Pollution and Environmental monitoring, Colbeck I, Draghici C., Perniu D. (eds), The Publishing House of the Romanian Academy, Bucharest, 87–95.

Simeonov L.I., Managadze G.G., Schmitt C. and Scheuermann K., 1997, Ecology screening of heavy metal pollution of the soil with laser mass spectrometry, Comp. rend. Acad. bulg. Sci., 51, 5–6, 29–32.

Managadze G. and Simeonov. L., 2006. Technological Transfer. Miniature Laser Mass Spectrometer for Express Analysis of Environmental Samples, in Chemicals as intentional and accidental global environmental threats, Borovetz, 2006, L. Simeonov, E. Chirila (eds). NATO Science Series C, Environmental Security, Springer, Dordrecht, 149–163.

Schwedt G., 2001, The essential guide to environmental chemistry, Wiley, 20.

Statescu Fl., Cotiusco-Zauca D., 2006, Heavy metal soil contamination, EEMJ, 5, 1205–1213.

WHO, 2004 Manganese and its compounds: environmental aspects http://www.inchem.org/documents/cicads/cicads/cicad63.htm#4.0.

# HEALTH-ECOLOGICAL INVESTIGATION OF SOIL POLLUTION WITH HEAVY METALS AND ARSENIC IN A METALLURGICAL AND ORE OUTPUT REGION

ALEXANDER SPASOV, ANNA DIMITROVA[*]
AND MOMTCHIL SIDJIMOV
*National Center of Public Health Protection*
*15 Acad. Ivan Geshov Blvd., 1431 Sofia, Bulgaria*

**Abstract.** The investigation is executed in a central Bulgarian region with both well developed agricultural and metallurgical activities. The agricultural land is approximately 38,700 decares. Geographically the region is estimated as closed valley with dominating winds into south-west and south-east direction. The survey of the heavy metal soil pollution is executed in the south-west part of the valley, characterized with high density of ore-dressing factories and landfills for industrial hazardous wastes. Executed is collection of soil samples and analysis of probes from 4,500 decares including agricultural land and three living areas, two of the last situated 1,000 m to the north-east (Chavdar village) and south (Benkovski village) from the landfill for industrial wastes. Determined are the following average concentrations for heavy metals: lead – 110 mg/kg soil; copper – 280 mg/kg; zinc – 130 mg/kg; cadmium – 0.75 mg/kg; arsenic – 18 mg/kg. Studied are also the heavy metals content in plants. The soil and plant pollution in the region is realized mainly through the aerosol pathway. Proved is the existence of possibilities for dispersion of dust particles, coming from the landfill for industrial wastes. Analyzing the morbidity rate for the region for groups of the population (children and adults), found are respiratory and allergic disturbances to be with higher frequency rate compared with the average data for the country. The general conclusion of the achieved results proves that the pollution of the soils comes primarily from the found 12 km in east direction non-ferrous ore-dressing and metallurgical plant. The correlation between the soil pollution and the morbidity of the population allows the creation of prophylactic programs including change in the land used for

---

[*] To whom correspondence should be addressed: a.dimitrova@ncphp.government.bg

agricultural purposes, sowing only definite plant cultures and some other measures aiming at limitation of the health risk.


**Keywords:** soil pollution, heavy metals, health risk


## 1. Introduction

Soil, as the major depot of heavy metals in the environment, plays an important role in global metal cycling. Consequently, anthropogenic contamination of soils has received much attention. Numerous investigations have been conducted in soils in the vicinity of smelters to study the heavy metal contamination of soils (Jauncein, 1995). These studies suggest that heavy metal concentrations are extremely high in the upper horizons of the soils close to the smelter and they decrease progressively with increasing distance.

This study was conducted in the district of Pirdop where the area located around the non-ferrous metallurgical complex is known as one of Bulgaria's most important "hot spots". In the area operates the biggest copper smelter complex in Bulgaria. It was built in 1957 and is currently operated by a Belgium firm. It has a treatment capacity of 500,000 tons per year of non-ferrous ores. This metallurgical operation, although still using mixture of contemporary and original technology installed before decades, plays a central role in the industrial base of the Pirdop area. The region is endowed with good soils and climate. However, because of the industrial activity of the metallurgical complex, the soils around it are highly contaminated by heavy metals such as cadmium, nickel, chromium, lead, and copper. The basic way for soil pollution is through dispersion of improperly fixed dust particles, containing mixture of heavy metals from the industrial landfill of the plant towards the area around it. Consuming vegetables and other crops grown in this area presents a serious health risk to those eating these products and especially for the most vulnerable part of the population – children and elderly.


## 2. Methods and Results

In October 2002, soil sampling on 50 private yards in Pirdop town was conducted on the surface and to a depth of 15 cm. Since the results of this sampling showed increased metal levels, the sampling depth was increased to 30 cm for the soil sampling in the same fields in April 2003. Efforts were made to sample all selected yards but in some cases conditions made it impossible to collect an appropriate soil sample. For example, sampling to the 30 cm depth was not possible on every property, as occasionally stony fill was encountered. Also, some yards were covered with gravel, asphalt, or debris (e.g., vehicles, construction material), which physically prevented the correct soil sampling. Generally, soil samples were not collected within 4 meters of driveways, walkways, building structures, fences and debris to reduce the likelihood of encountering local residential sources of contamination (e.g., driveway spills, eroded paint from painted surfaces). Each property was assessed for sampling. A back yard was usually used. Each soil sample was divided into two depth intervals (0–5 cm and 20–30 cm) and the different sections for each of the sample depths were placed in one labeled polyethylene bag. In Table 1 are presented some of the results from the investigation with participation of samples taken from all districts in the town of Pirdop.

TABLE 1. Heavy metals content in soil samples from private yards in the town of Pirdop

| Sample No | Lead mg/kg | Cadmium mg/kg | Arsenic mg/kg | Copper Mg/kg | Zinc mg/kg | Chromium mg/kg | Nickel mg/kg |
|---|---|---|---|---|---|---|---|
| 1* | 615 | 4.00 | 25.5 | 12,000 | 830 | 33.8 | 36.5 |
| 1-1 | 92.5 | 0.25 | 32.4 | 648 | 195 | 29.0 | 19.0 |
| 2* | 93.5 | 0.75 | 4.40 | 1,100 | 258 | 23.8 | 43.0 |
| 3* | 200 | 2.25 | 25.5 | 1,930 | 615 | 27.0 | 25.8 |
| 4* | 158 | 3.00 | 44.1 | 2,300 | 470 | 34.5 | 34.0 |
| 5* | 120 | 2.25 | 33.2 | 2,600 | 955 | 27.3 | 22.2 |
| 5-1 | 90.0 | 0.75 | 32.3 | 1,030 | 248 | 30.5 | 31.8 |
| 6* | 163 | 1.25 | 60.4 | 2,900 | 235 | 24.0 | 22.8 |
| 7* | 130 | 1.75 | 45.3 | 2,650 | 255 | 28.0 | 28.2 |
| 7-1 | 118 | 1.00 | 37.1 | 1,090 | 193 | 19.3 | 31.8 |
| 8* | 90.0 | 0.50 | 20.1 | 628 | 210 | 21.0 | 28.5 |
| 8-1 | 77.5 | 0.25 | 19.8 | 518 | 190 | 30.5 | 31.0 |
| 9* | 373 | 3.25 | 89.4 | 6,800 | 830 | 23.0 | 22.3 |
| 9-1 | 150 | 1.00 | 39.3 | 1,630 | 495 | 27.0 | 29.3 |
| 10* | 108 | 1.25 | 23.9 | 1,730 | 620 | 28.5 | 27.0 |
| 10-1 | 40.0 | 0.25 | 10.1 | 213 | 163 | 18.5 | 32.5 |
| 11* | 82.5 | 0.75 | 10.9 | 363 | 240 | 29.5 | 31.5 |
| 11-1 | 95.0 | 0.50 | 17.2 | 455 | 295 | 23.5 | 35.8 |
| 12* | 24.5 | 0.25 | 7.25 | 128 | 175 | 14.3 | 22.0 |
| 12-1 | 37.5 | <0.25 | 8.50 | 190 | 215 | 18.5 | 25.8 |

* – samples taken from the soil surface.
x-1 – samples taken from 30 cm depth.

The data from the atom absorption analysis prove that recent or ongoing atmospheric deposition results in accumulation of heavy metals in the upper most soil layers and decreasing quite abruptly with depth. Where atmospheric deposition is the only source of contamination, as it is in the Pirdop region, measurably elevated soil contaminant levels are usually confined to the top 20–30 cm of soil. For most of the private yards, the highest soil contaminant

levels tended to occur not right at the surface but between 5 and 10 cm, and
then fell rapidly to near background concentrations below about 25–30 cm. This
pattern, which has been observed in other regions that have been impacted by
historic industrial emissions, is consistent with deposition that was much higher
in the past and was much lower or abated entirely in more recent years.
Although an overall spatial pattern was evident for some contaminants, particularly lead, copper, cadmium and arsenic, in our investigation the variability of
the soil metal levels in the samples from the different yards makes it difficult to
judge the extent of the contamination. The results confirm that soil to a depth of
30 cm in the vicinity and downwind of the copper smelter is severely
contaminated with heavy metals. Based on the soil sampling data, the soil
maximum concentration limit for Pb, Cu, Cd, As, and Zn is exceeded several
times beyond 10 km into south-west and south-east direction from the plant
over an area greater than 120 km$^2$ (Figure 1).

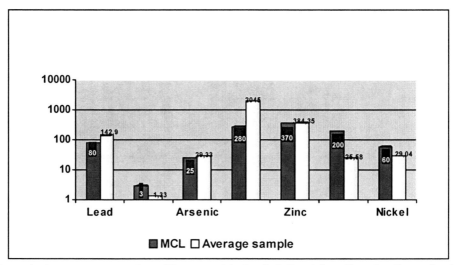

*Figure 1.* Average concentration of heavy metals (mg/kg) found in collected soil samples in
Pirdop region compared to maximum concentration limit (MCL) for agricultural land in Bulgaria.

The soil metal contamination in the Pirdop area is unquestionably source-
oriented, resulting from more than 40 years of atmospheric deposition from the
smelter plant. These heavy metals are very persistent in soil and potentially
hazard for the population's health through different ecological vectors. Since
the plant emissions lessened several years ago, further increases in soil metal
concentrations is not expected. Additionally is developed and implemented a
pilot project of a technology for treatment of polluted soil in that region where
the ions of heavy metals are fixed sustainably with the help of a proper

meliorant and are taken out from the biological circle with second effect of achieving improvement of the structure and fertility of the soil layer (Demnerova, 2001).

In April 2003 collection and analysis of plant samples was executed with additional analysis of approximately 200 samples of agricultural land around Pirdop. In Table 2 are presented the concentrations of the investigated heavy metals into several plant cultures from totally collected 50 samples. To assist in the interpretation of the plant bioavailability of the soil contaminant concentrations, soil pH was determined in distilled water using standard protocols.

TABLE 2. Heavy metals content in plant samples from private yards in the town of Pirdop

| Sample No | Lead mg/kg | Cadmium mg/kg | Arsenic mg/kg | Copper mg/kg | Zinc mg/kg | Chromium mg/kg | Nickel mg/kg |
|---|---|---|---|---|---|---|---|
| 1 | 1.40 | 0.51 | 2.15 | 120 | 65.5 | 0.97 | 1.23 |
| 2 | 1.50 | 0.58 | 3.55 | 167 | 68.3 | 4.43 | 2.50 |
| 4 | 0.77 | 0.33 | 2.41 | 114 | 96.7 | 2.53 | 1.20 |
| 5 | 0.57 | 0.23 | 0.77 | 44.0 | 47.3 | 0.43 | 0.45 |
| 6 | 0.40 | 0.22 | 1.00 | 30.7 | 48.7 | 0.23 | 0.53 |
| 7 | 0.25 | 0.21 | 0.64 | 41.7 | 43.7 | 0.47 | 0.93 |
| 8 | 0.55 | 0.27 | 0.60 | 41.3 | 52.3 | 1.03 | 0.93 |
| 9 | 0.50 | 0.26 | 0.84 | 39.0 | 117 | 0.53 | 0.60 |
| 10 | 1.40 | 0.43 | 2.95 | 197 | 123 | 6.57 | 3.97 |
| 12 | 0.53 | 0.10 | 0.80 | 39.0 | 59.3 | 1.80 | 1.43 |

The collected plant cultures determine that the practice of regularly tilling agricultural fields do not substantially reduce the soil contaminant burden. Sometimes the collection of surface soil samples only from undisturbed sites may over-estimate the severity and extent of contamination (Dumitru, 1992), particularly in the downwind direction, as this area is predominantly agricultural. Tilling tends to reduce the concentrations in the surface soil layers but increases the concentrations at depth, essentially spreading the contamination throughout the plow layer and subsequently into plant cultures. The difference between tilled and untilled sites is greatest farthest from the smelter complex, with the metal concentrations at surface being higher in the untilled sites. However, at tilled sites closer to the plant soil metal contamination exceed the maximum concentration limits (MCL) at depths greater than 25–30 cm. Therefore, tilling may exacerbate future remediation efforts in the region as the contamination has been distributed deeper into the soil profile.

## Conclusions

1. The soil metal contamination in Pirdop region is a potential threat to human health and the natural environment. Subsequent reductions in soil metal concentrations as a result of natural processes will be extremely gradual. The potential for toxicity can be reduced by adding a liming agent to the soil to raise the pH.
2. The soil and plant pollution in the region is realized mainly through the aerosol pathway. The investigation was successful in better defining the area of surface and 25–30 cm soil contamination in the zone where the concentration gradient was steep (within about 5–10 km of the smelter complex).
3. The investigation defines the extent of agricultural soil contamination and determines if soil contaminant levels had changed during the last years. The agricultural limitations would potentially include reduced yields of cereal crops on mineral soil and metal-enriched vegetable crops on organic soil.

## References

Jauncein M. et al., Public Health Implications of Soil Pollution, Draft, WHO, 1995
Demnerova K., Innovative Approaches to the On-Site Assessment and Remediation of Contaminated Sites, Proceedings of the NATO Advanced Study Institute, Prague, Czech Republic, 24 May–2 June 2001
Dumitru, M., Research for Establishing Measures to Ecologically Restore Soils Polluted with Heavy Metals in Zlatna Area, GHELASE, 1992

# QUANTITATIVE HEALTH RISK ASSESSMENT OF HYDROCARBON CONTAMINATION AT FORMER INNER TOWN BUS DEPOT (CASE STUDY)

GYULA DURA[1*], ZOLTÁN DEMETER[1]
AND PÉTER MENSÁROS[2]
[1]*National Institute of Environmental Health
Gyali ut 2–6.,1097 Budapest, Hungary*
[2]*GEOHIDROTERV Ltd.*
*Temesvári u. 20, Budapest 1116, Hungary*

**Abstract.** Hungary has a National Clean-up programme in line with policy and governmental decree for contaminated land that do specify explicit environmental and health risk assessment. This presentation provides a case study of risk assessment and modelling methods applied to a site contaminated by hydrocarbons at former inner town bus depot. A comprehensive health and environmental risk assessment framework was carried out for the evaluation of contaminated site. The goal of the study was to provide estimates of potential changes in exposure and risk that may occur either under a no-further action alternative (in present status after removing of contaminated soil layer) or following implementation of remedial actions (decontamination of heavily polluted groundwater). The presentation describes the danger that the hydrocarbon contamination source represents. The potential fugacity of pollutants to environmental media was evaluated by transfer and transport model according to RBCA rules. The calculated risk values for selected receptors (population, local and construction workers) showed marked hazard inside the site and descending off-site risk. The risk values were used in the selection of appropriate remedial action.

**Keywords:** contaminated site, human health risk assessment, remediation target concentration

---

* To whom correspondence should be addressed: duragy@okk.antsz.hu

L. Simeonov and V. Sargsyan (eds.),
*Soil Chemical Pollution, Risk Assessment, Remediation and Security.*
© Springer Science+Business Media B.V. 2008

The quantitative human health risk assessment of hydrocarbon contamination at former inner town bus depot was carried out. The risk assessment of contaminated soil and groundwater is essentially aimed at the determination of the remediation target concentration.

## 1.   The Aim of the Study

### 1.1.   ASSESSMENT OF THE HEALTH RISK OF HYDROCARBONS

- Total petroleum hydrocarbons (TPH) and BTEX (benzene, toluene, ethylbenzene, xylene and other alkylbenzene derivatives, e.g., cumene) contamination in soil and groundwater
- Halogenated aliphatic hydrocarbons (tetrachloroethylene, trichloroethylene, cis-1,2-dichloroethylene) contamination
- Polycyclic aromatic hydrocarbons (PAH) (fluorene, phenanthrene, anthracene, fluoranthene, pyrene, benz[a]anthracene, chrysene, benz[b]fluoranthene, benz[k]fluoranthene, benz[a]pyrene, indene[1,2,3-cd]pyrene, dibenz[a,h]anthracene, benz[g,h,i]perylene, naphtalenes) contamination

### 1.2.   SUMMING UP THE CHEMICAL RISKS OF THE SOIL
AND GROUNDWATER CONTAMINANTS

### 1.3.   DETERMINATION OF THE REMEDIATION TARGET CONCENTRATION
FOR THE CONTAMINATED SOIL AND GROUNDWATER

In our risk assessment we accomplished the complete evaluation of the possible exposure pathways from the source of contamination to the receptors.

The remediation target concentration based on the relevant risk can be determined, when the site specific health risk exceeds the acceptable level. In "those cases reverse risk assessment was carried out: according to this method, instead of calculating the risk posed by the contamination, the acceptable risk was taken into account when determining the residual level of the contaminants, the so-called "remediation target level", below which no health impact occurs.

## 2.  The Environmental Health Risk Assessment Method

The risk assessment was accomplished on the basis of the "The detailed quantitative risk assessment of contaminated sites" guidance, which was prepared by the Department of the Protection of Environmental Media, Ministry of Environment, Hungary. We have also taken into consideration the recommendations of the European Communities and the US EPA "Exposure Assessment Methodology" guidance.

### 2.1.  THE AVAILABLE DATA USED FOR THE ASSESSMENT OF THE EXPOSURE AND HEALTH RISK

In the report we used the available data, with which the Sponsor provided us. The data is presented in Tables 1a–1c and 2a–2e.

TABLE 1a. Inorganic contaminants in the groundwater (ug/L). The contaminant concentration values of the samples taken from the monitoring wells (Or5; Or14; Or20; Or23; Or24; Or26 and the one marked with "well") compared with the current Hungarian legal regulations

|  | A | B | C1 | C2 | C3 | Average | Max | Standard deviation (SD) | N | Upper 95% confidence limit |
|---|---|---|---|---|---|---|---|---|---|---|
| Silver | k* | 10 | 50 | 80 | 100 | 0.05 | 0.09 | 0.056 | 2 | 0.12 |
| Boron | 100 | 500 | 750 | 1,000 | 1,500 | 659.86 | 1,130 | 260.16 | 7 | 852.58 |
| Barium | 200 | 700 | 1,000 | 1,500 | 2,000 | 157.17 | 429 | 126.55 | 7 | 250.92 |
| Cadmium | 0.4 | 5 | 6 | 8 | 10 | 0.104 | 0.31 | 0.103 | 7 | 0.18 |
| Cobalt | 1 | 20 | 40 | 75 | 150 | 2.37 | 5.24 | 1,588 | 7 | 3.54 |
| Chrome | 1 | 50 | 100 | 150 | 200 | 2.45 | 5.47 | 1,826 | 7 | 3.80 |
| Copper | 10 | 200 | 300 | 500 | 1,000 | 6.71 | 32.9 | 11,618 | 7 | 15.32 |
| Molibden | 5 | 20 | 75 | 100 | 300 | 8.35 | 41.2 | 14,595 | 7 | 19.16 |
| Nickel | 5 | 20 | 50 | 75 | 100 | 7.55 | 11.9 | 2,683 | 7 | 9.54 |
| Lead | 3 | 10 | 40 | 75 | 100 | 1.13 | 1.59 | 0.643 | 2 | 2.02 |
| Selenium | 1 | 5 | 10 | 20 | 50 | 3,317 | 9.72 | 3,164 | 7 | 5.66 |
| Tin | 2 | 10 | 50 | 100 | 150 | 0.317 | 0,63 | 0.204 | 7 | 0.46 |
| Zink | 65 | 200 | 300 | 500 | 1,000 | 203.76 | 484 | 201.41 | 7 | 352.96 |

According to the the KöM-EüM-FvM-KHVM Joint Ministerial Decree no.10/2000 (VI.2.):

| | |
|---|---|
| A | is the background concentration |
| B | is the contamination limit value |
| C1, C2, C3 | are the intervention values |
| N | is the number of samples analysed |
| k* | is the smallest detectable concentration |

TABLE 1b. Total aliphatic TPH, BTEX and halogenated aliphatic hydrocarbon contaminants in the groundwater compared with the concentration values of the current Hungarian legal regulations (ug/L)

| | A | B | C1 | C2 | C3 | Average | Max | Standard deviation (SD) | N | Upper 95% confidence limit |
|---|---|---|---|---|---|---|---|---|---|---|
| Total aliphatic hydrocarbons (TPH) | 50 | 100 | 500 | 1000 | 2,000 | 2,618 | 22,100 | 5,442 | 29 | 4,598 |
| Benzene | 0.05 | 1 | 5 | 10 | 20 | 2.40 | 23.80 | 5.78 | 28 | 4.54 |
| Toluene | 0.05 | 20 | 30 | 50 | 80 | 1.49 | 26.10 | 5.02 | 28 | 3.35 |
| Ethylbenzene | 0.05 | 20 | 30 | 50 | 80 | 1.38 | 18.70 | 3.63 | 28 | 2.72 |
| Xylol | 0.05 | 20 | 30 | 50 | 80 | 2.86 | 26.30 | 6.15 | 28 | 5.14 |
| Other alkyl benzenes | | 20 | 30 | 50 | 80 | 85.60 | 683.00 | 157.69 | 28 | 144.01 |
| 1,2-dichloro-ethylene | 0.05 | 10 | 50 | 100 | 200 | 50.69 | 273.00 | 79.46 | 22 | 83.90 |
| Trichloroethylene | 0.05 | 10 | 40 | 150 | 400 | 7.81 | 31.00 | 8.53 | 24 | 11.22 |
| Tetrachloro-ethylene | 0.05 | 10 | 40 | 150 | 400 | 399.39 | 4820.00 | 978.25 | 28 | 761.73 |

TABLE 1c. PAH contaminants in the groundwater and regulatory values of the current Hungarian Ministerial order: (ug/L)

| | A | B | C1 | C2 | C3 | Aver. | Max | Standard deviation (SD) | N | Upper 95% confidence limit |
|---|---|---|---|---|---|---|---|---|---|---|
| Naphtalenes* | 0.1 | 2 | 5 | 20 | 70 | 19.55 | 166 | 41.21 | 29 | 34.55 |
| Acenaphthylene | 0.02 | 0.2 | 0.5 | 2 | 5 | 0.10 | 1.16 | 0.29 | 16 | 0.24 |
| Acenaphthene | 0.02 | 0.05 | 0.5 | 2 | 5 | 0.42 | 3.27 | 0.69 | 26 | 0.69 |
| Fluorene | 0.02 | 0.05 | 0.5 | 2 | 5 | 1.17 | 10.2 | 2.10 | 29 | 1.94 |
| Phenanthrene | 0.02 | 0.1 | 0.5 | 2 | 5 | 1.37 | 17.8 | 3.35 | 29 | 2.59 |
| Anthracene | 0.02 | 0.05 | 0.5 | 2 | 5 | 0.23 | 3.08 | 0.58 | 29 | 0.44 |
| Fluoranthene | 0.005 | 0.1 | 0.2 | 0.5 | 1 | 0.31 | 4.77 | 0.88 | 29 | 0.63 |
| Pyrene | 0.002 | 0.1 | 0.2 | 0.5 | 1 | 0.31 | 3.99 | 0.74 | 29 | 0.58 |
| Benz[a]anthracene | 0.002 | 0.02 | 0.1 | 0.2 | 0.5 | 0.09 | 1.65 | 0.30 | 29 | 0.20 |
| Chrysene | 0.002 | 0.02 | 0.05 | 0.1 | 0.2 | 0.12 | 1.82 | 0.33 | 29 | 0.24 |
| Benz[b]fluoranthene | 0.001 | 0.02 | 0.03 | 0.05 | 0.1 | 0.15 | 0.29 | 0.14 | 3 | 0.31 |
| Benz[k]fluoranthene | 0.001 | 0.02 | 0.03 | 0.05 | 0.1 | 0.10 | 2.03 | 0.37 | 29 | 0.23 |
| Benz[e]pyrene | 0.001 | 0.01 | 0.02 | 0.05 | 0.1 | | | | | |
| Benz[a]pyrene | 0.001 | 0.01 | 0.02 | 0.05 | 0.1 | 0.06 | 1.31 | 0.24 | 29 | 0.15 |
| Indene[1,2,3-cd]pyrene | 0.001 | 0.01 | 0.02 | 0.05 | 0.1 | 0.03 | 0.59 | 0.11 | 28 | 0.07 |
| Dibenz[a,h]anthracene | 0.02 | 0.01 | 0.02 | 0.05 | 0.1 | 0.01 | 0.13 | 0.03 | 24 | 0.02 |
| Benz[g,h,i]perylene | 0.002 | 0.01 | 0.02 | 0.05 | 0.1 | 0.04 | 0.53 | 0.12 | 28 | 0.09 |
| Total PAH without naphtalene(s) | 0.1 | 2 | 5 | 8 | 15 | 4.27 | 52.30 | 9.66 | 29 | 7.79 |

*Naphtalenes: Naphtalene + 1-Methyl-naphtalene + 2-Methyl-naphtalene

TABLE 2a. Inorganic contaminants in the soil (mg/kg): Level of contamination in the following samples collected from 0.5–1.0 m depth: Ort5; Ort37; Ort38; Ort39; Ort40; Ort42; Ort43; Ort44; Ort45; Ort46; Ort17; Ort21; Ort22; Ort24; Ort30; Or4; Or5; Or9; Or11; Or12; Or13; Or14; Or15; Or19; Or20; Or21; Or22; Or23; Or24; Or26 and the concentration limit values of the current Hungarian legal regulations

| | A | B | C1 | C2 | C3 | Average | Max | Standard deviation (SD) | N | Upper 95% confidence limit |
|---|---|---|---|---|---|---|---|---|---|---|
| Silver | 0.3 | 2 | 10 | 20 | 40 | 0.13 | 2.18 | 0.389142 | 30 | 0.27 |
| Boron | | | | | | 140.4 | 501 | 92.38925 | 30 | 173.50 |
| Barium | 150 | 250 | 300 | 500 | 700 | 0.25 | 0.82 | 0.145139 | 30 | 0.30 |
| Cadmium | 0,5 | 1 | 2 | 5 | 10 | 7.79 | 21 | 4.495893 | 30 | 9.40 |
| Cobalt | 15 | 30 | 100 | 200 | 300 | 20.85 | 91 | 14.1496 | 30 | 25.91 |
| Chrome (sum) | 30 | 75 | 150 | 400 | 800 | 37.36 | 648 | 116.2206 | 30 | 78.94 |
| Copper | 30 | 75 | 200 | 300 | 400 | 0.57 | 4.58 | 0.840647 | 30 | 0.87 |
| Molibden | 3 | 7 | 20 | 50 | 100 | 19.87 | 89.7 | 16.21246 | 30 | 25.67 |
| Nickel | 25 | 40 | 150 | 200 | 250 | 18.33 | 80.8 | 15.92542 | 30 | 24.03 |
| Lead | 25 | 100 | 150 | 500 | 600 | 0.93 | 2.41 | 0.59598 | 28 | 1.15 |
| Selen | 0.8 | 1 | 5 | 10 | 20 | 1.58 | 6.31 | 1.545154 | 30 | 2.13 |
| Tin | 5 | 30 | 50 | 100 | 300 | 119.32 | 1,280 | 225.6115 | 30 | 200.0 |

According to the KöM-EüM-FvM-KHV Joint Ministerial Regulation no.10/2000 (VI.2.):
A                is the background concentration
B                is the contamination limit value
C1, C2, C3     are the intervention values

TABLE 2b. Total aliphatic TPH, BTEX and halogenated aliphatic hydrocarbon contaminants in the soil and concentration values of the current Hungarian legal regulations (mg/kg)

| | A | B | C1 | C2 | C3 | Aver. | Max | Standard deviation (SD) | N | Upper 95% confidence limit |
|---|---|---|---|---|---|---|---|---|---|---|
| Total aliphatic CH (TPH) | 50 | 100 | 300 | **3,000** | 5,000 | 1,426 | 7,270 | 2557.85 | 12 | 2,873 |
| Benzene | 0.1 | 0.2 | 0,5 | **3** | 5 | 0.01 | <0.01 | | | |
| Toluene | 0.05 | 0.5 | 5 | **15** | 25 | 0.01 | <0.01 | | | |
| Ethylbenzene | 0.05 | 0.5 | 1 | **15** | 25 | 0.02 | <0.01 | | | |
| Xylol | 0.05 | 0.5 | 5 | **15** | 25 | 0.04 | <0.01 | | | |
| Other alkylbenzenes | 0.05 | 0.5 | 5 | **30** | 50 | 2.52 | 3,260 | | | |
| 1,2-dichloroethylene | 0.001 | | | | | 0.01 | <0.01 | | | |
| Trichloroethylene | 0.001 | | | | | 0.01 | <0.01 | | | |
| Tetrachloroethylene | 0.001 | | | | | 0.01 | <0.01 | | | |

Only those samples that were collected from the surface soil of the Ort 44 point could have been tested for BTEX, and only the "other alkyl-benzenes" were detectable with a result of 2.52 mg/kg. The BTEX contamination levels of the 3 and 4 m deep soil layers were detectable at the following sample points: ort17; ort18; ort39; or9; or12; or13; or26.

The level of soil TPH contamination at 0.5–0–7 m depth:

| Depth | Average | Maximum | Standard deviation (SD) | N | Upper 95% confidence limit |
|-------|---------|---------|-------------------------|---|----------------------------|
| 0.5 m | 1,426 | 7,270 | 2557.85 | 12 | 2,873 |
| 1.0 m | 1,572 | 20,500 | 4598.66 | 27 | 3306.2 |
| 2.0 m | 2,961 | 14,800 | 4559.91 | 25 | 4748.1 |
| 3.0 m | 3,210 | 16,700 | 4325.53 | 31 | 4,733 |
| 4.0 m | 2,005 | 22,900 | 4480.96 | 29 | 3635.7 |
| 5.0 m | 871 | 6,840 | 1843.53 | 15 | 1804.4 |
| 6.0 m | 1,134 | 4,500 | 2244.21 | 4 | 3,333 |
| 7.0 m | 26 | 26 | | 1 | |

TABLE 2c. The concentration values of the samples tested for PAH collected from the surface (0.5 meter) soil of the ORT-44/0.5 m; OR 21/0.5 m; ORT 23/0.5 m; ORT 25/0.5 m; ORT 12/0.5 m sampling points and the concentration limit values of the current Hungarian legal regulations (mg/kg)

| | A value | Average | Maximum | Standard deviation (SD) | N | Upper 95% confidence limit |
|---|---------|---------|---------|-------------------------|---|----------------------------|
| Naphthalene | | 0.0222 | 0.049 | 0.018833 | 5 | 0.038708 |
| 2-methylnaphthalene | | 0.0816 | 0.222 | 0.096308 | 5 | 0.166016 |
| 1-methylnaphthalene | | 0.0788 | 0.262 | 0.108792 | 5 | 0.174159 |
| Acenaphthylene | 0.03 | 0.0398 | 0.103 | 0.042646 | 5 | 0.07718 |
| Acenaphthene | 0.03 | 0.0302 | 0.097 | 0.037453 | 5 | 0.063028 |
| Fluorene | 0.04 | 0.0464 | 0.15 | 0.058453 | 5 | 0.097636 |
| Phenanthrene | 0.04 | 0.7846 | 2.18 | 0.815867 | 5 | 1.499726 |
| Anthracene | 0.04 | 0.4532 | 1.71 | 0.708556 | 5 | 1.074265 |
| Fluoranthene | 0.05 | 2.5408 | 6.96 | 2.683025 | 5 | 4.89253 |
| Pyrene | 0.02 | 3.805 | 13.9 | 5.718877 | 5 | 8.817721 |
| Benzo(a)anthracene | 0.02 | 1.5164 | 3.76 | 1.446885 | 5 | 2.784626 |
| Chrysene | 0.02 | 1.7868 | 4.83 | 1.866238 | 5 | 3.422599 |
| Benzo(b)fluoranthene+ | 0.02 | | | | | |
| Benzo(k)fluoranthene | 0.01 | 3.4654 | 11.4 | 4.594746 | 5 | 7.492796 |
| Benzo(a)pyrene | 0.02 | 2.2776 | 7.42 | 2.959582 | 5 | 4.871738 |
| Indeno(1,2,3-cd)pyrene | 0.025 | 1.4016 | 4.82 | 1.944057 | 5 | 3.105609 |
| Dibenzo(a,h)anthracene | 0.01 | 0.3272 | 1.04 | 0.411098 | 5 | 0.687537 |
| Benzo(g,h,i)perylene | 0.02 | 1.267 | 4.32 | 1.736424 | 5 | 2.789014 |
| Total naphthalenes | 0.03 | 0.1826 | 0.519 | 0.217652 | 5 | 0.373377 |
| Total PAH without naphthalenes | | 19.7422 | 62.7 | 24.84708 | 5 | 41.52121 |
| Total PAH | 0.5 | 19.932 | 63 | 24.87203 | 5 | 41.73288 |

According to the KöM-EüM-FvM-KHVM Joint Ministerial Regulation no.10/2000 (VI.2.):
A   is the background concentration
N   is the number of samples analysed

TABLE 2d. The average PAH concentrations of the samples collected from OR-26/4.0 m; ORT 17/4.0 m; ORT 18/4.0 m; ORT 39/4.0 m; ORT 23/4.0 m; ORT5/4.0 m; OR 9/3.0 m; OR-12/3.0 m; OR 13/3.0 m points; and their upper 95% confidence limit in the 3–4 m deep soil layer – without the outlier value of the OR3 sample – compared with the current Hungarian legal regulations (mg/kg)

| | A value | Average | Max | Standard deviation (SD) | N | Upper 95% confidence limit |
|---|---|---|---|---|---|---|
| Naphthalene | | 0.615125 | 2.72 | 0.952677 | 8 | 1.275284 |
| 2-methylnaphthalene | | 3.116375 | 11.4 | 4.64122 | 8 | 6.332515 |
| 1-methylnaphthalene | | 3.59375 | 12.2 | 5.101191 | 8 | 7.128628 |
| Acenaphthylene | 0.03 | 0.04225 | 0.153 | 0.056835 | 8 | 0.081634 |
| Acenaphthene | 0.03 | 0.08075 | 0.267 | 0.107262 | 8 | 0.155077 |
| Fluorene | 0.04 | 0.32375 | 1.11 | 0.426106 | 8 | 0.619021 |
| Phenanthrene | 0.04 | 0.724875 | 2.79 | 1.006615 | 8 | 1.42241 |
| Anthracene | 0.04 | 0.02425 | 0.072 | 0.023057 | 8 | 0.040228 |
| Fluoranthene | 0.05 | 0.017875 | 0.027 | 0.00799 | 8 | 0.023412 |
| Pyrene | 0.02 | 0.033 | 0.067 | 0.018632 | 8 | 0.045911 |
| Benzo(a)anthracene | 0.02 | 0.0105 | 0.038 | 0.011588 | 8 | 0.01853 |
| Chrysene | 0.02 | 0.063625 | 0.216 | 0.076189 | 8 | 0.116421 |
| benzo(b)fluoranthene+ | 0.02 | | | | | |
| benzo(k)fluoranthene | 0.01 | 0.01325 | 0.038 | 0.012326 | 8 | 0.021791 |
| benzo(a)pyrene | 0.02 | 0.006625 | 0.022 | 0.006886 | 8 | 0.011396 |
| Indeno(1,2,3-cd)pyrene | 0.025 | 0.002875 | 0.009 | 0.00285 | 8 | 0.00485 |
| Dibenzo(a,h)-anthracene | 0.01 | 0.001667 | 0.002 | 0.000577 | 3 | 0.00232 |
| benzo(g,h,i)perylene | 0.02 | 0.003125 | 0.01 | 0.003182 | 8 | 0.00533 |
| Total naphthalenes | 0.03 | 7.327625 | 24.9 | 10.57983 | 8 | 14.65894 |
| Total PAH without naphthalenes | | 1.347625 | 4.75 | 1.735044 | 8 | 2.549926 |
| Total PAH | 0.5 | 8,682 | 29.7 | 12.30093 | 8 | 17.20595 |

| | A | B | C1 | C2 | C3 |
|---|---|---|---|---|---|
| Total PAH | 0.5 | 1** | 5 | 25 | 40 |

According to the KöM-EüM-FvM-KHVM Joint Ministerial Regulation no.10/2000 (VI.2.):

| | |
|---|---|
| A | is the background concentration |
| B | is the contamination limit value |
| C1, C2, C3 | are the intervention concentrations |
| N | is the number of samples analysed |

TABLE 2e. Exceedingly high PAH contamination in the samples collected from the soil of the OR3 sampling point (mg/kg)

|  | The maximum contaminant value of the OR3 sampling point at 3–4 m depth |
|---|---|
| Naphthalene | 15.20 |
| 2-methylnaphthalene | 41.40 |
| 1-methylnaphthalene | 36.00 |
| Naphthalenes | 0.16 |
| Acenaphthylene | 0.48 |
| Acenaphthene | 2.49 |
| Fluorene | 27.80 |
| Phenanthrene | 1.47 |
| Anthracene | 1.49 |
| Fluoranthene | 2.01 |
| Pyrene | 0.09 |
| Benzo[a]anthracene | 0.36 |
| Benzo[b]fluoranthene | 0.10 |
| Benzo[k]fluoranthene | 0.05 |
| Benzo[a]pyrene | 0.02 |
| Indeno[1,2,3-cd]pyrene | 0.01 |
| Dibenzo[a,h]anthracene | 0.02 |
| Benzo[g,h,i]perylene | 92.60 |
| Total naphthalenes | 36.50 |
| Total PAH without naphtalenes | 129 |

## 2.2.  EVALUATION OF THE DATA

When determining the risk we had to aggregate the concentration data of the contaminants. The upper 95% confidence limits (UCL) of the average concentrations were calculated from the measured values of the contaminants, as shown below:

$$UCL = x + t \, (s/\sqrt{n})$$

where

UCL    is the upper 95% confidence limit of the average
x        is the average
s        is the standard deviation
t        is the t upper 95% confidence limit value of the Student probe
n        is the number of samples

In the risk assessment we applied both average concentrations and the upper 95% confidence limit values of the averages.

## 2.3.  THE DATA DETERMINING THE LEVEL OF THE EXPOSURE

With the help of exposure models we are able to determine the human intake of the contaminants from the different contaminated environmental media directly (ingestion) or indirectly (vapour inhalation of soil contaminants) depending on the land use, exposure time, receptors and human activity. The local conditions were taken into consideration when selecting the relevant exposure ways. The levels of the groundwater and soil contamination are considerable according to the results of the analytical study performed on the site.

### 2.3.1.  *Exposure Scenarios*

Exposure situations were selected to reflect the local conditions of the site, namely that we considered those exposure pathways which originated from the proposed landuse (buildings, park) and the exposure of the local workers and remediation workers as well as residents in the direct vicinity (0 m) and 200 m away from the source (Table 3).

According to the traditonal (conservative) scenario ground water use (as drinkig water consumption) was taken into consideration for the assessment of potential health risk.

A risk assessment scenario which excluded the consumption of groundwater as drinking water had also been carried out for assessment of actually existing health risk.

In the exposure modelling the RBCA software was used to accomplish the health risk assessment according to the ASTM PS-104 standard.

TABLE 3. The selected parameters of the exposure scenarios

| Landuse | Public area | | Residental area | |
|---|---|---|---|---|
| Receptor | Local workers | Remediation workers | Residents | Residents |
| Distance between the source of contaminant and the receptor | 0 m | 0 m | 0 m | 200 m |
| Duration of the exposure | 25 years 5 years * | 1 year | 30 years | 30 years |
| Considered exposure pathways | | | | |
| Outdoor air | 10 m³/day | 10 m³/day | 20 m³/day | 20 m³/day |
| Indoor air | 10 m³/day | 10 m³/day | 20 m³/day | 20 m³/day |
| Soil | 100 mg per day | 200 mg per day | 200 mg per day | none |
| Groundwater Consumption | 1 liter/day none * | 1 liter/day none * | 2 liter/day none * | 2 liter/day none * |

* According to the scenario which excludes the consumption of groundwater as drinking water.

The quantitative expression of the exposure and the risk, the calculation of the hazard quotient and the carcinogenic risk as well as the acceptable level of risk are described in Annex.

## 2.4.  SITE-SPECIFIC DATA

The Sponsor provided us with the data determining the soil and groundwater qualities (hydrogeological, geological) of the contaminated site.

## 3.  Results

We summarized the calculated values of the total health risk (see Table 4) resulting from the exposure of the aliphatic TPH, BTEX, halogenated aliphatic hydrocarbons (tetrachloroethylene, dichloroethylene, cis-1, 2-dichloroethylene) and PAH contaminants in case of each environmental media (air, soil, ground-water) and receptors (local workers, remediation workers, residents) with regard to the distance between the source of contamination and the receptors (0 m and 200 m).

TABLE 4. The health risk posed by total aliphatic TPH, BTEX, halogenated ethylenes and PAH exposure in the place and 200 meters away from the site. The duration of the exposure is 30 years. Risk values related to average contamination of the groundwater and soil surface (0.5 m)

|  | Likelihood of the CARCINOGENIC RISK (excess risk) | | | | The value of the HAZARD QUOTIENT (in excess of the toxicological threshold [RfD] value indicating health damage) | | | |
|---|---|---|---|---|---|---|---|---|
|  | Public area | | Residental area | | Public area | | Residental area | |
| Source-receptor distance | 0 m | 0 m | 0 m | 200 m | 0 m | 0 m | 0 m | 200 m |
| Receptors | Local workers | Remedi-ation workers | Residents | Residents | Local workers | Remedi-ation workers | Residents | Residents |
| Outdoor air | 1.29E-7 | 2.06E-8 | 1.97E-7 | 8.86E-8 | 0.081 | 0.088 | 0.103 | 0.048 |
| Indoor air | 1.15E-6 |  |  |  | 0.178 |  |  |  |
| Soil | 3.70E-5 | 1.32E-6 |  |  | 0.005 | 0.004 |  |  |
| Ground-water | 9.03E-5 |  | 3.03E-4 | 3.73E-6 | 1.984 |  | 5.556 | 0.068 |
| Total | 1.29E-4 | 1.34E-6 | 3.04E-4 | 3.82E-6 | 2.248 | 0.092 | 5.659 | 0.116 |
| Tolerable | 1.0 E-05 | | 1.0 E-6 | | 1.0 | | | |
| Classifica-tion | large | small | large | large | large | small | large | mode-rate |

The likelihood of the carcinogenic effects and the hazard quotient indicating systemic toxicity are shown in each case. Attention was paid to the expression of the health risk resulting from the different soil layers.

## 3.1. THE RISK OF THE CONTAMINATED SITE

### 3.1.1. *The Results of Health Risk Assuming Groundwater Consumption/Intake According to the Traditional (Conservative) Exposure Scenario*

The average values of the groundwater concentrations and the soil surface contamination indicate that the contaminants reaching the outdoor air through evaporation and dust formation pose only a negligible risk.

On the other hand, indoor air of the building standing on the site can be contaminated with the volatile substances that diffuse through the cracks of the building's foundation. The people staying eight hours a day in the building might be exposed to moderate risk (using default building parameters of the RBCA software). While the risk deriving from soil contamination is considered small, the health risk value of groundwater consumption is large.

TABLE 5. The health risk posed by the aliphatic TPH, BTEX, chlorinated ethylenes and PAH exposure on site and 200 m away. The duration of the exposure is 30 years. The carcinogenic risk and the hazard quotient, indicated as a function of the upper 95% confidence limit of groundwater and the maximum level of contamination in the soil surface (0.5 m)

| | Likelihood of the CARCINOGENIC RISK (excess risk) | | | | The value of the HAZARD QUOTIENT (in excess of the toxicological threshold [RfD] value indicating health damage) | | | |
|---|---|---|---|---|---|---|---|---|
| | Public area | | Residental area | | Public area | | Residental area | |
| Source-receptor distance | 0 m | 0 m | 0 m | 200 m | 0 m | 0 m | 0 m | 200 m |
| Receptors | Local workers | Reme-diation workers | Residents | Residents | Local workers | Reme-diation workers | Residents | Residents |
| Outdoor air | 2.07E-8 | 4.42E-9 | 3.16E-8 | 8.79E-9 | 2.143 | 38.522 | 2.501 | 1.163 |
| Indoor air | 1.95E-6 | | | | 52.578 | | | |
| Soil | 1.11E-4 | 3.96E-6 | | | 1.210 | 0.898 | | |
| Ground-water | 1.47E-4 | | 4.94E-4 | 6.08E-6 | 1847.83 | | 5173.92 | 63.598 |
| Total | 2.60E-4 | 3.96E-6 | 4.94E-4 | 6.08E-6 | 1903.76 | 39.420 | 5176.42 | 64.761 |
| Tolerable | 1.0 E-05 | | 1.0 E-6 | | 1.0 | | | |
| Classification | large | small | large | large | extremely large | very large | extremely large | very large |

The carcinogenic risk resulting from the maximum concentrations of the soil surface contamination and the upper 95% confidence limit of average values of contaminants in groundwater are considered large.

The hazard quotient indicating the potential health damage is extremely large. This exceeding value might be the result of the high contamination in the groundwater. The indoor air sould also be regarded as a potential source of risk.

TABLE 6. The health risk posed by the total aliphatic TPH, BTEX, chlorinated ethylenes and PAH exposure on site and 200 m away. The duration of the exposure is 30 years. The carcinogenic risk and the hazard quotient, indicated as a function of the average groundwater concentration and the average soil contamination at 3–4 m depth

| | Likelihood of the CARCINOGENIC RISK (excess risk) | | | | The value of the HAZARD QUOTIENT (in excess of the toxicological threshold [RfD] value indicating health damage) | | | |
| --- | --- | --- | --- | --- | --- | --- | --- | --- |
| | Public area | | Residental area | | Public area | | Residental area | |
| Source-receptor distance | 0 m | 0 m | 0 m | 200 m | 0 m | 0 m | 0 m | 200 m |
| Receptors | Local workers | Reme-diation workers | Residents | Residents | Local workers | Reme-diation workers | Residents | Residents |
| Outdoor air | 4.84E-8 | 1.59E-8 | 7.02E-8 | 2.95E-8 | 0.103 | 0.472 | 0.129 | 0.060 |
| Indoor air | 1.77E-6 | | | | 0.738 | | | |
| Soil | 1.89E-6 | 6.69E-8 | | | 0.021 | 0.017 | | |
| Ground-water | 8.43E-5 | | 2.83E-4 | 3.48E-6 | 20.479 | | 57.341 | 0.705 |
| Total | 8.80E-5 | 8.28E-8 | 2.83E-4 | 3.51E-6 | 21.341 | 0.488 | 57.470 | 0.765 |
| Tolerable | 1.0 E-05 | | 1.0 E-6 | | 1.0 | | | |
| Classification | large | Negligible | large | large | very large | moderate | very large | moderate |

The concentration values of the average groundwater contamination and average soil contamination at 3–4 m depth exceed the acceptable level of risk, although they are two orders of magnitude less than the calculated risk resulting from the maximum level of contamination.

### 3.1.2. *The Results of the Health Risk According to the Scenario which Excludes the Consumption of Groundwater as Drinking Water*

In the modified version of the exposure scenario direct contact with the groundwater has been excluded. It seems reasonable not to use the exposure deriving from the groundwater consumption in the calculations, because the groundwater is not utilized either as drinking water or for other purposes on a contaminated site. At the same time, groundwater should be regarded as a kind of receptor with regard to the underground water base protection efforts, especially in case of the inhabited areas.

TABLE 7. The total values of the health risk deriving from the aliphatic TPH, BTEX and chlorinated ethylenes exposure on site and 200 m away. The duration of the exposure is 30 years. Summary of the risk values according to the exposure scenario excluding groundwater consumption

| | Likelihood of the CARCINOGENIC RISK (excess risk) | | | | The value of the HAZARD QUOTIENT (in excess of the toxicological threshold [RfD] value indicating health damage) | | | |
|---|---|---|---|---|---|---|---|---|
| | Public area | | Residental area | | Public area | | Residental area | |
| Source-receptor distance | 0 m | 0 m | 0 m | 200 m | 0 m | 0 m | 0 m | 200 m |
| Receptors | Local workers | Remediation workers | Residents | Residents | Local workers | Remediation workers | Residents | Residents |
| Groundwater average, Soil surface average | 3.8E-5 | 1.34E-6 | 1.9E-7 | 8.86E-8 | 0.263 | 0.092 | 0.103 | 0.048 |
| Groundwater UCL, Soil surface maximum | 1.11E-4 | 3.91E-6 | 3.16E-8 | 8.79E-9 | 55.932 | 39.420 | 2.501 | 1.163 |
| Groundwater average Soil (3–4 m) | 3.7E-6 | 8.28E-8 | 7.0E-8 | 2.95E-8 | 0.862 | 0.488 | 0.129 | 0.060 |

Table 7 summarizes the health risk values of the exposure scenario excluding groundwater consumption. In other words these values point at the consequencies of exposure to inhaled contaminants evaporated from groundwater or soil and to the direct soil contact. It can be concluded, that the carcinogenic risk for local workers, deriving from the average contamination of the surface soil and the groundwater, slightly exceeds the tolerable value ($10^{-5}$), and is acceptable considering other types of receptors. The hazard quotient values indicate small-moderate risk.

The 95% UCL concentrations in groundwater and the maximum values of soil contamination result a significantly high level of risk through direct contact with soil and diffusion to the building. Its level can be determined with indoor air pollution measures, while models indicate the probability of volatilization.

Using average concentrations, the risk posed by indoor air is moderate, its confidence strongly depends on the building parameters (ventilation, thickness of the foundation, surface/volume ratio).

TABLE 8. The summed values of the health risk resulting from the soil exposure of the aliphatic TPH, BTEX, chlorinated ethylenes and PAH contaminants

|  | Likelihood of the CARCINOGENIC RISK (excess risk) | | | | The value of the HAZARD QUOTIENT (in excess of the toxicological threshold [RfD] value indicating health damage) | | | |
|---|---|---|---|---|---|---|---|---|
|  | Public area | | Residental area | | Public area | | Residental area | |
| Source-receptor distance | 0 m | 0 m | 0 m | 200 m | 0 m | 0 m | 0 m | 200 m |
| Receptors | Local workers | Remedi-ation workers | Residents | Residents | Local workers | Remedi-ation workers | Residents | Residents |
| Soil surface average | 3.7E-5 | 1.32E-6 | – | – | 0.005 | 0.004 | – | – |
| Soil surface maximum | 1.11E-4 | 3.96E-6 | – | – | 1.210 | 0.898 | – | – |
| Soil (3–4m) average | 1.89E-6 | 6.69E-8 | – | – | 0.021 | 0.017 | – | – |

Table 8 compares the different levels of risk deriving from contamination measured in the different soil layers as well as risks calculated from the contamination indicated by the average and the 95% UCL values.

We can conclude from the results of the hazard quotient calculations that the risk posed by the average contamination of the soil surface-assuming 100 mg soil or dust ingestion per day for local workers and 200 mg for reclamation workers – is negligible, but the carcinogenic risk deriving from the PAH contamination exceeds the tolerable level ($3.7 . 10^{-5}$).

The health risk deriving from the maximum soil surface contamination for the exposed people on the site – due to the PAH contaminants – exceeds the tolerable level ($1.11 . 10^{-4}$) in one order of magnitude. Health risks calculated for remediation wokers are whitin the accepteale limits. The hazard quotient also indicates a risk to the human health when using the concentration values of the maximum contamination level in the calculations.

The direct risk to the health resulting from the soil – measured at 3–4 m depth – is considered moderate.

Direct contact with the contaminated soil as an exposure route was not considered relevant for the local population.

## 4.  Determination of the Remediation Target Concentration

The table seen below shows the contamination levels of the soil and groundwater compared with the appropiate C2 intervention values of the site in question [according to the specifications in the KöM-EüM-FvM-KHVM Joint Ministerial Decree no.10/2000 (VI.2.) on "The limit values for the protection of the groundwater and soil quality" issued for the execution of the Government Regulation no. 33/2000. (III.17.) on "The issues related to the activities affecting the quality of the groundwater bodies".

It can be concluded from Tables 9, 10 and 11 that the levels of TPH, BTEX and chlorinated ethylenes and PAH in the soil and groundwater exceed the C2 intervention values; the presence of the inorganic (metal) components is negligible.

Using the reverse risk assessment method we recalculated the acceptable concentrations of the contaminants from the acceptable health risk values (hazard quotient <1, carcinogenic excess risk <1 . $10^{-5}$ regarding to the local workers and <1 . $10^{-6}$ to the residents).

The health risk values presented in Tables 4–8. exceeded the tolerable level, therefore the target concentrations of contamination that do not result excess risk had to be recalculated from the acceptable health risk values. We performed the reverse risk assessment according to both the traditional scenario and the modified version that excludes groundwater consumption.

TABLE 9. Data indicating whether the intervention values according to the KöM-EüM-FvM-KHVM Joint Ministerial Decree no. 10/2000 (VI.2.) on soil and groundwaters are exceeded based on the highest measured concentrations

| Contaminant | Soil | Groundwater |
|---|---|---|
| Inorganic contaminants, metals | - | - |
| Total-aliphatic TPH | + | + |
| Benzene | - | + |
| Toluene | - | - |
| Ethylbenzene | - | - |
| Xylenes | - | - |
| Other alkylbenzenes | + | + |
| Cis-1,2-dichloroethylene | - | + |
| Trichloroethylene | - | - |
| Tetrachlorothylene | - | + |
| Total PAH without naphthalenes | | + |
| Total PAH | + | |

+   The intervention value (C2) is exceeded (C2)
-   The intervention value (C2) is not exceeded

TABLE 10. The groundwater contamination compared to the intervention values (µg/l)

|                                           | Average (ug/l) | Maximum (ug/l) | C2 (ug/l) |
|-------------------------------------------|----------------|----------------|-----------|
| Total aliphatic TPH                       | 2,618          | 22,100         | 1,000     |
| Benzene                                   | 2.40           | 23.80          | 10        |
| Toluene                                   | 1.49           | 26.10          | 50        |
| Ethylbenzene                              | 1.38           | 18.70          | 50        |
| Xylene isomers                            | 2.86           | 26.30          | 50        |
| Cumene, as other aliphatic benzene derivative | 85.60      | 683.00         | 50        |
| 1,2-dichloroethylene                      | 50.69          | 273.00         | 100       |
| Trichloroethylene                         | 7.81           | 31.00          | 150       |
| Tetrachloroethylene                       | 399.39         | 4820.00        | 150       |
| Total naphthalenes                        | 19.55          | 166.00         | 20        |
| Total PAH without naphthalenes            | 4.27           | 52.30          | 8         |

TABLE 11. Presents the soil contamination compared to the intevention values (mg/kg)

|                                           | Average (mg/kg) | Maximum (mg/kg) | C2 (mg/kg) |
|-------------------------------------------|-----------------|-----------------|------------|
| Total aliphatic TPH                       | 1,426           | 7,270           | 3,000      |
| Benzene                                   | 0.01            | <0.01           | 3          |
| Toluene                                   | 0.01            | <0.01           | 15         |
| Ethylbenzene                              | 0.02            | <0.01           | 15         |
| Xylene isomers                            | 0.04            | <0.01           | 15         |
| Cumene, as other aliphatic benzene derivative | 2.52        | 3,260           | 30         |
| 1,2-dichloroethylene                      | 0.01            | <0.01           |            |
| Trichloroethylene                         | 0.01            | <0.01           |            |
| Tetrachloroethylene                       | 0.01            | <0.01           |            |
| Naphthalene                               | 0.1826          | 0.519           |            |
| Total PAH                                 | 19.932          | 63              | 25         |

Table 12 presents the remediation D limit values, which have been recalculated according to the traditional exposure scenario (presumes groundwater consumption, applied to the use of the public area), are aiming at the prevention of the potential health risk that is likely to occur without appropiate intervention.

The soil and groundwater concentrations presented in Tables 12 and 13 should be regarded as the remediation target concentrations (*target values*), which ensure that any exposure after the remediation would not damage the health, and the health risk does not exceed the acceptable level. The D values which refer to the potential health risk (presuming groundwater consumption) were calculated according to the traditional exposure scenario.

TABLE 12. The determined remediation target concentrations of the contaminated groundwater applying the reverse risk assessment method

| Contaminant | D-value for groundwater (ug/L) |
|---|---|
| Total aliphatic TPH | 1,500 |
| Benzene | 15 |
| Toluene | 26.1* |
| Ethylbenzene | 18.7* |
| Xylene somers | 26.3* |
| Cumene, as other aliphatic benzene derivative | 85 |
| 1,2-dichloroethylene | 50 |
| Trichloroethylene | 31* |
| Tetrachloroethylene | 40 |
| Total naphthalenes | 19.5 |
| Total PAH without naphthalenes | 5 |

* The present maximum concentration of the contamination is considered equivalent to the D value, because the health risk calculated from the concentration does not exceed the acceptable level. An increase in the level of contamination cannot be accepted.

TABLE 13. The remediation target concentrations determined by reverse risk assessment method for the contaminated soil layer in depth 2–4 meter (except for the outlier value of the Or 3 saple point)

| Contaminant | D-value for soil (mg/kg) |
|---|---|
| Total aliphatic TPH | 3,000* |
| Benzene | n.a. |
| Toluene | n.a. |
| Ethylbenzene | n.a. |
| Xylol isomers | n.a. |
| Cumene, as other aliphatic benzene derivative | n.a. |
| 1,2-dichloroethylene | n.a. |
| Trichloroethylene | n.a. |
| Tetrachloroethylene | n.a. |
| Naphthalene | 14.6** |
| Total PAH | 17.2** |

* The present average concentration of the contamination is considered equivalent to the D value, because the health risk calculated from the concentration does not exceed the acceptable level.
** The present 95% UCL concentration of the contamination is considered equivalent to the D value, because the health risk calculated from the concentration does not exceed the acceptable level.
An increase in the level of contamination cannot be accepted.
n.a. not available due to lack of data.

The remediation D values, aiming at the prevention of the health risk that is actually exists, have been recalculated according to the adapted site specific exposure scenario (excluding groundwater consumption, taking into consideration soil surface) are presented in Tables 14 and 15.

TABLE 14. Remediation D limit values, which have been calculated by the reverse risk assessment from maximum groundwater contamination according to the modified exposure scenario (excluding groundwater consumption) in order to prevent the possible health risk

| Contaminant | D value for groundwater (ug/L) |
|---|---|
| Total aliphatic TPH | 2,500 |
| Benzene | 23.8* |
| Toluene | 26.1* |
| Ethylbenzene | 18.7* |
| Xylenes (mixture of isomers) | 26.3* |
| Cumene, as an aliphatic benzene derivative) | 100 |
| 1,2-dichloroethylene | 50 |
| Trichloroethylene | 31* |
| Tetrachloroethylene | 200 |
| Total naphtalenes | 34.5** |
| Total PAH without naphtalenes | 7.8** |

* The current maximum concentration of the contaminant can be considered a D value at the same time, because the health risk, which is calculated from the concentration value itself, does not exceed the acceptable level.
** The current 95% upper confidence concentrations (UCL) of the average can be considered a D value at the same time, because the health risk, which is calculated from the concentration value itself, does not exceed the acceptable level.
An increase in the level of contamination is inadmissible.

TABLE 15. The remediation D limit values which have been calculated by the reverse risk assessment method and relate to the soil surface

| Contaminant | D value for soil (mg/kg) |
|---|---|
| Total aliphatic TPH | 1,400* |
| Benzene | n.sz. |
| Toluene | n.sz. |
| Ethylbenzene | n.sz. |
| Xylenes (mixture of isomers) | n.sz. |
| Cumene, as an aliphatic benzene derivative) | n.sz. |
| 1,2-dichloroethylene | n.sz. |
| Trichloroethylene | n.sz. |
| Tetrachloroethylene | n.sz. |
| Naphtalene | 0,2** |
| Total PAH | 20** |

* The current average concentration of the contaminant can be considered a D value at the same time, because the health risk, which is calculated from the concentration value itself, does not exceed the acceptable level.
** The current 95% UCL concentration of the contaminant can be considered a D value at the same time, because the health risk, which is calculated from the concentration value itself, does not exceed the acceptable level.
An increase in the level of contamination is inadmissible.
n.a. not calculable due to lack of data.

The concentrations given by the remediation D limit values (target values) presented in Tables 14 and 15 ensure that the possible exposure would not lead to health damage and the risk does not exceed the acceptable level. The D values, which have been calculated by the exposure model excluding ground-water consumption, are based on toxicological admissibility and refer to an actually existing health risk that is likely to occur.

## 5.   Comparsion with the Organoleptic Threshold Values

It might be useful to make comparisons between the concentrations of the oil-derivatives and the organoleptic (odour, taste) threshold values. The TPH concentrations which elicit an odour or have a disagreeable taste are smaller than the toxicologically tolerable daily doses. An unpleasant odour or taste might induce a feeling of discomfort. Ethylbenzene, toluene and xylol isomers elicit an odour effect above 20 µg/l concentration. The aliphatic TPH and alkylbenzene concentrations in the groundwater of the study area exceed the organoleptic threshold values. However, this comparison does not refer to the likelihood of an adverse health impact, it rather indicates the presence of the unpleasant organoleptic effects.

It should also be mentioned, that part C of the appendix 1. (water quality parameters and threshold values) of the Government Decree no. 201/2001. (X.25) on "The requirements and monitoring of the drinking water quality" established a threshold value of 50 µg/l for the oil-derivatives.

## Conclusions

Contamination of *groundwater* and soil with aliphatic TPH, alkylbenzene, and PAH on the former bus dapot exceed the C2 intervention values. *The detailed quantitative risk assessment results express concerns about the exposure deriving from the groundwater and soil contamination.*

The health risk calculated according to the traditional-conservative (worst case) exposure scenario (assuming groundwater consumption) from the average contamination of the soil surface and the groundwater exceeds the acceptable level for the local workers. That is only a slight risk for the remediation workers, staying only for a short interval of time (1 year) on the site. The level of risk for the exposed population living in the vicinity of the area exceeds the acceptable level. The contaminated groundwater pose risk not only on-site but off-site if it consumed or used for other purposes, i.e. irrigation.

The hazard quotient, calculated from the maximum concentrations of the surface soil contamination and the 95% upper confidence concentrations (UCL) of the average groundwater contamination, indicates an extremely high value, which can be attributed to the high level of contamination in the groundwater. The indoor air should be also regarded as a posssible source of risk. It also can be stated that human health risk from contaminated soil can take place in case of opening up the deep soil layers.

According to the scenario which excludes groundwater consumption the average contamination level based health risk values for on-site workers slightly exceed the tolerable level ($10^{-5}$) and do not indicate risk to the other receptors. The risk values express a small to moderate level of risk.

The *remediation target concentration* of the former bus depot was determined applying the reverse risk assessment method. These remediation target values ensure that the exposure deriving from the residue contamination after the clean-up would not lead to adverse health effects, and the health risk would not exceed the acceptable level.

The remediation D limit values, recalculated on the basis of the traditional-conservative scenario (including groundwater consumption) and aiming at the prevention of the potential health risk are as follows:

TABLE 16. The remediation D values are as follows, when presuming the multifunctional use of the soil and the groundwater

| Contaminants | D-value for groundwater (ug/L) | D-value for soil (mg/kg) |
|---|---|---|
| Total aliphatic TPH | 1,500 | 3,000* |
| Benzene | 15 | n.sz. |
| Toluene | 26.1** | n.sz. |
| Ethylbenzene | 18.7** | n.sz. |
| Xylene isomers | 26.3** | n.sz. |
| Cumene, as other aliphatic benzene derivative | 85 | n.sz. |
| 1,2-dichloroethylene | 50 | n.sz. |
| Trichloroethylene | 31** | n.sz. |
| Tetrachloroethylene | 40 | n.sz. |
| Total naphthalenes | 19.5 | 14.6** |
| Total PAH without naphthalenes | 5 | 17.2** |

* The present average concentration of the contamination (n = 5) is considered equivalent to the D value, because the health risk calculated from the concentration does not exceed the acceptable level.
** The present maximum concentration of the contamination (95% UCL PAH concentration in soil) is considered equivalent to the D value, because the health risk calculated from the concentration does not exceed the acceptable level. An increase in the level of contamination *cannot* be accepted.

The D values, recalculated from the site specific exposure scenario (excluding groundwater consumption) and aiming at the prevention of the actual health risk, are as follows:

TABLE 17. The remediation D limit values of the soil and the groundwater

| Contaminants | D-value for groundwater (ug/L) | D-value for soil (mg/kg) |
|---|---|---|
| Total aliphatic TPH | 2,500 | 1,400*** |
| Benzene | 23.8* | n.sz. |
| Toluene | 26.1* | n.sz. |
| Ethylbenzene | 18.7* | n.sz. |
| Xylene isomers | 26.3* | n.sz. |
| Cumene, as other aliphatic benzene derivative | 100 | n.sz. |
| 1,2-dichloroethylene | 50 | n.sz. |
| Trichloroethylene | 31* | n.sz. |
| Tetrachloroethylene | 200 | n.sz. |
| Total naphthalenes | 34.5** | 0.2** |
| Total PAH without naphthalenes | 7.8** | 20** |

* The current maximum concentration of the contaminant can be considered a D value at the same time, because the health risk, which is calculated from the concentration value itself, does not exceed the acceptable level. An increase in the level of contamination is inadmissible.
** The current 95% UCL of average contaminant concentration can be considered a D value at the same time, because the health risk, which is calculated from the concentration value itself, does not exceed the acceptable level. An increase in the level of contamination is inadmissible.
*** The current average concentration of the contaminant can be considered a D value at the same time, because the health risk, which is calculated from the concentration value itself, does not exceed the acceptable level. An increase in the level of contamination is inadmissible.
n.c. not calculable due to lack of data.

The difference between the potential and actually existing health risk values and the obtained remediation D limit values presented in Tables 16 and 17 can be attributed to the multifunctional use of the soil and the groundwater and the site specific (public area, park) landuse.

Considering the local conditions, the application of the reasonable (without groundwater consumption) remediation D value is recommended.

Appropriate measures (monitoring, clean-up) have to be taken on the site for total aliphatic TPH, other alkilbenzenes, cis-1, 2-dichloroethylene and tetrachloroethylene in groundwater.

The following facts should be considered in the interpretation of the obtained results from public health perspectives:

1. The evaluation is based on the assumption that the concentrations of the contaminants in the model calculations reflect exactly that level of contamination that the population is exposed to on the site and in the vicinity of the area during the given exposure time.

2. The conclusions presented in the risk assessment report are based on the available that the Sponsor provided us with, and relate to the total aliphatic TPH, BTEX, chlorinated ethylenes and PAH contaminants.

3. The actual exposure of the population – as a function of the daily activity-varies with their participation in the exposure situations (time spent on site, groundwater consumption, contaminant transport to the off-site receptors).

4. The consumption/intake of the contaminated groundwater reflects a worst-case scenario. Although the risk is exaggerated from the perspective of the healthy adult population, but the protection and interest of the vulnerable groups (children, pregnant women, elderly people and those suffering from illnesses) make it necassary.

5. The conservative, worst-case approach also counterweights the variability deriving from the uncertainity of the model-evaluation.

6. Special attention was paid in our model for the distribution of the contaminating substances between the different environmental media. Therefore the comparison of the model evaluation results with analytical measurements is considered essential during the supervision works.

## References

Szennyezett területek részletes mennyiségi kockázatfelmérése. Kármentesítési kézikönyv. KöM, 2001

Detailed quantitative risk assessment of contaminated sites. Technical guide. Ministry of Environmental Protection, 2001

Technikai útmutató az újonnan bejelentett anyagok kockázatbecsléséről szóló 93/67/EGK számú Bizottsági irányelv és a meglévő anyagok kockázatbecsléséről szóló (EK) 1488/94 számú Bizottsági rendelet alkalmazásához. Környezetvédelmi Minisztérium, 2002

Technical guide to the Council Directive no. 93/67/EGK on "Risk assessment of recently registrated substances" and the Council Directive no. (EK) 1488/94 on "The existing substances". Ministry of Environmental Protection, 2002

Risk Assessment Guidance for Superfund. Vol.1. Human Health Evaluation Manual (Part A.) US EPA Washington, 1989

IRIS (US EPA's Integrated Risk Information System). 1999

IARC monographs on the evaluation of carcinogenic risks to humans. Overall evaluation of carcinogenicity: An updating of IARC Monographs Vol. 1–69 (a total of 836 agents, mixtures and exposures). Lyon, 1997

Linders J.: Update of the role of modeling in risk assessment. In: Environmental and human health risk assessment for agrochemicals. 10–11 July, 1996. London, IBC UK Conference.

Bálint P.: Orvosi élettan I. - II. Medicina könyvkiadó, Budapest, 1979

ASTM American Society for Testing and Materials: Standard guide for risk-based corrective action. ASTM PS-104, Philadelphia, PA, USA 1998

IUCLID: International Uniform Chemical Information Database, European Commission Joint Research Centre, European Chemicals Bureau, 2001

# Appendix

## AP.1. QUANTITATIVE EXPRESSION OF THE EXPOSURE AND THE RISK

The exposure can be characterized by the contact with the contaminant, its extent, duration and frequency.

The amount of the contaminant absorbed by the body of the exposed organism – through ingestion, inhalation or skin contact – in other words the *intake* determines the average daily dose *(ADD)*, which can be calculated using the following formula:

*ADD* (mg/kg/day) = Cc . AA . EF . EH / TT

where

Cc          is the concentration of the contaminant in the different environmental media (soil, groundwater, air)

AA          is the absorbed amount (through ingestion, inhalation, skin contact) expressed in mg/day

EF          is the exposure frequency

The unit of exposure is mg/kg body weight/day

*ADD* (mg/kg/day) = Cc . BM . EF . EH / TT

The calculated averade daily doses are then compared with the toxicologically acceptacle (reference) doses or concentrations. The reference doses (RfD) and concentrations (RfC), which do not cause health effects, can be found in toxicological databases supervised by international organizations.

## AP.2. THE CALCULATION OF THE HAZARD QUOTIENT AND THE CARCINOGENIC RISK

From the aspect of risk characterization it is essential to draw distinctions between the deterministic and stohastic impact of the chemical substances on human health.

In the former case the dose-response relation can be described with a threshold dose (concentration), below which a damage in the organs would not occur. The latter means a probability, namely the potential of a chemical for causing damages in the DNA, the genetic material of a cell, and is not determined by a threshold dose.

The hazard posed by the ingestion of the chemicals which could be determined by a threshold dose is defined as follows:

Hazard ratio = Average Daily Dose/ Reference Dose

When dealing with inhalation exposure the inhaled concentration/ reference concentration ratio is calculated.

*The health risk can be indicated with descriptive factors [6.] and is divisible into chategories based on the the average daily dose (ADD) of the body and the Reference dose/concentration (which does not damage the health) ratio. (The acceptable, tolerable and reference dose expressions are used as synonims).*
*The health risk can be expressed as follows:*

| Hazard ratio   | The level of risk |
|----------------|-------------------|
| Less than 0.01 | Negligible        |
| 0.01–0.1       | Small             |
| 0.1–1          | Moderate          |
| 1–10           | Large             |
| more than 10   | extreme large     |

*The potential of a chemical substance for inducing carcinogenic (genotoxic) risk can be determined by the slope factor of the dose-response relationship. It is generally accepted, that the carcinogenic risk is zero at no exposure, and the dose-response relationshop is linear in a low dose-domain. If a chemical substance is carcinogenic by ingestion, then the likelihood of tumor formation can be calculated by using the following formula:*

Carcinogenic Riskat = $1 - e^{-(\text{Oral slope factor} * \text{Average Daily Dose})}$

If a chemical is carcinogenic by inhalation, the measure of risk is:

Carcinogenic Risk = $1 - e^{-(\text{Unit Risk} * \text{Concentration})}$

The incidence of cancer exceeds the limit value, it indicates a theoretical excess risk. The oral slope factor and the unit riska can be looked up in the relevant databases.

AP.3. THE ACCEPTABLE LEVEL OF RISK

The health risk is to be considered acceptable, when the average daily dose of the inhaled, ingested or dermally absorbed contaminant does not exceed the acceptable dose, namely that the value of the hazard ratio is smaller than one.

If a contaminant poses a carcinogenic risk, then the incidence of cancer, which is attributable to the exposure – above the limit value, the excess risk – must not exceed the socially accepted 1,0E - 6 ($1.0 . 10^{-6}$) level of risk. According to this condition, the incidence of cancer due to a certain exposure is likely to happen to one exposed individual out of a million (although cancer might develop because of a completely different type of exposure).

In working areas the $1.0 . 10^{-5}$ risk level value is considered acceptable, according to the ESZCSM Decree no. 12/2002. (XI.16.) based on the Government Decree no. 26/2000. (IX.30.) amendment on "Health impairment prevention".

*kármentesítési célállapot (D)-határérték*

AP.4. The MAPPING OF RISK

Mapping for hazard quotients and excess cancer probability for contaminants.

Iso-risk curves for Tetrachloroehylene, PAH and Cumene.

# EXTRACTION WITH EDTA TO ASSESS THE GLOBAL RISK PRESENTED BY HEAVY METALS (Cd, Ni, Pb) IN SOILS AND SEDIMENTS

MONIKA KEDZIOREK[*] AND ALAIN BOURG
*Environmental Hydro Geochemistry Laboratory (LHGE),
Department of Earth Sciences, University of Pau,
BP 1155, 64013 Pau Cedex, France*

**Abstract.** Only part of the heavy metal load of pristine and polluted soils and sediments poses a risk for Man and the Environment. This fraction is usually estimated using either single or sequential chemical extraction procedures. Here we compare, for six solids collected in oxidizing environments, the sequential chemical extraction protocol of Tessier et al. (1979) and single extractions with $HNO_3$ (at pH 2) or 0.01 M EDTA. The latter, a simple technique that is, moreover, independent of pH, provides results similar to the sum of all of the extractable metals in the sequential chemical scheme (total content minus the load associated with the residual phase) and is therefore proposed as a simple, easily implemented technique for evaluating the total potentially soluble fraction of metals in solids.

**Keywords:** heavy metals, sediment, soil, EDTA, solubility, geochemical availability

## 1. Introduction

Natural solids are major sinks for most heavy metals in aquatic cycles and, as a result, can contain large concentrations of toxic heavy metals. For environmental risk assessment, it is necessary to determine how permanent their trapping of heavy metals actually is, in other words—whether these metals are permanently fixed under solid forms or whether they could reenter the aquatic cycle as solutes. Heavy metals are associated with natural solids as

---

[*] To whom correspondence should be addressed: monika.kedziorek@univ-pau.fr

L. Simeonov and V. Sargsyan (eds.),
*Soil Chemical Pollution, Risk Assessment, Remediation and Security.*
© Springer Science+Business Media B.V. 2008

adsorbed (surface) species, precipitates or incorporated in mineral amorphous or crystalline structures.

The determination of only the total concentration of toxic heavy metals in solids is therefore not a satisfactory criterion to evaluate their risk of remobilization, and thus their potential dispersion and bioavailability.

Chemical extraction is the classical procedure for determining whether metals might possibly be solubilized from solids. A variety of procedures or "recipes" have been developed. These fall into two categories:

1. Multiple chemical extractions in which the solid is subjected to either a sequential series of extractions (e.g., Tessier et al., 1979; Davidson et al., 1994; Simeonova and Simeonov, 2006) or simultaneous extractions (e.g., Jeanroy et al., 1984; Heron et al., 1994). These techniques are expensive and their uncertainty, due to solid sampling and extract analysis, increases with the number of extraction steps. Sequential techniques have been highly criticized for the lack of selectivity of each extraction step (e.g., Rapin et al., 1985; Nirel and Morel, 1990; Whalley and Grant, 1994).
2. Extractions with a single chemical reactant (e.g., EDTA, DTPA, ...). These have been used in agronomy to evaluate the availability of nutrients in soils (e.g., Baghdady and Sippola, 1984; Li and Shuman, 1996) and have been extended to assess the bioavailability of contaminants (e.g., Ure et al., 1993; Gyori et al., 1996) and to evaluate the feasibility of chemical decontamination of polluted solids (e.g., Barona and Romero, 1996; Huang et al., 1997). They are easy to implement (one extraction step) but are usually only empirical and therefore not easily extrapolable to other situations.

EDTA as a single extracting agent has been the object of many studies. EDTA-extractable metal concentrations have been reported to be well correlated with metal concentrations in plants (e.g., Pickering, 1981; Banjoko et al., 1991). Attempts have been made to use it to assess mobile metal pollutant pools in soils and sediments (e.g., Banjoko et al., 1991; Quevauviller et al., 1994; Bezvedova, 1995). However, all of these studies have been empirical and cannot fully explain the solubilizing phenomena involved. Sufficient data to enable a phenomenological interpretation are usually lacking. For example, two of the major parameters controlling the solubilization processes—pH and the ratio of available EDTA to potentially extractable metals—are not usually sufficiently controlled or determined.

The objective of this study was to develop a simple, inexpensive, rapid method that could be used for a large number of samples, but still capable of determining the potential solubility of heavy metals associated with natural solids in order to evaluate their bioavailability and their potential mobility in hydrological cycles. This is not a phase-specific approach but rather a global determination of the risk of remobilization of metal pollutants from natural or polluted solids. Extraction with a single chemical ($HNO_3$ at pH 2, or EDTA at a sufficient concentration) is compared with the extraction of one of the most widely used chemical sequential extraction procedures.

Due to the scarcity of widely available polluted reference material for selective extraction methods, the present study used only true natural samples.

## 2. Materials and Methods

Samples were collected from 6 different locations in France—four specifically for this study and two for a similar study (Crouzet and Altmann, 1996). The first sample, a sediment (upper 20 cm) from the Deûle River near Lille, was sampled from a small boat with a dredge in an area where barges dock to load and unload mining ores. It contains a large fraction of ore grains washed off nearby stock piles by rain, which explains the extremely high metal concentrations measured in this pseudo-sediment (Table 1). All of the other sediments were collected with an aluminum scoop (upper 3–5 cm). The second solid was collected from the bank of the Lot River (southwestern Massif Central) downstream from a historical pollution source (smelting activities). The third solid was collected in Le Palais Creek, a tributary of the Vienne near Limoges, downstream from a copper electro refinery (Bordas, 1998). The forth and fifth solids were collected (upper 5 cm) from the Garonne River and the Gironde Estuary, respectively (Crouzet and Altmann, 1996). The sixth solid is a topsoil (upper 30 cm) collected near Lille, in the vicinity of smelting operations. The very fine-textured sediments from the Garonne River and Gironde Estuary were not sieved and were stored wet at 4°C until used (Crouzet and Altmann, 1996). The other samples were air-dried at 40°C and passed through a 2 mm nylon sieve. Air-drying enables satisfactory preservation of sediments collected in oxidized environments (e.g., Bordas and Bourg, 1998). A small evolution of metal fractions extracted by the sequential protocol used in this study was observed. This was most likely due to the contact with atmospheric oxygen, which favors the formation of Fe and Mn oxides and subsequent associations of other trace metals with these geochemical fractions (e.g., Bordas and Bourg, 1998), but does not perturb our study.

The geochemical composition of the solids was determined by multielement ICP (Jobin Yvon) and the organic carbon content using a Shimadzu Total Organic Analyzer: model TOC-5000 (Table 1).

TABLE 1. Selected characteristics of the solids studied

| Solid sample | $SiO_2$ % | $Al_2O_3$ % | Ca % | Fe % | Mn % |
|---|---|---|---|---|---|
| Sediment Deûle canal | 50.4 | 7.4 | 6.5 | 2.0 | 0.05 |
| Sediment Lot River | 65.3 | 12.1 | 1.0 | 2.4 | 0.07 |
| Sediment Le Palais Creek | 46.0 | 15.5 | <1.4 | 3.4 | 0.13 |
| Sediment Garonne River | 56.4 | 18.6 | 3.7 | 2.9 | 0.06 |
| Sediment Gironde Estuary | 57.4 | 19.4 | 3.8 | 2.1 | 0.05 |
| Soil Lille area | 77.8 | 7.4 | 0.9 | 1.74 | 0.04 |

| Solid sample | Org C % | Cd µg/g | Ni µg/g | Pb µg/g | Source |
|---|---|---|---|---|---|
| Sediment Deûle canal | 4.00 | 350 | 36 | 8,570 | (a) |
| Sediment Lot River | 3.05 | 4.1 | 21 | 50 | (a) |
| Sediment Le Palais Creek | 4.70 | 13 | n.d.* | 177 | (b) |
| Sediment Garonne River | 0.67 | 1.5 | 45 | 50 | (c) |
| Sediment Gironde Estuary | 0.72 | 2.6 | 35 | 51 | (c) |
| Soil Lille area | 1.85 | 9 | 16.5 | 350 | (a) |

* Not determined.
(a) This study; (b) Bordas (1998); (c) Crouzet and Altmann (1996).

From amongst the numerous sequential selective extraction methods, we selected the procedure of Tessier et al. (1979) because it is the most commonly used and it has largely inspired many of the other procedures. It is a series of 5 consecutive extractions that permits, according to its authors, the identification of 5 operationally defined phases (Table 2).

For the single extractions (acid or EDTA), batch experiments were performed at different pH values in both the absence and presence of EDTA. The ionic strength was maintained constant using $NaNO_3$ as the supporting electrolyte. The concentrations of the various solids in the reactor ranged from 1 to 10 g/L (10 g/L for the Deûle canal and Lot River sediments and the Lille topsoil, and 1 g/L for the other samples) and the EDTA concentrations from $10^{-5}$ to $10^{-2}$ M. The pH was adjusted using Suprapure $HNO_3$ or NaOH. After 24 h of stirring (48 h for the sediments from the Garonne River and the Gironde Estuary), time sufficient to reach equilibrium, the pH was measured in the suspension and the sample was centrifuged at 3,000 rpm for 15 min. The supernatant liquid was separated by filtration through a 0.45 µm pore size filter and acidified with Suprapure $HNO_3$.

Dissolved concentrations of Cd, Cu, Ni and Pb were determined by atomic absorption (flame or Zeeman graphite furnace) or anodic stripping voltammetry.

TABLE 2. Sequential extraction protocol of Tessier et al. (1979)

| Extraction sequence | Operationally defined fraction | Extraction conditions |
| --- | --- | --- |
| F1 | Exchangeable | 1 M $MgCl_2$; pH 7.0; 1 h; 25°C |
| F2 | Bound to carbonates | 1 M NaOAc; pH 5 (HOAc); 5 h., 25°C |
| F3 | Bound to Fe and Mn oxyhydroxides | 0.04 M $NH_2OH.HCl$ in 25% (v/v) HOAc; 5 h; 96°C |
| F4 | Bound to organic matter and sulfides | 30% $H_2O_2$; pH 2 ($HNO_3$); 5 h; 85°C |
| F5 | Residual | $HNO_3/HClO_4$ (5/1); 80°C |

## 3.  Results

The sums of metals extracted during steps 1–4 of the protocol of Tessier et al. (1979), assuming that this value represents the maximum potentially available metal load, were compared with values resulting from extractions at pH 2 in the absence of EDTA and at any pH in the presence of EDTA.

### 3.1. SEQUENTIAL EXTRACTION (TESSIER ET AL., 1979) AND SOLUBILITY (TABLE 3)

Only 5% of the cadmium is associated with the residual phase or fraction (F5 in Table 2). This is not surprising since relatively high levels of cadmium in natural solids are mainly of anthropogenic origin (e.g., Förstner and Wittman, 1979). There is little cadmium in crystalline mineral structures and, in contaminated solids, much in potentially remobilizable fractions (i.e., surface species).

Lead is only weakly associated with the residual phase for the Deûle canal sediment (17%) and the Lille topsoil (6%)–the two solids most contaminated by this metal. This seems surprising since this metal can be naturally present in significant quantities in the crystalline structure of solids (e.g., Förstner and Wittman, 1979), but is readily explained if we consider relative amounts. For the less contaminated Le Palais Creek and Lot River sediments (the typical Pb content of uncontaminated river sediments ranges from 25 to 70 µg/g (Pereira-Ramos, 1988; Förstner and Müller, 1974)), the relative value of the residual fraction is accordingly higher (15–25%).

TABLE 3. Typical results of the extraction by the three methods (in μg extracted per g of solid) (average relative uncertainty 10%)

Extractability of cadmium

| Solid sample | Sequential extraction total of steps 1–4 | HNO3 pH 2 | EDTA 0.01 M |
|---|---|---|---|
| Sediment Deûle canal | 340 | 55 | 300 |
| Sediment Lot River | 3.8 | 4.0 | 4.0 |
| Sediment Le Palais Creek | 13 | 12 | 11* |
| Soil Lille area | 8.2 | 9 | 9 |

Extractability of lead

| Solid sample | Sequential extraction total of steps 1–4 | HNO3 pH 2 | EDTA 0.01 M |
|---|---|---|---|
| Sediment Deûle canal | 7,150 | 3,350 | 6,700 |
| Sediment Lot River | 37 | 26 | 43 |
| Sediment Le Palais Creek | 150 | 160 | 140* |
| Soil Lille area | 330 | 320 | 320 |

Extractability of nickel

| Solid sample | Sequential extraction total of steps 1–4 | HNO3 pH 2 | EDTA 0.01 M |
|---|---|---|---|
| Sediment Deûle canal | 18 | 29 | 19 |
| Sediment Lot River | 6.3 | 3.6 | 5.4 |
| Sediment Garonne River | 8.1 | 4.0 (pH3) | 8.3** |
| Sediment Gironde Estuary | 6.9 | 3.5 (pH3) | 7.2** |
| Soil Lille area | 7 | 6 | 6 |

Le Palais Creek data from Bordas (1998).
Garonne and Gironde data from Crouzet and Altmann (1996).
* At pH 7.2.
** At pH 5.8.

Nickel is the metal most generally associated with the residual phase, from 40% to 50% for the most polluted solids (Deûle canal sediment and Lille topsoil) to 80% for the solids collected in the Garonne-Gironde system.

## 3.2. ACID EXTRACTION AND SOLUBILITY (TABLE 3)

The concentration of solubilized metal increases as pH decreases for all of the heavy metals and solids studied (Figure 1). For cadmium in the Lot River sediments and the Lille topsoil, the solubilization is almost total at pH 2 (Table 3). For Cd, Cu and Pb in the Le Palais sediment, about 90% of the total load is solubilized at pH 2. For Ni, at most 55% of the metal is extracted from the various solids. The Deûle canal sediment releases only small fractions of its metal load at pH 2 (1%, 15% and 20% of the Ni, Cd and Pb, respectively). This is probably due to the high metal load of this solid (Table 1) and also to its peculiar nature (mining ore grains deposited at the bottom of the canal, Marot 1995).

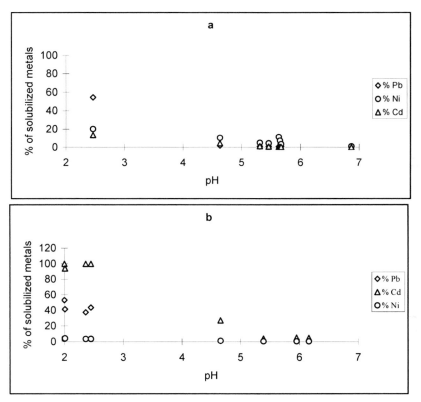

*Figure 1.* Release of heavy metals from solids as a function of pH in the absence of EDTA ((a) sediment Deûle canal; (b) soil Lille area).

The increased release of heavy metals to the aqueous phase with decreasing pH can be explained by increased dissolution of one or more minerals (oxide, hydroxides, carbonates, sulfides or mixed solids) and/or by increased desorption from solid surfaces due to the substitution of surface cationic species by the increasing number of protons present in the solution.

## 3.3. EDTA EXTRACTION AND SOLUBILITY

Metal solubilization increases with the EDTA concentration (Figure 2). As the chelate concentration increases, the remobilization of heavy metals behaves like what might be called an *EDTA concentration solubilization edge* (by analogy with the *pH desorption edge* concept). As the EDTA concentration increases one hundred-fold, the metal solubility increases sharply (from little to almost total release to the solution). The range of EDTA concentrations for which this occurs depends on the metal, the ease of release decreasing in the order cadmium > lead > nickel.

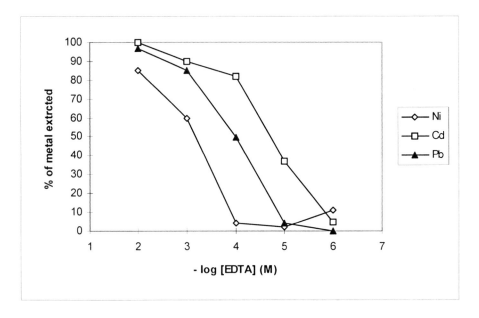

*Figure 2.* Release of heavy metals from the Lille soil as a function of EDTA concentration at pH 7 (The fraction released is calculated on the basis of the extractable load and not for the total concentration in the solid. The data represented at -log [EDTA] = 6 corresponds, in fact, to [EDTA] = 0.).

For low EDTA concentrations (<0.001 M), the solubility of metals is greatly affected by the acidity of the solution (with remobilization increasing with acidity) (Figure 3). For 0.01 M EDTA, it is independent of pH (with the exception of the Deûle canal sediment for which the metal solubility decreases slightly with increasing pH (Figure 4).

For Pb in the *Deûle canal sediment*, the extraction efficiency at pH 2 increases from 60% to 80% as the EDTA concentration increases from 0.001 to 0.01 M. No significant increase in solubilization is observed for Cd and Ni. The extraction at pH 2 and 0.01 M EDTA is always slightly lower (by 5–15% of the

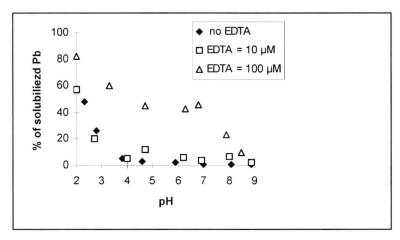

*Figure 3.* Release of Pb from the Lille soil as a function of pH for low EDTA concentrations (<0.1 mM).

total metal content in the solid) than that predicted by the addition of steps 1–4 of the chemical extraction procedure (Table 3). The lower value given by EDTA extraction and the slight decrease in solubility with increasing pH in the presence of EDTA are surely due to the peculiar nature of this sample. The amount of EDTA available for extracting the metals of interest might be limited because of the richness of this sample in many potentially extractible metals (9.4 mg/g Zn, 8.6 mg/g Pb, 0.32 mg/g Cu and 6.5% Ca). For example, the quantity of Pb extracted uses up 3% of the total EDTA. An alternative or additional explanation is the nature of this sediment, which might be rich in sulfides as it contains a large fraction of mining ore grains (Marot, 1995). To our knowledge, no extraction by EDTA of heavy metals from sulfide minerals has been reported.

For the *Lot River sediment*, the increase in added EDTA from 0.001 M to 0.01 M at pH 2 does not change Cd solubility and only slightly increases that of Pb and Ni. For the 0.01 M EDTA experiment, the solubility pattern is almost completely independent of pH (Figure 4).

For the *Le Palais sediment*, Cd and Pb solubility do not change as the EDTA concentration is increased ten-fold.

For the *Lille topsoil*, the ten-fold increase in added EDTA does not affect the solubilization of the metals. As for the Lot River sediment, the fraction of metal solubilized varies little with pH.

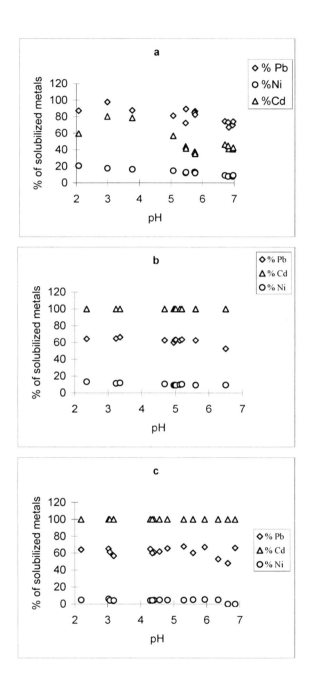

*Figure 4*. Release of heavy metals from solids as a function of pH in the presence of 0.01 M EDTA ((a) sediment Deûle canal; (b) sediment Lot River; (c) soil Lille area).

For the *Garonne River and Gironde Estuary sediments*, for which only Ni was investigated, the EDTA extractions agree with the non-residual phase of the protocol of Tessier et al. (1979).

For all of the solids, the quantities of metals extracted by 0.01 M EDTA are, within experimental errors, the same as the sum of steps 1–4 of the chemical extraction procedure.

## Discussion

Since the selectivity of the so-called selective extraction methods has been questioned (see Introduction), we have investigated the potential availability of metals from natural solids using a different approach. Instead of making vain attempts at determining the geochemical phases to which the metals are associated, our objective was simply to determine the total available metals or, in other words, all of the metal fraction that could possibly be released to the aqueous phase (i.e., not associated with the "residual" fraction of the selective extraction methods). This is a global approach. It does not give detailed information concerning the geochemical speciation of metal pollutants in solids, but it does provide quantitative data on the global risk of remobilization of metals in natural and polluted soils and sediments.

For low EDTA concentrations ($<10^{-3}$ M), the extraction pH is important because the greater the pH, the lower the extraction yield. At low pH, extraction is aided by the presence of the protons supplied by the acid. The assumption that the complexing capacity of EDTA, increasing with the availability of the complex former (i.e., the fully deprotonated EDTA form), should increase with pH is too simplistic. A sufficient concentration of EDTA enables a pH-independent extraction. The maximum concentration of EDTA to be used depends on its solubility. If this is the limiting factor, smaller concentrations of solids might be used.

Even though the solids used in this study contained significant amounts of both major metals (such as calcium) and minor but strongly complexable metals (such as iron), there seems to be enough EDTA in the 0.01 M EDTA extracting solution to remobilize all of the geochemical forms that are not bound in the mineral structure (the residual phase) in the sequential chemical extraction protocol). The Deûle sediment, rich in a variety of extractable metals, is an exception, the available EDTA possibly being a limiting factor.

Acidification to pH 2 is not appropriate for determining the potentially soluble fraction of metals. Although this simple leaching method does indeed quite effectively solubilize all of the extractable Cd (except in the Deûle canal sediment), in many samples, it does not solubilize all of the extractable Pb and Ni (Table 3).

For all of the metals studied here and all of our samples (with the exception of the Deûle canal sediment), chemical extraction with 0.01 M EDTA (single extraction, independent of pH) is shown to be an easy, inexpensive, rapid method that furnishes valuable and highly accurate information concerning potentially remobilizable metal load in polluted solids (especially for field sites where strong heterogeneities require the treatment of a large number of samples). More investigations on a larger number of samples of more diverse origin are needed, however, before this method can be proposed as a standard.

## Acknowledgements

This work was partially financed by the French Ministry of the Environment Program GESSOL and the European Union (ERDF).

## References

Baghdady NH, Sippola J. Extractability of polluting elements Cd, Cr, Ni and Pb of solids by three methods. Acta Agr Scand 1984; 34: 345–348.

Banjoko VA, McGrath SP. Studies of the distribution and bioavailability of soil zinc fractions. J Sci Food Agric 1991; 57: 325–334.

Barona A, Romero F. Fractionation of lead in soils and its influence on the extractive cleaning with EDTA. Environ Technol 1996; 17: 63–70.

Bezvedova B. Mobility of heavy metals in soils in the Krusne Hory and Slavkovasky mountains. Environ Monitoring Assess 1995; 34: 163–166.

Bordas F. Remobilisation de Micro-polluants Métalliques à partir de Sédiments de Rivière naturellement Pollués en Présence de Complexants Organiques Synthétiques Doctoral Dissertation, Univ. of Limoges, France, 1998.

Bordas F, Bourg ACM. A critical evaluation of sample pretreatment for storage of contaminated sediments to be investigated for the potential mobility of their heavy metal load. Water, Air, Soil Pollut 1998; 103: 137–149.

Crouzet C, Altmann RS. Comparaison de diverses méthodologies pour estimer le pouvoir toxique des sédiments 1996. BRGM Report R 38 894, Orléans, France, 40 pp.

Davidson CM, Thomas RP, Mc Vey SE, Perala R, Littlejohn D, Ure AM. Evaluation of a sequential extraction procedure for the speciation of heavy metals in sediments. Anal Chim Acta 1994; 291: 277–286.

Förstner U, Müller G. Schwermetalle in Flüssen und Seen als Ausdruck der Umweltverschmutzung, Springer, Berlin, 1974, 225 pp.

Förstner U, Wittman GTW. Metal Pollution in the Aquatic Environment, Springer, Berlin, 1979, 486 pp.

Heron G, Crouzet C, Bourg ACM, Christensen TH. Speciation of Fe(II) and Fe(III) in contaminated aquifer sediments usning chemical extraction techniques. Environ Sci Technol 1994; 28: 1698–1705.

Huang JW, Chen J, Berti WR, Cunningham SD. Phytoremediation of lead-contaminated soils: role of synthetic chelates in lead phytoextraction. Environ Sci Technol 1997; 31: 800–805.

Jeanroy E, Guillet B, Ortiz R. Applications pédogénétiques de l'étude des formes du fer par les réactifs d'extraction: cas des sols brunifiés et podzolisés sur roches cristallines. Sci Sol 1984; 3: 199–211.

Gyori Z, Goulding K, Blake L, Prokisch J. Changes in the heavy metal contents of soil from the Park Grass Experiment at Rothamsted Experimental Station. Fresenius J Anal Chem 1996; 354: 699–702.

Li Z, Shuman LM. Extractability of zinc, cadmium, and nickel in soils amended with EDTA. Soil Sci 1996; 161: 226–232.

Marot F. Etude et traitement de boues résiduaires, de vases de dragages contenant des polluants métalliques et des matières organiques 1995. BRGM Report no 1 to ADEME, Orléans, France, 34 pp.

Nirel PMV, Morel FMM. Pitfalls of sequential extraction. Water Res 1990; 24: 1055–1056.

Pereira-Ramos L. Etude et exploitation critique de résultats d'nalyses de métaux sur sédiments. Campagnes sur les grandes rivières du bassin Seine-Normandie de 1981 à 1986 1988. AFBSN-IHC Report, 59 pp.

Pickering WF. Selective chemical extraction of soil components and bound metal species. CRC Crit Rev Analyt Chem 1981; 12: 233–266.

Quevauviller Ph, Rauret G, Muntau H, Ure A, Rubio R, Lopez-Sanchez JF, Fiedler HD Griepink B. Evaluation of a sequential extraction procedure for the determination of extractable trace metal contents in sediments. Fresenius J Anal Chem 1994; 349: 808–814.

Rapin F, Tessier A, Campbell PGC, Carignan R. Potential artifacts in the determination of metal partitioning in the sediments by a sequential extraction procedure. Environ Sci Technol 1986; 20: 836–840.

Simeonova B, Simeonov L. An application of a phytoremediation technology in Bulgaria. The Kremikovtzi Steel Works experiment Remediation Journal, Spring edition 2006; Wiley Periodicals, New York, pp. 113–123.

Tessier A, Campbell PGC, Bisson M. Sequential extraction procedure for the speciation of particulate trace metals. Anal Chem 1979; 51: 844–851.

Ure AM, Quevauviller Ph, Muntau H Griepink B. Speciation of heavy metals in soils and sediments. An account of the improvement and harmonization of extraction techniques undertaken under the auspices of the BCR of the Commission of European Communities. Intern J Environ Anal Chem 1993; 51: 135–151.

Whalley C, Grant A. Assessment of the phase selectivity of the European Community Bureau of Reference (BCR) sequential extraction procedure for metals in sediment., Anal Chim Acta 1994; 291: 287–295.

# THE IMPACT OF SOIL POLLUTANTS ON SOIL MICROBIAL ACTIVITY

RADKA DONKOVA[*] AND NADYA KALOYANOVA
*Institute of Soil Science "N. Poushkarov"
7, Shosse Bankya Str., 1080 Sofia, Bulgaria*

**Abstract.** The functioning of the soil as a vital system and the support on its biological productivity depends to a higher extent on the soil microflora activity. That is why in the assessing of anthropogenic soil pollution it is necessary to take into account the changes in the the size, composition and activity of the soil microbial community, variation in loss of the normal bands and the appearance of new bands compared with the unpolluted soil. There is not yet a commonly accepted system of bio-indication on soil pollution. As bio-indicators are used pure cultures of microorganisms, sensitive to determined type of pollutant; the number and ratio of the main taxonomic and ecologic trophic groups of microorganisms; bacterial community tolerance; intensity of the microbiological processes – soil respiration, fixation of nitrogen, cellulose decay; soil enzyme activity and so on. With higher sensibility are distinguished the indexes, reflecting more narrowly the special processes, which are implemented by the limited number of microorganisms. The changes in soil microbial equilibrium can serve as an "early warning" for negative alterations in the soil conditions long before they could be detected by classical chemical methods and before they could become irreversible. The complex investigations on the soil biological activity should be conducted in assessing ecological risk of soil pollutants. The reliable microbiological indicator must be established and used.

**Keywords:** soil pollutants, risk assessment, soil biological activity soil microflora, enzyme activity, bio-indicators

---

[*] To whom correspondence should be addressed: rada_donkova@yahoo.com

## 1. Introduction

Soil ecological status is determined by natural and anthropogenic factors, which at given combination can lead to destruction of the ecological balance, formed during the natural soil-formation process.

One of the most important soil characteristics is the ability to filtrate, adsorb and precipitate the substances falling on its surface. The presence of significant concentrations of different highly biologically active chemical substances (pesticides, heavy metals, polycyclic carbohydrates, polychlorinated biphenyls and furans, dioxins, petroleum products, etc.) influence the status and functioning of the soil microbial biocenosis, soil fertility and human health.

The functioning of the soil as a vital system and the support on its biological productivity depends to a higher extent on the soil microflora activity. The soil is a habitat for a vast, complex and interactive community of soil organisms whose activities largely determine its chemical and physical properties. In a fertile soil the soil biota may have a biomass exceeding 20 t.ha$^{-1}$, with life forms ranging from microscopic bacteria to the largest of earthworms which may be 1 m in length. In most terrestrial ecosystems the soil contains by far the greatest diversity of organisms, in some cases it is estimated that one gram of soil can contain several thousand genotypes.

The microorganisms can be active destructors of the pollutants, but on the other side the pollutants can destroy the microbial succession, can suppress or kill some varieties of microorganisms and activate the development of others, which will lead to change in the resistance of the soil ecosystem as a whole.

That is why in the assessing of anthropogenic soil pollution it is necessary to take into account the changes in the the size, composition and activity of the soil microbial community, variation in loss of the normal bands and the appearance of new bands compared with the unpolluted soil.

## 2. The Impact of the Main Soil Pollutants on Soil Microflora

In monitoring of the pesticides of extreme importance are those with the largest persistence, the largest practical use and with potential risk of harmful influence on the bio-ecosystems. On the first place concerning the persistence is the chlorinated organic pesticides count in so called "dirty dozen" – DDT, dieldrin, eldrin and others.

According to the data of Balinova (1998), DDT has a period of half-decay 10–13 years, and in particular places in the country (intensive agricultural regions, stores, sides of preparing solutions for agricultural aviation, burial plots for unusable pesticides) are detected significant concentrations as of DDT, as well as of its metabolite product DDE, which surpass the Limit Concentrations in many of the samples taken for analysis in 1996 and 1998. The higher values of DDT, as compared to those of DDE indicate a new introduction of this substance, regardless of the prohibition for its use as early as in 1970. The studies of Shegunova (2001) confirm the fact that these pesticides are still in use as a mixture of trade products and even after 30 years their remains are present in the soil surface layers. This necessitates the determination of the pesticides' influence on the micro flora of the polluted sides and the need to apply remediation techniques.

The DDT has a negative influence on the growth and the activity of soil microorganisms. It decreases significantly the amount of bacteria, actinomycetes (Ko & Lockwood, 1968), totally inhibits *Alternaria humicola, Trichoderma viride, Botryotrichum sp., Helmintosporium sativum, Sepedonium sp, Mucor sp, Rhizoctonia solani, Botrytis* (Varshney & Gaur, 1972), the growth of *Azotobacter* (Callao & Montoya, 1956), nodulation (Braithwaite et al., 1958; Masefield, 1955). It does not influence soil respiration (Tate, 1974; Salonius, 1972). In concentrations higher then 5 ppb DDT inhibits *Nitrobacter, Nitrosomonas* (Gaur & Pareek, 1971) and it also inhibits the soil protozoa (MacRae & Vinckx, 1973) and has a negative effect on *Collembola* and *Acarina* and on some useful worms (Sheals, 1955, 1956; Hartenstein, 1960).

When applied in the recommended dozes the Aldrin does not have such an influence on the amount and activity of the soil microorganisms, which could influence soil fertility (Fletcher & Bollen, 1954; Eno, 1957; Martin et al., 1959; Shaw, 1960). Negative influence is not observed also with the nematodes (French et al., 1959). At higher insecticide concentrations a negative effect is observed as it is especially strongly expressed with the nitrifying microorganisms, the microorganisms oxidizing sulfur to sulfates, and it totally inhibits cultures of *Nitrobacter agilis*. A strongly expressed negative effect is found to exist in *Acarina, Collumbola, Edwards, wire worm* (Wilkinson et al., 1964).

The Dieldrin is subjected to microbial transformation. In experiments with sterile and non-sterile soils, treated with 50 ppb Dieldrin, Yagnow and Haider (1972) isolated 177 bacterial strains, which produce water-soluble metabolites of the dieldrin, as the most active *Nocardia, Corinobacterium, Micrococcus spp.,* incubated with 0.4 ppm Dieldrin for 5 weeks in aerated cultures transformed 0.06–0.11% and 0.14–0.2% in stationary cultures. *Rhizoctonia solani,* as well as some actinomycetes also transformed the molecules of the Dieldrin (Ko & Lockwood, 1968). Mutsumura et al. (1968) describe 4 possible ways of Dieldrin transformation from Pseudomonas sp. As a whole-the preparation is stable to the biological and non-biotic destruction (C. Friedel et al., 2000). The Dieldrin not only is transformed by the soil microorganisms, but it influences their growth and activity. Chandra (1967) found that the Dieldrin has a negative influence on the nitrifying organisms. This influence is with different continuity, depending on the soil type, soil temperature and wetness. Bollen and Tu (1971), studying the influence of the Endrin on the soil microorganisms, found out that in amounts three times higher then the used in the practice the fungicide does not have a negative influence on the amount of the soil microorganisms, or on the activity of ammonifying, nitrifying or oxidizing of sulfur microorganisms. It was found that the rate of organic matter decay is increased.

Heptachlor does not have a negative effect on the formation of nodule formation (Shamiyen & Johanson, 1973) but it inhibits to a bigger extent the growth of some soil microorganisms. Added in concentrations 100 mg/l in agar preparations it retards the growth of 89% bacteria, 81% of actinomycetes, and 50% of fungi.

The investigations in relation with different polycyclic aromatic hydrocarbons (PAH) are directed first of all to the possibility to be actively metabolized by the microorganisms. It is accepted that the microbial destruction is more important for the destruction of these organic substances, because it leads to their full mineralization, while, at abiotic decay medial products are formed, which can be more dangerous for the environment as compared to the initial materials. The available references concerning the influence of PAH on the soil micro flora are very scarce, which justify this type of studies.

The polychlorinated biphenyls (PCPs) influence negatively some soil microorganisms. Dusek (1995) in experiments with increasing dozes of PCPs found a strong decrease of the nitrite-oxidizing bacteria.

High concentrations of heavy metals have been shown to adversely affect the size, diversity, and activity of microbial populations in soil. The risk of heavy metals presence in soil is related to the fact that they can accumulate everlastingly in toxic doses, and cannot be degraded. The most harmful are Cd, Pb, As, Cu, Ni, etc. (McGrath, et al., 1995; Chen et al., 2000). Significant negative correlation between the influence of the heavy metals on the soil microflora and their available concentration was established. Thus, the factors influencing the bioavailability of the heavy metals in the soil can influence the toxicity of the metals on soil microbial community. Among the most important factors are: soil organic matter amount, soil structure, soil pH. The metal pollution of soil provokes a significant decrease of the soil microbial biomass. In most of the cases this decrease is observed at low metal concentrations and is kept during some years (Khan & Scullion, 2000). This suggests that the metal concentrations in the soil that are near the EC (Exciding Concentrations-determined limit values) probably will lead to a significant inhibition of the microbial biomass with a durable effect on the soil productivity. The critical toxic level of the metals affecting the microbial biomass is difficult to be determined yet. This is due mainly to the difference of the polluting metals, to the physical and chemical properties of the soils, to the difference in the used methodology and the experimental conditions. The relative toxicity of the metals is relatively constant and they can be ranged by it as follows: Cd > Cu > Zn > Pb.

The separate metals exert different by strength influence on the different microorganisms. For example: the negative influence of cadmium increases in the range: fungi < actinomycete < bacteria (Hiroki, 1992).

Results of our studies confirmed the existing reference data about the disturbing of the soil microbiological balance due the activity of heavy metals. This disturbing to a significant extent is determined by the soil type, soil pH and the amount of the pollutants.

Thus the data from a pot experiment on the effect of cadmium contamination, applied in three doses (5, 10 and 15 mg/kg soil) in the conditions of alluvial soil (Fluvosol) and leached chernozem (Eutric Vertisol) on microbiological properties revealed that all applied doses had strong toxic effect on the main groups of microorganisms and their activities. During the incubation period this effect decreased, but it still remained quite strong within a period of 90 days (Tables 1 and 2). The studied parameters were more influenced in the alluvial soil (Kaloyanova, 2007).

TABLE 1. Dynamics of the examined microorganisms

| Variants | Bacteria | | | Actinomicetes | | | Fungi | | |
|---|---|---|---|---|---|---|---|---|---|
| | 15 day | 45 day | 90 day | 15 day | 45 day | 90 day | 15 day | 45 day | 90 day |
| | x $10^6$ g/soil | | | | | | x $10^3$ g/soil | | |
| Fulvous | | | | | | | | | |
| Control | 6.5 | 4.3 | 4.8 | 0.9 | 1.4 | 1.2 | 5 | 3 | 2 |
| Soil + Cd$_1$ | 0.9 | 2.2 | 3.3 | 0.3 | 1.2 | 1.1 | 2 | 2 | 1 |
| Soil + Cd$_2$ | 0.8 | 1.1 | 2.3 | 0.2 | 1.0 | 0.9 | 2 | 2 | 1 |
| Soil + Cd$_3$ | 0.4 | 0.8 | 2.0 | 0.1 | 0.8 | 0.5 | 1 | 1 | 1 |
| LSD (P = 0.05) | 1.9 | 2.1 | 1.4 | 0.3 | 0.5 | 0.3 | 1.2 | 1.0 | 0.5 |
| Eutric Vertisol | | | | | | | | | |
| Control | 1.6 | 8.7 | 4.4 | 0.7 | 2.4 | 1.1 | 14 | 4 | 2 |
| Soil + Cd$_1$ | 0.7 | 2.6 | 2.1 | 0.3 | 1.5 | 0.3 | 7 | 3 | 1 |
| Soil + Cd$_2$ | 0.5 | 2.2 | 1.3 | 0.2 | 1.0 | 0.2 | 6 | 1 | 1 |
| Soil + Cd$_3$ | 0.4 | 1.2 | 1.2 | 0.1 | 0.7 | 0.2 | 5 | 1 | 1 |
| LSD (P = 0.05) | 2.3 | 1.2 | 1.2 | 0.1 | 0.5 | 0.2 | 2.6 | 1.1 | 0.7 |

Cd$_1$ - 5mg Cd/kg soil, Cd$_2$ - 10mg Cd/kg soil, Cd$_3$ - 15mg Cd/kg soil

TABLE 2. Total bilogical activity, microbial biomass and specific respiratory activity

| Variants | Total biological activity | | | Biomass C | | | Specific respiration activity | | |
|---|---|---|---|---|---|---|---|---|---|
| | 15 day | 45 day | 90 day | 15 day | 45 day | 90 day | 15 day | 45 day | 90 day |
| | mg $CO_2$/g soil | | | mg C/g soil | | | mg $CO_2$/biomass C | | |
| Fluvosol | | | | | | | | | |
| Control | 6.4 | 5.9 | 4.9 | 6.2 | 8.0 | 7.4 | 1.0 | 0.7 | 0.7 |
| Soil + $Cd_1$ | 0.6 | 4.6 | 4.3 | 4.5 | 7.4 | 6.3 | 0.1 | 0.6 | 0.7 |
| Soil + $Cd_2$ | 0.5 | 3.5 | 3.3 | 3.8 | 5.4 | 5.1 | 0.1 | 0.6 | 0.6 |
| Soil + $Cd_3$ | 0.2 | 2.2 | 2.1 | 2.6 | 4.8 | 4.6 | 0.1 | 0.8 | 0.5 |
| LSD (P = 0.05) | 2.1 | 0.8 | 0.7 | 1.4 | 1.1 | 1.2 | | | |
| Eutric Vertisol | | | | | | | | | |
| Control | 7.3 | 5.7 | 4.4 | 6.1 | 6.2 | 4.1 | 1.2 | 0.9 | 1.1 |
| Soil + $Cd_1$ | 0.5 | 2.9 | 2.2 | 4.9 | 4.5 | 3.2 | 0.1 | 0.6 | 0.7 |
| Soil + $Cd_2$ | 0.4 | 1.3 | 1.1 | 2.9 | 3.1 | 2.2 | 0.1 | 0.6 | 0.5 |
| Soil + $Cd_3$ | 0.3 | 1.2 | 0.9 | 1.9 | 2.5 | 1.6 | 0.7 | 0.5 | 0.6 |
| LSD (P = 0.05) | 2.2 | 0.9 | 0.7 | 2.0 | 1.1 | 0.8 | | | |

$Cd_1$, 5mg Cd/kg soil, $Cd_2$, 10mg Cd/kg soil, $Cd_3$, 15mg Cd/kg soil

Our investigations (Figure 1) revealed that cadmium had negative effect on some of the investigated group of microorganisms in the condition of calcareous chernozem (Typical chernozems) even though, in concentration lower than

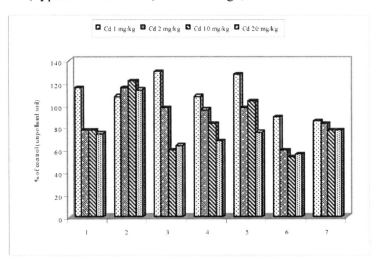

*Figure 1*. Influence of Cd on soil microflora: 1. Amonifying bacteria 2. N utilizing bacteria 3. Actynomicetes 4. Fungi 5. Nitrogenfixing bacteria 6. Cellulose decomposing microorganisms 7. Total biological activity.

maximum permitted concentration for the soil (Petkova & Donkova, 2006). The microorganisms taking part in nitrogen transformation of organic matter were the most sensitive.

In laboratory and pot experiments carried out with alluvial-meadow soil (Fluvisol) treated with urea, $CuSO_4$ and two types of composts we established that Cu in dose of 750 mg/kg$^{-1}$ soil inhibited the development of ammonifying bacteria with 69% and increased the number of fungi (Kaloyanova & Kostov, 2004). The addition of composts decreases the inhibition of the development of bacteria and actinomycetes. Bacteria were more sensitive to Cu toxicity compared to actinomycetes.

Copper tolerance coefficients varied at fungi from 20% to 90%, at actinomycetes from 0.5% to 4% and at bacteria from 4% to 15%. Composts treatments showed weak growth. The growth inhibition at vine branches compost was 60–71% but grape pruning, husks and seeds compost showed inhibition of 33–65% (Tables 3, 4 and 5).

TABLE 3. Total quantity of amonifying bacteria, microscopical fungi and Cu tolerant fungi

| Variants | Amonifying bacteria | Fungi | Cu tolerant fungi | Coefficient of tolerance |
|---|---|---|---|---|
| | x 10$^9$ g$^{-1}$ soil | | | % |
| 1. Soil + N | 21.7 ± 3.7 | 0.06 ± 0.02 | 0.04 ± 0.02 | 66 |
| 2. Soil + N + Cu | 6.7 ± 1.3 | 0.30 ± 0.50 | 0.10 ±0.01 | 33 |
| 3. Soil + N + Cu + compost₁ | 15.2 ± 0.9 | 0.40 ± 0.30 | 0.04 ± 0.01 | 11 |
| 4. Soil + N + Cu + compost₂ | 13.6 ± 2.8 | 0.30 ± 0.05 | 0.05 ± 0.01 | 16 |
| 5. Soil + N + Rhizobium meliloti 116 | 11.5 ± 2.5 | 0.05 ± 0.02 | 0.04 ± 0.02 | 80 |
| 6. Soil + N + Cu + Rhizobium meliloti 116 | 6.8 ± 0.8 | 0.30 ± 0.03 | 0.09 ± 0.01 | 30 |
| 7. Soil + N + Cu + Rhizobium meliloti 116 + compost₁ | 14.2 ± 1.9 | 0.30 ± 0.06 | 0.06 ± 0.01 | 20 |
| 8. Soil + N + Cu + Rhizobium meliloti 116 + compost₂ | 13.9 ± 0.6 | 0.30 ± 0.04 | 0.06 ± 0.01 | 20 |
| LSD (P = 0.05) | 1.0 | 0.1 | 0.06 | |

[a], 50 mg urea kg$^{-1}$ soil; [b], 750 mg Cu kg$^{-1}$ soil; compost₁, 10% (v/v) vine branches; compost₂, 10% (v/v) grape pruning

TABLE 4. Total quantity of Cu tolerant bacteria, actinomycetes and fungi grown on SAA

| Variants | Bacteria | Actinomycetes | Fungi |
|---|---|---|---|
| | CFU x $10^9$ $g^{-1}$ soil | | |
| 1. Soil + N | 0.20 ± 0.05 | 0.50 ± 0.07 | 0.09 ± 0.01 |
| 2. Soil + N + Cu | 0.06 ± 0.02 | 0.04 ± 0.01 | 0.20 ± 0.03 |
| 3. Soil + N + Cu + compost₁ | 0.04 ± 0.01 | 0.02 ± 0.00 | 0.10 ± 0.00 |
| 4. Soil + N + Cu + compost₂ | 0.07 ± 0.01 | 0.20 ± 0.06 | 0.20 ± 0.05 |
| 5. Soil + N + *Rhizobium meliloti* 116 | 0.20 ± 0.08 | 0.30 ± 0.09 | 0.10 ± 0.005 |
| 6. Soil + N + Cu + *Rhizobium meliloti* 116 | 0.07 ± 0.02 | 0.03 ± 0.04 | 0.30 ± 0.02 |
| 7. Soil + N + Cu + *Rhizobium meliloti* 116 + compost₁ | 0.03 ± 0.005 | 0.02 ± 0.00 | 0.20 ± 0.01 |
| 8. Soil + N + Cu + *Rhizobium meliloti* 116 + compost₂ | 0.07 ± 0.02 | 0.03 ± 0.01 | 0.30 ± 0.03 |
| LSD (P = 0.05) | 0.01 | 0.04 | 0.01 |

[a], 50 mg urea $kg^{-1}$ soil; [b], 750 mg Cu $kg^{-1}$ soil; compost₁, 10% (v/v) vine branches; compost₂, 10% (v/v) grape pruning

TABLE 5. Coefficient of the bacteria tolerant, actinomycetes and fungi grown on SAA

| Variants | Bacteria | Actinomycetes | Fungi |
|---|---|---|---|
| | Tolerance (subsistence) % | | |
| 1. Soil + N | 9 | 4 | 90 |
| 2. Soil + N + Cu | 15 | 1 | 33 |
| 3. Soil + N + Cu + compost₁ | 8 | 1 | 20 |
| 4. Soil + N + Cu + compost₂ | 9 | 2 | 50 |
| 5. Soil + N + *Rhizobium meliloti* 116 | 4 | 2 | 50 |
| 6. Soil + N + Cu + *Rhizobium meliloti* 116 | 14 | 1 | 75 |
| 7. Soil + N + Cu + *Rhizobium meliloti* 116 + compost₁ | 6 | 1 | 28 |

[a], 50 mg urea $kg^{-1}$ soil; [b], 750 mg Cu $kg^{-1}$ soil; compost₁, 10% (v/v) vine branches; compost₂, 10% (v/v) grape pruning

In model experiment carried out with leached chernozem (Luvic phacozem) and light gray forest soil (Orthic Luvisols) we established that lead in concentration equal to Maximum Permitted Concentration and twice higher had short-term negative effects on soil microflora and its activity (Donkova & Petkova, 2003). They were observed immediately after simulated lead pollution. The term of the disappearing of negative effects is different for investigated groups of microorganisms. The fungi were influence most strongly (Figure 2).

The knowledge, concerning the metal influence on the soil microorganisms and influence of the soil properties on their manifestation are still not complete and justify the future studies. In natural conditions the pollution is never due only to one metal. Just the opposite – the most frequent case is a combination of several metals together with other pollutants. This necessitates a more detailed study of the complex pollution influence on the soil microflora.

In the study carried out in the region of non-ferrous metals factory, Plovdiv (Donkova & Dinev, 2006) was established that the main pollutants in the region are Cd, Pb, and Zn, which amounts in some of the studied points surpass, with several orders, the Maximum Permitted Concentrations (Table 6).

Leached chernozem

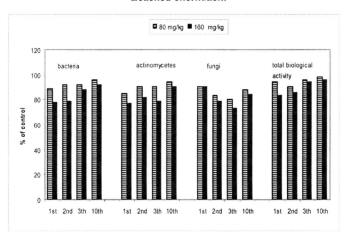

Light gray forest soil

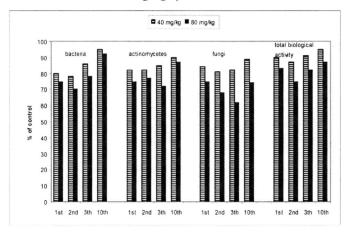

*Figure 2.* Lead influence on amount of microflora and total biological activity.

TABLE 6. Content of Cd, Pb, Zn and Cu in soil samples in the region of non-ferrous metals factory, Plovdiv (mg/kg)

| Site | Cd | Pb | Zn | Cu |
|------|------|-------|-------|------|
| 4 | 1.5 | 70 | 185 | 45.5 |
| 11 | 40.5 | 1,935 | 3,100 | 203 |
| 21 | 103 | 6,295 | 6,400 | 520 |
| 23 | 8.5 | 513 | 755 | 142 |
| 32 | 5 | 217 | 500 | 86.5 |
| 33 | 2.5 | 207 | 365 | 97 |

In the strongly polluted areas a change was established in the amount of the soil microorganisms, which lead to change in their relation, i.e. to upset the balance system. The amount of bacteria and of the cellulose destructing microorganisms decrease to 80%, that of the actinomycetes – to 50%, and at fungi was observed a stimulating, which is in accordance with the established range of sensitivity of the soil microorganisms to the influence of the heavy metals by Hiroki (1992). It is established that the decrease of the intensity of soil respiration which shows shrinkage of the microbiological spectrum of activity, decrease of functional activity of the microorganisms and probably also a immediate death of part of the microorganisms (Table 7).

TABLE 7. Soil microflora in the region of non-ferrous metals factory in Plovdiv, Bulgaria

| Microorganisms | Site | | | | | |
|---|---|---|---|---|---|---|
| | 4 | 11 | 21 | 23 | 32 | 33 |
| Aerobic heterotrofic bacteria [1] LSD P<0.05 = 26.73 | 219.33 ±13.78 | 96.33 ± 3.53 | 49.33 ± 5.82 | 156.00 ± 8.32 | 106.00 ± 3.46 | 128.00 ± 6.43 |
| Anaerobic heterotrofic bacteria [2] | $0.10^5$ | $0.10^4$ | $0.10^3$ | | | |
| Fungi[1] LSD P<0.05 = 0.0092 | 0.056 ± 0.001 | 0.077 ± 0.001 | 0.104 ± 0.005 | 0.054 ± 0.004 | 0.054 ± 0.002 | 0.054 ± 0.003 |
| Actinomycetes[1] LSD P<0.05 = 0.3564 | 3.300 ± 0.404 | 2.800 ± 0.208 | 1.876 ± 0.120 | 3.467 ± 0.240 | 1.967 ± 0.176 | 2.333 ± 0.333 |
| Denitrifying bacteria[2] | $0.10^4$ | $0.10^4$ | $0.10^4$ | | | |
| Nitrifying bacteria[2] | $0.10^5$ | $0.10^4$ | $0.10^4$ | | | |
| Utilizing min N bacteria[1] LSD P<0.05 = 2.6543 | 25.2677 ± 0.6173 | 18.0000 ± 0.5859 | 10.7000 ± 0.4163 | 19.2048 ± 0.7234 | 16.3667 ± 0.6960 | 24.6678 ± 1.8269 |
| Sporeforming bacteria[1] LSD P<0.05 = 0.0126 | 0.0387 ± 0.0035 | 0.0340 ± 0.0076 | 0.0107 ± 0.0078 | | | |
| Cellulose decomposing microorganisms[1] LSD P<0.05 = 0.0160 | 0.0927 ± 0.0049 | 0.0267 ± 0.0014 | 0.0213 ± 0.0152 | 0.0337 ± 0.0152 | 0.0913 ± 0.0.115 | 0.0763 ± 0.0143 |
| Total biological activity[3] LSD P<0.05 = 3.686 | 27.90 ± 1.29 | 13.75 ± 1.49 | 10.84 ± 0.70 | 23.64 ± 4.35 | 27.86 ± 0.75 | 26.86 ± 0.98 |

[1] – CFU $.10^6$ g soil; [2] – cells/g soil; [3] – mg $CO_2$/100 g soil

The combined effects were dependent on the type and dose of pollutants, soil condition, season variation, test parameters, on the addition of concentration ratios to soils and differ from antagonistic, no effect to synergistic.

In model experiments we studied the combined effect of lead and herbicide relay (900 g/l acetochlor) on soil microflora with four soils Leached Smolnitsa (Haplic Vertisol, FAO), Alluvial-Meadow soil (Fluvisol, FAO), Leached Chernozem and Light Gray Forest Soil. A lead pollution of the studied soils was simulated – application of concentrations equal to twice of the Maximum Permitted Concentrations. Two herbicide concentrations recommended for practical use and twice higher were investigated. It was established that Relay has a negative effect on the soil microflora, which occurred immediately after herbicide application, better expressed in soils with lower sorption capacity. The negative effect was most strongly reflected at the variants with higher herbicide concentrations. The presence of lead in the studied soils did not determine a significant change in the herbicide effect (Donkova & Petkova, 2005; Donkova, 2007).

Wang et al. (2006) assessing the combined effects of cadmium (Cd, 10 mg/kg of soil) and butachlor (5, 10 and 50 mg/kg of soil) on enzyme activities and microbial community structure established that combined effects of Cd and butachlor on soil urease and phosphatase activities depend largely on the addition concentration ratios to soils. Phosphatase activities were decreased in soils with Cd (10 mg/kg of soil) alone, whereas urease activities were unaffected by Cd. Urease and phosphatase activities were significantly reduced by high butachlor concentration (50 mg/kg of soil). When Cd and butachlor concentrations in soils were added at milligram ratio of 2:1 or 1:2, urease and phosphatase activities were decreased, while enzyme activities were greatly improved at the ratio of 1:5.

The results obtained from Maliszewska-Kordybach (2002) indicate that combined effect of PAH (flourene, anthracene, pyrene and chrysene) and $Zn^{2+}$, $Pb^{2+}$ and $Cd^{2+}$ on soil microorganisms activity can be stronger than in soils amended with HM or PAH separately. The reaction of the tested organisms was related to soil properties, PAH concentration.

## 3.  Bioindicators of Soil Pollution

There is not yet a commonly accepted system of bio-indication on soil pollution. The review of the references data shows that as bio-indicators are used pure cultures of microorganisms, sensitive to determined type of pollutant; the number and ratio of the main taxonomic and ecologic trophic groups of microorganisms; bacterial community tolerance; intensity of the microbiological processes – soil respiration, fixation of nitrogen, cellulose decay; soil enzyme activity and so on.

With higher sensibility are distinguished the indexes, reflecting more narrowly the special processes, which are implemented by the limited number of microorganisms – nitrification, nitrogen fixation, cellulose decomposition.

From the integral indexes about the soil microorganisms' activity, the Total biological activity, (determined trough the $CO_2$ production) is unsuitable for determining the effect of the pollutants, due to the significant variation of this index, from the soil conditions. Apart from this, the production of $CO_2$ is a result of the vital activity of all organisms in the soil and is suppressed only at very high levels of pollution (Tyler, 981). With higher sensibility are distinguished the indexes, reflecting more narrowly the special processes, which are implemented by the limited number of microorganisms – nitrification, nitrogen fixation, cellulose decomposition.

From the integral indexes the microbial biomass is studied most broadly. It represents from 1% to 4% of the soil organic matter, but with its activity is realized the transformation of the whole organic matter, entering in the soil. It is established that the amount of the microbial biomass is changed a lot faster, then the soil organic matter, depending on the applied agro-technical activities and anthropogenic influence, due to which it is a reliable index for early changes in the soil ecological status (Powlson et al., 1987; Insam et al., 1989). The microbial biomass is a component of two main biological indexes – the ratio of the biomass Carbon to the organic Carbon and the metabolite coefficient. The last represents the ratio of the total biological activity, expressed with the production of $CO_2$ to the microbial biomass and according to a huge number of researchers it reflects the physiological status of the soil microorganisms and the stability of the microbial communities at the unfavorable conditions (Wardle & Parkinson, 1990; Insam et al., 1991; Blagodatskaya et al., 1995). Many authors established the decrease of the microbial biomass and increase of the metabolite coefficient, which shows ineffective use of the nutrients by the soil microorganisms, at soil heavy metals pollution (Chander & Brookes, 1991, 1991a; Bargett & Saggta, 1994; Kuperman & Careiro, 1997; Aceves et al., 1999; Smejkalova et al., 2003). Similar unfavorable tendencies concerning the stated indexes are received by Dahlin et al. (1997), and at soil pollution with Cd, Cu, Pb and Zn in amounts lower than the allowed limit concentrations, accepted in the European Union.

The biological nitrogen fixation by obligate endophytic diazotroph bacteria (e.g., Rhizobium, Azotobacter, Azospirillum, etc.) is the major alternative to the use of commercial nitrogen fertilizer in agriculture. This especially applies to "Rhizobium-legumes" symbiotic system that has priority in the process of nitrogen fixation. The symbiotic nitrogen fixation is an exceptionally complex biological process, in which macro as well as micro-symbionts, play specific roles. Many authors report about the suppressing influence of the heavy metals pollution on the growth and activity of free living and symbiotic nitrogen fixing organisms (Castro, 2000; Martyaniuk et al., 2003; Zviagintsev et al., 1997; Simon, 1999). This influence depends to a significant extend on the kind of the nitrogen fixing bacteria. Thus, Tong and Sadowsky (1994), announce, that the Bradyrhizobium strains are more stable to heavy metals pollution as compared to Rhizobium strains.

A.

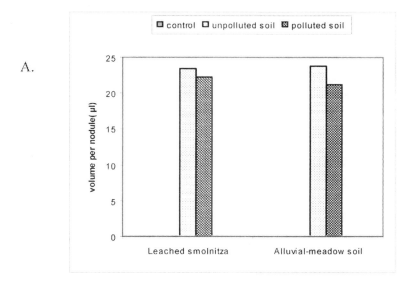

B.

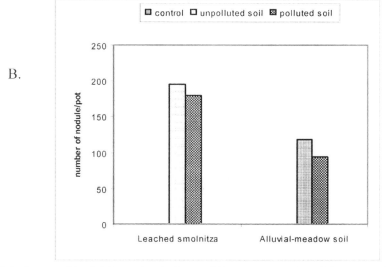

*Figure 3.* Influence of lead (Pb) on the virulence of *Br.japonicum*: A. nodule volume; B. nodule number.

In our studies we established that the influence of the pollutants on the symbiotic fixation of atmosphere Nitrogen depends on the type and the concentration of the pollutants, on soil properties, on the strain of the nitrogen-fixating bacteria.

Thus in pot experiment with two soils (Leached Smolnitsa-Pelic Vertisol and Alluvial-Meadow soil – Fluvisol) and soybean variety "Daniela" the obtained results showed, that lead in amount two times the Permissible Level Content decreased the virulence (Figure 3), symbiotically nitrogen fixing activity (Figure 4) and the efficiency (Figure 5) of the Bradyrhizobium japonicum strain 646 (Donkova, 2006). The negative influence was more clearly pronounced in the Alluvial–Meadow soil. The different degree of the lead influence can be explained with the higher cation exchange capacity of the Haplic Vertisol. As a result the bigger part of the lead is bound in the soil absorption complex, and thus has lower toxicity (Figure 3).

In the condition of pot experiment, carried out with three soils and soyaben sort "Boryana" (Figure 6) it was established that the influence of the escort herbicide on main properties of Br. japonicum depended on the soil characteristics. It was most strongly manifested on calcaric chernozem (Donkova & Chanova, 2002).

The investigation conducted in a vegetable pot experiment with carbonate chernozem, herbicides alachlor and acetochlor and four strains Br. japonicum (Figure 7) show that the herbicide effect depended on the nature of the strain and differ from stimulation, no effect to completely inhibition at high concentration (Donkova, 1998).

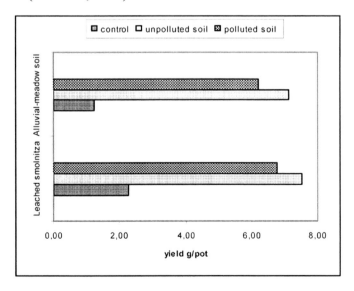

*Figure 4.* Fixed nitrogen amount.

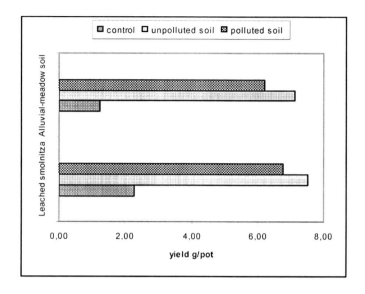

*Figure 5.* Influence of lead on the effectiveness of *Br. japonicum.*

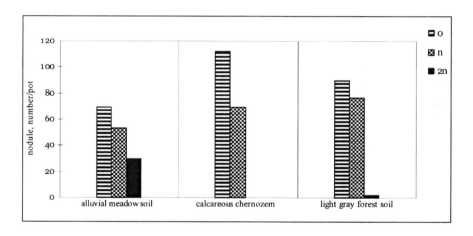

*Figure 6.* Influence of escort on the virulence of *Br. japonicum.*

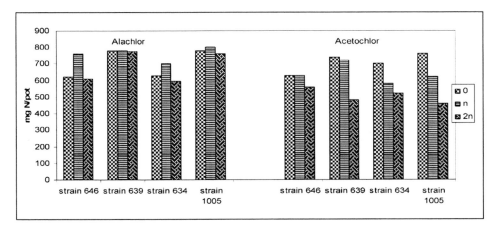

*Figure 7.* Influence of herbicides on *Br. japonicum* strain N₂ fixation activity.

Soil enzyme activities are responsible for soil organic matter decomposition and are involved in the supply of nutrients to crops (Nannipieri, 1994). Soil enzyme activity is also sensitive to soil heavy metal and pesticide pollution and it can therefore be used as a tool for monitoring modifications that occur in the soil environment caused by anthropogenic activities (Gianfreda & Bollag, 1996). Moreno et al. (2003a) observed that phosphatase activity was the most sensitive soil enzyme to evaluate soil contamination by heavy metals. According to these authors, the other analyzed enzymes were in the following sequence: urease > glicosidase > protease. Arylsulphatase, cellulose, catalase, dehydrogenase activities also are used in assessing impact of soil pollutants. The influence of dose, soil properties and available concentration of pollutants were observed.

The data available concerning use of the mycorrhiza fungi as bio indicators for the soil pollution is contradictive. Arbuscular mycorrhizal fungi (AMF) are soil microorganisms that establish mutual symbioses with the majority of higher plants, providing a direct physical link between soil and plant roots. The AMF occur in almost all habitats and climates. The activity of the heavy metals depends on the type of the mycorrhizal infection and the amount of the soil pollutant. For example, some mycorrhizal fungi can protect the plant root, from the activity of the heavy metals, due to which they live in symbiosis (Colpaert et al., 1997), others have increased sensibility to heavy metals pollution, and are suggested as bio indicators of this cases (Weissenhorn et al., 1995). Based on the data obtained from their study Del Val et al. (1999) concluded that the size and diversity of AMF populations were modified in metal-polluted soils, even in those with metal concentrations that were below the upper limits accepted by the European Union for agricultural soils. The PAH manifested strongly effects on AMF. Alarcon et al. (2006) found that Phenantrene reduced spore germination of Gigaspora margarita more than 90%, whereas Benzol[a]pyrene at 42%. In both PAH germinated spores had greater hyphal elongation.

## Conclusion

The data represented shows that due to high ecological resilience of the soil microorganisms they respond quickly to natural and anthropogenic impacts, adapting to the environmental conditions. Therefore, the changes in soil microbial equilibrium can serve as an "early warning" for negative alterations in the soil conditions long before they could be detected by classical chemical methods and before they could become irreversible. That is why the complex investigations on the soil biological activity should be conducted in assessing ecological risk of soil pollutants.The reliable microbiological indicator must be established and use.

## References

Aceves, M., Grace, C., Ansorena, J., Dendooven, L. & Brookes, P. (1999) Soil Microbial Biomass and Organic C in a Gradient of Zn Concentrations in Soils around a Mine Spoil Tip. Soil Biology and Biochemistry, 31(6), 867–876.

Alarcon, A., Delgadillo-Martines, J., Franko-Ramirez, A., Davies, F. T. & Ferrera-Cerrato, R. (2006) Influence of two polycyclic aromatic hydrocarbons on spore germination and phytoremediation potential of Gigaspora margarita – Edynochloa polystachya symbiosis in benzo[a]pyrene polluted substrate. Revista International Embiental, 22(1), 39–47.

Balinova, A. (1998) Environmental risk from point sources of pesticides in soil. Agricultural Science, 4, 51–54.

Bargett, R. & Saggta, S. (1994) Effect of Heavy Metal Contamination on the Short-term Decomposition of Labeled $^{14}$C Glucose in a Pasture Soil. Soil Biology and Biochemistry, 26(6), 727–733.

Blagodatskaya, Ye. V., Ananeva, N. D. & Miakshina, T. N. (1995) Characteristics of soil microbial community by the metabolic coefficient. *Pochvovedenie* (Russian Soil Science), 2, 205–210.

Bollen, W. L. & Tu, C. M. (1971) Influence of endrin on soil microial population and their activity. Res. Paper, US Forest Service PNW-114, 4 pp.

Braithwaite, B. M., Jane, A. & Swain, F. G. (1958) Effect of insecticides on sod sown sub clover. J. Aust. Inst. Agric. Sci., 24, 155–157.

Callao, V. & Montoya, E. (1956) Action de certains insecticides sur la croissance d'azotobacter dans le sol. VI Cong. Int. Sci. Sol. Rapp. C, 327–329.

Castro, I. (2000) Exotoxicological Effects of Heavy Metals in the Biolological Fixing of Nitrogen in Industrially Contaminated Soils. Silva-Lusitanaq 8(2), 165–194.

Chander, K. & Brookes, P. (1991) Microbial Biomass Dynamics during the Decompopsition of Glucose and Maize in Metal-contaminated and Non-contaminated Soils. Soil Biology and Biochemistry, 23(10), 917–925.

Chander, K. & Brookes, P. (1991a) Effect of Heavy Metals from Past Application of Sewage Sludge on Microbial Biomass and Organic Matter Accumulation in a Sandy Loam and Silty Loam. Soil Biology and Biochemistry, 23(10), 927–932.

Chandra, P. (1967) Effect of two chlorinated insecticides on soil microflora and nitrification process as influenced by different soil temperatute and textures. In: Progress in soil Biology, 320–330.

Chen, H. M., Zheng, C. R., Wang, S. Q. & Tu, C. (2000) Combined pollution and pollution index of heavy metals in red soil. Pedosphere, 10(2), 117–124.

Colpaert, J. & Assche, J. (1987) Heavy Metal Tolerance in Some Ectomycorhizal Fungi. Functional Ecology, 1, 415–421.

Colpaert, J. V., van Tichelen, K. K., Vangronsveld, J. (2000) Ectomycorrhizal fungi can protect their host trees against heavy metal toxicity. In: Proceedings of the InterCOST Workshop on Bioremediation, Sorrento, 15–18 November 2000, 72–74.

Dahlin, S., Witter, E., Mart, A., Turnew, A. & Baath, E. (1997) Where is the Limit? Changes in the Microbilogical Properties of Agricultural Soils at Low Level of Metal Contamination. Soil Biology and Biochemistry, 22(9–10), 1405–1415.

Del Val, C., Barea, J. M. & Azcón-Aguilar, C. (1999) Diversity of arbuscular mycorrhizal fungus populations in heavy-metal-contaminated soils. 1: Appl Environ Microbiol. Feb; 65(2), 718–23.

Donkova, R. & Petkova, D. (2003) Influence of lead on microbial activity of leached chernozem and light gray forest soil. Proceding Int. Scientific conference "50 Years University of Foresty. Session "Ecology and Environment protection", 11–13.

Donkova, R., Chanova, D., Petkova, D. & Markova, A. (1998) Influence of the herbicides alachlor and acetochlor on some properties of the strains Br. japonicum and their detoxication in the soil. Agricultural Science, 4, 55–56.

Donkova, R. & Chanova, D. (2002) Influence of Escort herbicide on the Br. japonocum in relation with soil characteristics. Soil Science, Agrochemistry and Ecology, XXXVII, 1–3.

Donkova, R. & Petkova, D. (2005) Influence of acetochlor on the microbial activity and its detoxication in the lead polluted soils. Proceedings National Conferencewith International Participattion: Management, Used and Protection of Soil Resources, 245–248.

Donkova, R. (2006) Lead impact on the basic properties of Bradyrhizobium japonicum. Bul. J. of Agricultural science, 12(5), 683–689.

Donkova, R. & Dinev, N. (2006) Microbiological Characteristic of soils in the area of non-ferrous metals factory, town of Plovdiv, Bulgaria. Eleventh Congress of the Microbiologists in Bulgaria,Varna, 2006.

Donkova, R. (2007) Influence of Rilay on microbiological activity of lead polluted soils. Proceedings International conference 60-years Institute of Soil Science – Soil Science – base for sustainable agriculture and environment protection. 13–17 May 2007, Sofia, 572–575.

Dusek, L. (1995) Activity of nitrifying populations in grass-land soil polluted by polychlorinated biphenils (PCPs). Plant and Soil, 176(2), 273–282.

Eno, C. F. (1957) Field accumulation of insecticide residues in soil. Exp. Sta. Rep. 142.

Fletcher, D. W. & Bollen, W. B. (1954) The effect of aldrin on soil microorganisms and some of their activities related to soil fertility. Appl. Microbiology, 2, 349–354.

French, N., Lichtenstain, E. P. & Thorne, G. (1959) Effect of some chlorinated hydrocarbon insecticides on nematode populations in soils. J. ecn. Ent, 52, 861–865.

Gaur, A. C. & Pareek, R. P. (1971) Tolerance of nitrification to DDT. Indian Journal of Entomology, 33(3), 368–370.

Gianfreda L. & Bollag J. M. (1996) Influence of natural and antropogenic factors on enzyme activity in soil. In: Stotzky, G. & Bollag J. M. Soil biochemistry. New York: Marcel Dekker, 9, 123–193.

Insam H., Parkinson, D. & Domsch K. (1989) Influence of Macroclimate on Soil Microbial Biomass. Soil Biology and Biochemistry, 21, 211–221.

Insam H., Mitchell C. & Dormaar J. (1991) Relationship of Soil Microbial Biomass and Activity with Fertilization Practice and Crop Yield of Three Ultisoils. Soil Biology and Biochemistry, 23(5), 459–464.

Hartenstein, R. C. (1960) The effect of DDT and malation upon forest soil microarthropods. J. eco. Ent., 53, 357–362.

Hiroki, M. (1992) Effect of heavy metal contamination on soil microbial population. Soil. Sci. Plant. Nutr., 38(1), 141.

Kaloyanova, N. & Kostov, O. (2004) Soilmicrobial characteristic and growth of lucerne at Cu contaminated soil. Soil Science and Ecology, 39(1), 26–31.

Kaloyanova, N. (2007) Effect of cadmium contamination on the microbiological properties of two soils and the yield of Lucerne. Proceedings International conference 60-years Institute of Soil Science – Soil Science – base for sustainable agriculture and environment protection. 13–17 May 2007, Sofia, 581–584.

Khan, M. & Scullion, J. (2000) Effect of metal (Cd, Cu, Ni, Pb or Zn) enrichment of sewage-sludge on soil microorganisms and their activities. Applied Soil Ecology, 20, 145–155.

Ko, W. H. & Lockwood, J. L. (1968) Convertion of DDT to DDD in soil and the effect of these compounds on soil microorganisms. Can. J. Microbiol. 14, 1069–1073.

Ko, W. H. & Lockwood, J. L. (1968) Accumulation and concentration of chlorinated hydrocarbon pesticides by microorganisms in soil. Can. J. Microbiol. 14, 1075–1078.

Kuperman R. & Carreiro M. (1997) Soil Heavy Metal Concentrations, Microbial Biomass and Enzyme Activities in a Contaminated Grassland Ecosystem. Soil Biology and Biochemistry, 29(2), 179–190.

McGrath, S. P., Chaudri A. M. & Giller K. E. (1995) Long-term effects of metals in sewage sludge on soils, microorganisms and plants. Journal of Industrial Microbiology and Biotechnology, 14(2), 94–104.

MacRae, I. C. & Vinckx, E. (1973) Effect of Lindane and DDT on population of protozoa in a garden soil. Soil Biol. Biochem., 5(2), 245–247.

Maliszewska-Kordybach, B. & Smreczak, B. (2002) Habitat function of agricultural soils as affected by heavy metals and polycyclic aromatic hydrocarbons contamination. Soil Science and Land Reclamation Department, Institute of Soil Science and Plant Cultivation, ul Czartoryskich, Pulawy, Poland, 8, 24–100.

Martin J. P., Harding, R. B. & Cannell, G. H. et al. (1959) Influence of fine annual field application of organic insecticides on soil biologycal and physical properties. Soil Sci, 87, 334–338.

Martyniuk, S., Wozniakowska A., Tujka A. & Martyniuk M. (2003) Microbial and biochemical characteristics of two soils treated with heavy metals and a reclaiming material. Pamietnik-Pulawski, 133, 115–121.

Masefield, G. B. (1955) Condition affecting the nodulation of leguminous crop in the field. Emp. J. exp. Agric., 23, 17–24.

Moreno, J. L., García, C. & Hernández, T. (2003) Toxic effect of cadmium and nickel on soil enzymes and the influence of adding sewage sludge. European Journal of Soil Science 54(2), 377–386.

Mutsumura, F., Boush, G. M. & Tai, A. (1968) Breakdown of dieldrin in the soil by a microorganisms. Nature. L., 219, 965–967.

Nannipieri, P. (1994) The potential use of soil enzymes as indicators of productivity, sustainability and pollution. In: Pankhurst C. E., Double, B. M., Gupta V. V. S. R., Grace P. R. (ed.). Soil biota management in sustainable farming systems. Melbourne: CSIRO, 238–244.

Petkova, G. & Donkova, R. (2006) Cadmium influence on microbiological activity of calcareous chrnozem. Eleventh Congress of the Microbiologists in Bulgaria, Varna'2006 (submitted for publication).

Powlson, D., Brokes, P. & Christensen, B. (1987) Measurement of Soil Microbial Biomass Provides an Early Indicator of Changes in Total Organic Matter due to Straw Incorporation. Soil Biology and Biochemisty, 19(1), 159–164.

Salonius, P. O. (1972) Effect of DDT and fenitrothion on forest soil microflora. J. of Entomology, 65(4), 1089–1090.

Simon, T. (1999) The effect of increasing rates of Ni and As on the growth of radish and soil microflora. *Rostlinna-Vyroda-UZPI*, 45 (9): 421–430.

Shamiyen, N. B. & Johanson, R. F. (1973) Effect of heptachlor on number of bacteria, actinomycetes and fungi in soil, Soil Biol and Biochem., 5, 3, 309–314.

Shaw, W. M. (1960) Pesticides Effects in soils on nitrification and plant growth. Soil Sci., 90, 320–323.

Sheals, J. G. (1955) The effect of DDT and BHC on soil Collembola and Acarina. Soil Zoology, 241–252.

Sheals, J. G. (1956) Soil population studies. Bull. ent. Res., 4, 803–822.

Shegunova, P., Teruze, K. & Atanasov, I. (2001) Priority organic pollutants in soils of Bulgaria. Assessment of the quality of contaninated soil and sites in Central and Eastern European Countries (CEEC) and New Independent States. Int. Workshop. Sept. – Oct. 3, Sofia, Bulgaria. Proceedings, 202–208.

Smejkalova, M., Mikanova, O. & Borunka, L. (2003) Effect of Heavy Metal Concentrations on Biological Activity of Soil Microorganisms.Plant, Soil and Environment-UZPI, 49, 7, 321–326.

Tate, K. R. (1974) Influence of four pesticide formulation on microbial processes in a New Zealand pasture soil. New Zealand J. of Agric. Res., 17, 1, 1–7.

Tong, Z. & Sadowsky, M. J. (1994) A selective medium for the isolation and quantification of *Bradyrhizobium japonicum* and *Bradyrhizobium elkanii* strains from soils and inoculants. Appl. Environ Microbiol., 60: 581–586.

Tyler G. (1981) Heavy Metals in Soil Biology and Biochemistry. In: Paul E., Ladd J. (eds.) Soil Biochemistry, vol. 5. Marcel-Dekker, New York, 371–414.

Varshney, T. N. & Gaur, A. C. (1972) Effect of DDT and Sevin on soil fungi. Acta Microbiol. Acad. Scient. Hungaricae., 19, 2, 97–102.

Wardle, D. & Parkinson D. (1990) Interactions between Microclimate Variables and the Soil Microbial Biomass. Biology and Fertility of Soils, 9, 273–280.

Weissenhorn, I., Merich, M. & Leyval, C. (1995) Bioavailability of Heavy Metals and Arbuscular Mycorhiza in Sewage-Sludge Amended Sandy Soil. Soil Biology and Biochemistry, 27, 287–296.

Wilkinson, A. T. S., Finlayson, D. G. & Morley, H. V. (1964) Toxic residues in soil 9 years after treatment with aldrin and heptachlor, Science, 143, 681–682.

Yagnow, G. & Haider, K. (1972) Evolution of $14CO_2$ from soil incubated with dieldrin-14C and the action of soil bacteria on labelled dieldrin. Soil Biol. Biochem., 4, 1, 43–49.

Zviagintsev, D., Kurakov, A., Umarov, M. & Z., Philip. (1997) Microbiological and biochemical indexes of polluted with lead Podsolic soil. *Pochvovedenie* (Russian Soil Science), 9: 1124–1131.

# ASSESSMENT OF INORGANIC PRIORITY POLLUTANTS IN CONTAMINATED SOILS: HARMONIZATION OF ANALYTICAL PROTOCOLS FOR HEAVY METAL EXTRACTION: ANALYTICAL SPECIATION

ROSER RUBIO[*] AND JOSÉ FERMIN LÓPEZ-SÁNCHEZ
*Universitat de Barcelona*
*Department of Analytical Chemistry*
*Martí I Franquès 1-11, E-08028 Barcelona, Spain*

**Abstract.** In environmental studies heavy metals as well as some metalloids, such as arsenic, are considered as priority pollutants. The development and use of extraction schemes aims the evaluation of the metal fractions available to plants and the evaluation of the mobility of element under changing environmental parameters. The extracting agents are not selective and the "extractable metal" is defined under operational conditions. The lack of harmonized procedures did not allow the results to be compared worldwide. The first part of the current chapter deals on the studies carried out within the Standards Measurements and Testing Program (formerly BCR) of the European Union and the proposals of extraction systems for both soils and sediments, as well as the production of soils Certified Reference Materials for extractable metals, in such Programs our research group participated actively. A second strategy is based on analytical speciation. Such approach permits the extraction, separation and quantification of the chemical species present in the soil. The knowledge of the chemical species is a valuable tool for further studies on metal transport and on their translocation in plants. The main goal of that strategy is to achieve the separation and quantification of the chemical species of the element of interest with the highest possible selectivity and sensitivity while maintaining its integrity. The second part of the present chapter deals mainly on the studies carried out in our group on arsenic speciation in soils by emphasizing on some critical analytical aspects such as the stability of the species during the analysis as well as their recovery. Some of our recent studies

---

[*] To whom correspondence should be addressed: roser.rubio@ub.edu

L. Simeonov and V. Sargsyan (eds.),
*Soil Chemical Pollution, Risk Assessment, Remediation and Security.*
© Springer Science+Business Media B.V. 2008

on selenium speciation in soils and some new results on the most recent studies on antimony speciation are also reported.

**Keywords:** analytical protocols, analytical speciations, heavy metals, inorganic pollutants, priority pollutants, risk assessment, contaminated soil

## 1.  Selective Extraction Strategies

Trace elements in soils can appear in different chemical forms or ways of binding. In unpolluted soils, trace elements exist mainly as relative immobile species in silicates, aluminates and other primary minerals but as a result of weathering these can be gradually mobilised to soil solution and became available to plants. In polluted soils, trace elements are mainly in more labile forms (sorbed, complexed, co-precipitated, etc.) and have and important contribution to the pool of potentially available metals. The most relevant trace elements are arsenic, boron, cadmium, chromium, cobalt, copper, lead, molybdenum, nickel, selenium, titanium, vanadium, and zinc. Some of these trace metals such us chromium, nickel or zinc are essential to plant growth but they have toxic effects at high levels. Others, such cadmium or lead, are non-essential and potentially toxic.

In environmental, agricultural and geochemical studies the determination of the ways of metal binding in soils and sediments provides more useful information on element mobility and availability than the determination of the total content. However, the determination of the different ways of binding is quite difficult due to the complexity of the analysed matrix and often impossible. Different analytical approaches are used, many of them focused on element desorption from the solid phase using chemical reagents; others are focused on the element adsorption from a solution by the solid phase or in the use of instrumental techniques such as X-ray (Kersten et al., 1989). Among them the approaches based on extraction/leaching procedures are the most widely accepted and used. During the last decades, extraction procedures for extractable heavy metals in soils have been developed and modified. In this respect, two groups of tests should be considered: the single reagent extraction test, one extraction solution on one soil sample, and the sequential extraction procedures, several extraction solutions are used sequentially on the same sample, although this last type of extraction is still in development for soils. Both types of extractions are applied, using not only different extracting schemes but also under different experimental conditions. This leads to the use of a great deal of extraction procedures and to the production of operationally defined results that make difficult data comparison and validation. As the results obtained by applying leaching procedures is procedure dependent, there is no fully satisfactory method available, and attempts are being made to improve existing methods or adapt them to specific problems or circumstances. As a result a large number of procedures have emerged, using not only different types of extracting solutions, including different pH values for the same extractant, but also a variety of laboratory conditions. The main variables which are modified may be summarised as follows: solid/liquid ratio, time of extraction, temperature, shaking intensity and type of shaking, the method for liquid-solid separation and the use or not of inert atmosphere. As all these factors affect the results obtained, leaching procedures in general cannot be compared directly.

## 1.1. SINGLE EXTRACTION PROCEDURES

Extraction procedures by use of a single extractant are widely used in soil science. These procedures are designed to dissolve element contents correlated with the availability of the element to the plants. This approach is well established for major components and nutrients and it is commonly applied in studies of fertility and quality of crops. The approach is also applied to predict the plant uptake of essential elements, to determine element deficiency or excess in a soil, to study the physical-chemical behaviour of elements in soils or in survey purposes. They are also applied, in a lesser extent, to elements considered as pollutants such as heavy metals. The application of extraction procedures to polluted or naturally contaminated soils is mainly focussed to ascertain the potential availability and mobility of the pollutants and its migration in a soil profile, which is usually connected with groundwater problems (Van der Sloot et al., 1997). Single extraction procedures are always restricted to a reduced group of elements and they are applied to a particular type of soil, siliceous, carbonated or organic. In a regulatory context, two applications for leaching tests can be recognised: the assessment or prediction of the environmental effects of a pollutant concentration in the environment and the promulgation of guidelines or objectives for soil quality as for example for land application of sewage sludge or dredged sediments. The data obtained when applying these tests are used for decision-makers in topics such as land use of soil or in countermeasures application (Rauret, 1998). Table 1 shows a summary of the most common leaching test used in soil analysis.

From the table it can be observed that single extraction approaches includes a large spectrum of extractants. It embraces from very strong acids, such as *aqua regia*, nitric acid or hydrochloric acid, to neutral non buffered salt solutions, mainly $CaCl_2$ or $NaNO_3$. Other extractants such as buffered salt solutions or complexing agents, because of their ability to form very stable water-soluble complexes with a wide range of cations, are frequently applied. For boron, hot water is also used. Basic extraction by using sodium hydroxide is used to assess the influence of the dissolved organic carbon in the release of heavy metals from soils. Information and details about a large number of extractants was extensively reviewed (Pickering, 1986; Lebourg et al., 1996).

TABLE 1. Extraction tests used in soil analysis

| Reagent group | Type and solution concentration |
|---|---|
| Unbuffered salt solution | $CaCl_2$ 0.1 mol $l^{-1}$ (Novozamski, 1993) <br> $CaCl_2$ 0.05 mol $l^{-1}$ (Novozamski, 1993) <br> $CaCl_2$ 0.01 mol $l^{-1}$ (Novozamski, 1993) <br> $NaNO_3$ 0.1 mol $l^{-1}$ (Gupta, 1993) <br> $NH_4NO_3$ 1 mol $l^{-1}$ (Novozamski, 1993) <br> $AlCl_3$ 0.3 mol $l^{-1}$ (Hughes, 1991) <br> $BaCl_2$ 0.1 mol $l^{-1}$ (Juste, 1988) |
| Buffered salt solution | $NH_4$-acetate, acetic acid buffer pH =7, 1 mol $l^{-1}$ (Ure, 1993b) <br> $NH_4$-acetate, acetic acid buffer pH = 4.8, 1 mol $l^{-1}$ (Novozamski, 1993) |
| Chelating agents | EDTA 0.01–0.05 mols $l^{-1}$ at different pH (Novozamski, 1993) <br> DTPA 0.005 mol $l^{-1}$ +TEA 0.1 mol $l^{-1}$ + $CaCl_2$ 0.01 mol $l^{-1}$ (Lindsay, 1978) <br> Melich 3: <br> $CH_3COOH$ 0.02 mol $l^{-1}$ + $NH_4F$ 0.015 mol.$l^{-1}$ + $HNO_3$ 0.013 mols $l^{-1}$ + EDTA 0.001 mol $l^{-1}$ (Melich, 1984) |
| Acid extraction | $HNO_3$ 0.43–2 mol.$l^{-1}$ (Novozamski, 1993) <br> Aqua regia (Colinet, 1983) <br> HCl 0.1$^{-1}$ mol.$l^{-1}$ (Novozamski, 1993) <br> $CH_3COOH$ 0.1 mol.$l^{-1}$ (Ure, 1993a) <br> Melich 1: HCl 0.05 mol.$l^{-1}$ +$H_2SO_4$ 0.0125 mol.$l^{-1}$ (Mulchi, 1992) |

The increasing performance of the analytical techniques used for element determination in an extract, together with the increasing evidence that exchangeable metals better correlate with plant uptake, has lead extraction methods to evolve towards the use of less and less aggressive solutions (Gupta et al., 1993). These solutions are sometimes called soft extractants and are based on non-buffered salt solutions although diluted acids and complexant agents are also included. Neutral salts dissolve mainly the cation exchangeable fraction although in some cases the complexing ability of the anion can play a certain role. Diluted acids solubilise partially trace elements associated to different fractions such as exchangeable, carbonates, iron and manganese oxides and organic matter. Complexing agents solubilise not only exchangeable element fraction but also the element fraction forming organic matter complexes and the element fraction fixed on the soil hydroxides. Nowadays, it is generally accepted that extractants are not selective and that minor variations in analytical procedures have significant effect on the results. According to Lebourg et al. (1996) some of these methods have been adopted officially or its adoption is under study in different countries with different objectives. An account of these methods is given on Table 2.

TABLE 2. Extraction methods normalised or proposed for normalisation in several European countries

| Country | Method | Objective |
|---|---|---|
| Germany | 1 mol·l$^{-1}$ NH$_4$NO$_3$ | Mobile trace element determination (DIN, 1993) |
| France | 0.01 mol·l$^{-1}$ Na$_2$-EDTA + 1 mol·l$^{-1}$ CH$_3$COONH$_4$ at pH = 7 DTPA 0.005 mol·l$^{-1}$ + TEA 0.1 mol·l$^{-1}$ + CaCl$_2$ 0.01 mol·l$^{-1}$ at pH = 7.3 | Available Cu, Zn and Mn evaluation for fertilisation purposes (AFNOR, 1994) |
| Italy | 0.02 mol·l$^{-1}$ EDTA + 0.5 mol·l$^{-1}$ CH$_3$COONH$_4$ at pH=4.6 DTPA 0.005 mol·l$^{-1}$ + TEA 0.1 mol·l$^{-1}$ + CaCl$_2$ 0.01 mol·l$^{-1}$ at pH = 7.3 | Available Cu, Zn, Fe and Mn evaluation in soils (UNICHIM, 1991) |
| Netherlands | CaCl$_2$ 0.1 mol·l$^{-1}$ | Availability and mobility of heavy metals in polluted soils evaluation (Houba, 1990) |
| Switzerland | NaNO$_3$ 0.1 mol·l$^{-1}$ | Soluble heavy metal (Cu, Zn, Cd, Pb and Ni) determination and ecotoxicity risk evaluation (VSBo, 1986) |
| United Kingdom | EDTA 0.05 mol·l$^{-1}$ at pH = 4 | Cu availabitity evaluation (MAFF, 1981) |

## 1.2. SEQUENTIAL EXTRACTION PROCEDURES

These procedures are widely applied for sediment analysis and are focused on differentiating among the several association forms of metals in the solid phases. To do so, several extracting reagents are applied sequentially to the sample according the following order: non buffered salts, weak acids or buffered salts, reducing agents, oxidising agents, and strong acids.

During the last decade, there was increasing interest on applying sequential extraction to study trace metal partitioning in soils, although first studies on nutrient element fractionation were already reported some decades ago (Williams et al., 1971a, b; Sommers et al., 1972). Most of the published literature is based on the work of Tessier et al. (1979), but new approaches, improvements and/or modifications have been also proposed (Arunachalam et al., 1996; Flores et al.,

1997; Maiz et al., 1997; Thöming et al., 1998; Narwal et al., 1999), i.e. the modification of the procedure to allow multi elemental determination by ICP-AES (Li et al., 1995) or in order to analyse a calcareous matrix (Orsini et al., 1993). In this way, the work carried out in Europe to develop a harmonised sequential extraction procedure is remarkable, i.e. the so called BCR procedure (Ure et al., 1993a; Sahuquillo et al., 1999; Rauret et al., 1999), that gained acceptance among the scientists using such procedures (Barona et al., 1999; Szakova et al., 1999; Díaz-Barrientoset et al., 1999; Sutherland et al., 2000) during the last decade. There are also studies dealing on problems of sequential extraction schemes already reported for sediments, as redistribution of metal fractions during the extraction process (Lo et al., 1998; Kim et al., 1991; Bunzl et al., 1999) or lack of selectivity when dissolving the different soil phases (Benitez et al., 1999).

However, most of the work carried out is focused to the use of sequential extraction as a tool to evaluate availability of metals to plants (Chlopecka, 1993, 1996c; Qian et al., 1996; Planquart et al., 1999; Aumada et al., 1999; García-Sánchez et al., 1999; Maiz et al., 2000; Simeonova and Simeonov, 2006) or to study metal distribution and/or mobility in polluted, forest and agricultural soils (Jeng et al., 1993; Ramos et al., 2000). In the first type of studies, most of the results show that some correlations exist between the exchangeable and the acid-soluble fractions and the plant uptake, although the relationship seems to be dependent on the type of soil and the type of plant. These results indicate that sequential extraction procedures, though operationally defined, may provide complementary and valuable data in order to predict metal availability to plants. In relation to the mobility and/or pollution studies, the main conclusion than can be drawn is that sequential extraction is useful to determine contamination problems, because metals from anthropogenic sources are retained by non residual fractions and, consequently, are more mobile than those from soil parent materials. On the other hand, Cd, Cu and Zn appear as mobile metals, whereas Cr, Ni and Pb are more strongly retained.

Moreover, there is literature about other uses of sequential extractions in soil science. In this way, there are studies on the heavy metal retention by silty soils (Cabral et al., 1998), the relationships between metal sorption and chemical forms (Schalscha et al., 1999), or the evaluation of the efficacy of restoration strategies using sequential extraction (Barona et al., 1996; Mulligan et al., 1999).

## 1.3. THE BCR APPROACH

In 1987 the European Commission launched a programme (in the framework of the BCR) aimed at harmonising extraction procedures, single and sequential, for the determination of trace metals in soils and sediments. The program started with a literature survey and a consultation with European experts (Ure et al., 1992). They identified the main requirements for procedures that could be widely accepted and able to be used with regulation purposes. These needs were summarised as single extraction for soil and sequential extraction for sediments. The pollutants considered to have priority were Cd, Cr, Cu, Ni, Pb and Zn and the need to prepare certified reference materials (CRMs) for their extractable trace metal content to be use in method validation in control laboratories was pointed out. As a complement to this study, interlaboratory exercises with the participation of expert laboratories took place from 1987 to 1990 and the outcome of the survey and the results of these trials were discussed in a workshop in 1992 (Ure et al., 1993a). The exercises showed the need for a better definition of the protocols used in order to obtain results that would be reproducible among laboratories. Moreover, the participants agreed on the need to move together in the direction of harmonising the extraction protocols among European countries. As a starting point well described single extraction protocols for soil, based on acetic acid and EDTA, and for sediment, based on a three-step sequential extraction protocol, were adopted and the need to prepare reference materials for extractable contents based on these procedures was again stressed.

In the third European Commission Framework Programme a project was developed (ETMESS) for validating the extraction procedures and for the preparation of CRMs. As a result of this project, the first CRMs for their extractable content of trace metals from soils and sediments were prepared. Four CRMs, three soils and one sediment, prepared within this project were available through the Institute for Reference Materials and Measurements. The three soil reference materials are CRM 483 and CRM 484, two sewage sludge-amended loamy soils with low organic matter content, and CRM 600, a calcareous soil and the sediment reference material is CRM 601. The success in the application of validated procedures for soils and sediments and the widespread use of the CRM already prepared for extractable contents highlighted the need to develop additional CRMs. Organic soils were considered to be as the most appropriate for certification of extractable heavy metal content in single extraction methods. A new CRM for sediment would need to be prepared for sequential chemical extraction after refining the procedure. It was also decided to assess whether the sequential extraction scheme could be applied to soil samples.

In the Fourth European Commission Framework Programme a new project was presented and approved for preparing new CRMs for their extractable trace metal content in soils and sediments (TRAMES). The new materials proposed were a soil enriched in organic matter, BCR-700, for certifying extractable heavy metal content by using acetic acid and EDTA, and a sediment, BCR-701, for the application of an improved three-step sequential chemical extraction. Moreover prior to certification it was proposed to carry out a study to overcome the sources of error in applying the three-step sequential extraction to sediment and to apply it as a feasibility study to a soil material (Sahuquillo et al., 1999). For this purpose it was chosen the already certified sediment material CRM 601 and the sewage sludge amended soil certified for its extractable contents by applying single extraction, CRM 483. In this context two new materials were prepared BCR-700, a soil enriched in organic matter, and BCR-701, a new sediment sample. In addition, after applying the modified procedure to the soil and the sediment samples, new information has become available for CRM 483 and CRM 601.

Table 3 summarises the reference materials produced and gives an indication on the certified or indicative values obtained for each material. The details on the preparation of materials, extraction procedures and certification campaigns have been published elsewere (Quevauviller et al., 1996, 1997a, b, c; Rauret et al., 2000a, b, 2001; Pueyo et al., 2001).

Figure 1 shows as flow diagrams the single and sequential extraction procedures as detailed at the end of the various certification campaigns. Note that the sequential extraction procedure finishes with a fourth step on the residue of the third step. This was added as an internal check on the procedure in order to carry out a mass balance between element extracted along the sequential extraction and the pseudo-total content extracted by *aqua regia* digestion.

In Table 4 the pseudo-total results obtained from the BCR-701 original sample are compared with the sum of the extracted metals from the three steps plus residual ($\Sigma$3steps+*aqua regia* extractable from residue). No significant differences were observed between the total metal extracted following the *aqua regia* protocol and the sum of extracted metals following the sequential extraction procedure, which indicates the good quality of the results obtained.

## 1.4. RECENT APPLICATIONS OF BCR METHODOLOGIES AND MATERIALS

After the development of the different procedures, these have been applied by numerous researchers as can be observed by inspecting to some literature databases as that from Chemical Abstracts Service. In this section some examples of application are mentioned.

TABLE 3. Soil and sediment certified reference materials for extractable trace elements produced in the framework of the BCR programme of the European Commission

| Procedure | Certified values | Indicative values |
|---|---|---|
| **BCR-483,** sewage sludge amended soil, Northampton (United Kingdom) | | |
| CH$_3$COOH 0.43 mol l$^{-1}$ | Cd, Cr, Cu, Ni, Pb, Zn | – |
| EDTA 0.05 mol l$^{-1}$ pH 7 | Cd, Cr, Cu, Ni, Pb, Zn | – |
| CaCl$_2$ 0.01 mol l$^{-1}$ | – | Cd, Cr, Cu, Ni, Pb, Zn |
| NaNO$_3$ 0.1 mol l$^{-1}$ | – | Cd, Cr, Cu, Ni, Pb, Zn |
| NH$_4$NO$_3$ 1 mol l$^{-1}$ | – | Cd, Cr, Cu, Ni, Pb, Zn |
| **BCR-483,** sewage sludge amended soil, Northampton (United Kingdom) | | |
| Modified sequential extraction | – | In all three steps: Cd, Cr, Cu, Ni, Pb, Zn |
| **BCR-484,** Sewage sludge amended soil, Catalonia (Spain) | | |
| CH$_3$COOH 0.43 mol l$^{-1}$ | Cd, Cu, Ni, Pb, Zn | Cr |
| EDTA 0.05 mol l$^{-1}$ pH 7 | Cd, Cu, Ni, Pb, Zn | Cr |
| CaCl$_2$ 0.01 mol l$^{-1}$ | – | Cd, Cr, Cu, Ni, Pb, Zn |
| NaNO$_3$ 0.1 mol l$^{-1}$ | – | Cd, Cr, Cu, Ni, Pb, Zn |
| NH$_4$NO$_3$ 1 mol l$^{-1}$ | – | Cd, Cr, Cu, Ni, Pb, Zn |
| **BCR-600,** Calcareous soil, San Pellegrino Parmense (Italy) | | |
| EDTA 0.05 mol l$^{-1}$ pH 7 | Cd, Cr, Ni, Pb, Zn | Cu |
| DTPA | Cd, Ni | Cr, Cu, Pb, Zn |
| **BCR-601,** Lacustrine Sediment, Lago Maggiore (Varese, Italy) | | |
| Original sequential extraction | 1st: Cd, Cr, Ni, Pb, Zn  2nd: Cd, Ni, Zn  3rd: Cd, Ni, Pb | 1st: Cu  2nd: Pb  – |
| Modified sequential extraction | – | In all three steps: Cd, Cr, Cu, Ni, Pb, Zn |
| **BCR-700,** Organic rich soil, Hagen (Germany) | | |
| CH$_3$COOH 0.43 mol l$^{-1}$ | Cd, Cr, Cu, Ni, Pb, Zn | – |
| EDTA 0.05 mol l$^{-1}$; pH 7 | Cd, Cr, Cu, Ni, Pb, Zn | |
| **BCR-701,** Lacustrine Sediment, Lago Orta (Piemonte, Italy) | | |
| Modified sequential extraction | In all three steps: Cd, Cr, Cu, Ni, Pb, Zn | – |

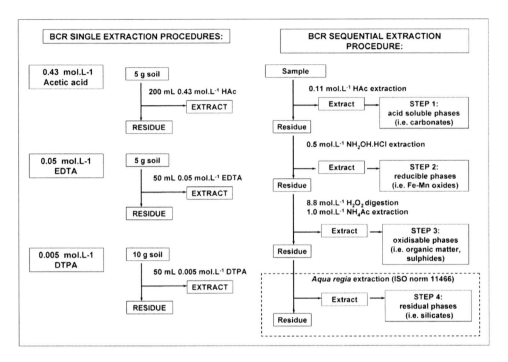

*Figure 1.* Single and sequential extraction procedures developed in the framework of the BCR Programme.

TABLE 4. Mass balance in sequential extraction. Comparison of the sum of three steps plus residual with the pseudo-total *aqua regia* extracted metal

| Element | Σ(3 steps + residual) (mg/kg) | ± | S.D. (mg/kg) | *Aqua regia* extraction (mg/kg) | ± | S.D. (mg/kg) |
|---------|---------|---|------|----------|---|------|
| Cd | 11.5 | ± | 0.46 | 11.7 | ± | 1.0 |
| Cr | 253 | ± | 10 | 272 | ± | 20 |
| Cu | 267 | ± | 12 | 275 | ± | 13 |
| Ni | 98.7 | ± | 4.4 | 103 | ± | 4.0 |
| Pb | 149 | ± | 6.3 | 143 | ± | 6.0 |
| Zn | 459 | ± | 15 | 454 | ± | 19 |

The assessment of element mobility in soils and the availability to plants has been the main application of the single and sequential extraction procedures (Schramel et al., 2000; Kowalska et al., 2002; Ayoub et al., 2003; Geebelen et al., 2003; Kaasalainen et al., 2003; Niesiobedzka, 2004; Tokalioglu et al., 2004; De Gregori et al., 2004; Kubova et al., 2005; Adamo et al., 2006; Alvarez et al., 2006; Pérez de Mora et al., 2006a, b; Davidson et al., 2007; Zemberyova et al., 2007). In some works the comparison of the results obtained by single, sequential extraction and other leaching tests (Sutherland et al., 2002; Tokalioglu et al., 2003; Margui et al., 2004; Kubova et al., 2004; Guevara-Riba et al., 2004; Kazi et al., 2005; Krasnodebska-Ostrega et al., 2006; Nakazatoet al., 2006) or electrochemical techniques (Kowalska et al., 2002) is also reported in order to determine the relevance of the environmental conclusions that can be drawn from the data obtained. Moreover, the use of microwave or sonication has been tested to speed up the extraction (Ipolyi et al., 2002; Canepari et al., 2005; Pazos-Capeans et al., 2005; Dutta et al., 2005). In this way some methodology for continuous extraction has been proposed (Chomochoei et al., 2002; Fedotov et al., 2005; Tongtavee et al., 2005; Song et al., 2006) in order to replace the classical batch experiment.

In relation to the certified reference materials sediments (BCR-601 and BCR-701) have been widely used, whereas the soil BCR-483 is the soil material with a better acceptance, probably due to the fact that indicative values for the extractable trace element contents with mild extraction procedures are also reported. The reference materials have been used to validation, quality control and the development of new methodological approaches. These materials are available from IRMM (IRMM web page: http://www.irmm.jrc.be/ html/homepage.htm).

Moreover the Reference Materials for Quality Control (QCM) for the routine use in the laboratories is highly recommended. Information on some of these materials can be obtained from Mat Control, a laboratory located in the Faculty of Chemistry of the University of Barcelona, together with some QCM soils for total and extractable metals: http://www.ub.es/dqa/grups/webquestram/ questramcat/serveis/unimaterials/english/01.htm.

## 2. Analytical Speciation Strategy

Analytical speciation permits to obtain information about chemical species of the elements of interest, such as oxidized or reduced forms, organometallic compounds, small complexes or macromolecules. In soils this knowledge is considered an important tool to determine the interaction of the species not only with the soil components but with the transport to the soil solution and to the plants growing in it. Moreover the information on the chemical species can be valuable to evaluate soil pollution levels and any possibility of remediation.

In this kind of studies a suitable procedure has to be established in such a way that the recovery studies have to perform by means of mass balance calculations. For this the determination of total metal content in both the sample and in the extracts has to be assessed as well as the quantification of the element species detected. For the determination of the total element in soils *aqua regia* digestion is recommended (ISO 11466 NORM 1995) but in some cases the main components of the soil matrix and their presence in the resulting solution could interfere seriously in the final measurement. For metal species determination the most critical point is their extraction form the original sample, since the system applied has to provide good recovery by maintaining the integrity and the stability of the species during the process. Several extraction systems have been proposed with the aim to provide good recovery but preserving the integrity of the species during the process (Hudson et al., 2004; Kahakachchi et al., 2004; Száková et al., 2005). For this physicochemical processes such as oxidation, volatilization, inter-conversion or adsorption have to be controlled. Additionally checking the stability of the species during the period from extraction to the final measurement is highly recommended. Pre treatment of the sample prior extraction could be also determinant, thus in some cases occasional increases of temperature during grinding can cause alterations of some organometallic species. Recommendations on the above mentioned points are reported in the literature (Rosenberg et al., 2001; Cornelis et al., 2003) mainly dealing on biological material and food, but no specific protocols for element analytical speciation in soils are reported. Nowadays it is widely accepted that the determination of chemical species in soils is a difficult task. This subject will be discussed through three selected elements such as arsenic, selenium and antimony, due to their environmental interest as well as the relevant role of selenium in agricultural soils. The occurrence of these elements in soils is briefly discussed and some results from the experience of our research group, dealing with speciation of such elements in these materials are reported here.

## 2.1. ARSENIC IN SOILS

As far as arsenic is concerned the occurrence of this element in soils is related to the use of pesticides and herbicides in agricultural practices as well as to the irrigation with water with high arsenic content. This last situation occurs in many parts of the world, among them Bangladesh is the one most representative region, where irrigation waters are naturally polluted and the concentration levels of arsenic can be very high. In these countries natural contamination of groundwater receives the greatest attention and more recently the contamination of soils, attributed to their role as a sink for arsenic, is a matter of interest due to the transfer of this element to the crops and consequently to the introduction into the food chain (Naidu et al., 2006).

Moreover industrial and mining residues can contribute to soil pollution. The knowledge of the arsenic species in soils is necessary for the studies on their interactions with living organisms and with the environment. The arsenic species identified in soils are mainly inorganic, arsenite and arsenate, but some methylated species generated under reactions microbially mediated, can be detected at lower amounts (Kuehnelt et al., 2003; Quaghebeur et al., 2005; Yehl et al., 2001). For total arsenic determination in soils digestion of the sample with *aqua regia* is a suitable choice and further derivatization step consisting of hydride generation is the most widely used. With this technique matrix interferences are avoided and high selectivity and sensitivity can be obtained. For the final detection atomic spectroscopic techniques, mainly ICP/AES or AFS are used and more recently ICPMS. Regarding the quantification of arsenic species coupled techniques have to be used. Among them Liquid Chromatography coupled with atomic spectroscopic techniques or to ICPMS are the most widely used and all of them can incorporate the hydride generation step before the detection, as a part of the overall coupling.

From the proposals for the extraction of arsenic species from soils, a mixture of phosphoric acid ascorbic acid is revealed as a suitable extractant (Ruiz-Chancho et al., 2005). In a study on arsenic contaminated soils from an old mining zone in the Pyrenees arsenite and arsenate have been measured in the extracts obtained by using this acidic mixture (Ruiz-Chancho et al., 2007). The properties of this reagent for solving arsenic in that soils was attributed on one hand to the ability to leaching the arsenate generated through the wheathering of the original arsenopyrite and further adsorption on the surface of the iron-oxides particles (Dove et al., 1985) and on the other hand to the generation of arsenite when the extractant used reacted with the remaining non wheathered arsenopyrite (Yu et al., 2004). Moreover and in relation to the quantification of the species the main problems are originated by the instability of them in the soil extracts and on the individual recovery studies of the species. Thus, in a study on polluted soils after an episode in Aznalcollar (Spain), we detected small amounts of methylarsonate (MA) and dimethylarsinate (DMA) and we realized that the stability of these organic species in the extracts was altered with time (Garcia-Manyes et al., 2002). Even in some cases oxidation is observed, in some extent, during soil grinding as the first pre-treatment of the sample. To our experience, purge of the extracts with an inert gas is recommended to avoid oxidation of arsenite to arsenate during the period of time from extraction to the measurement. Some examples of these behaviours can be observed in Figure 2.

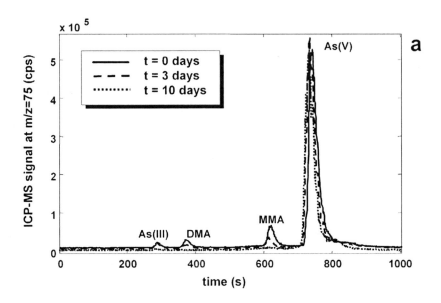

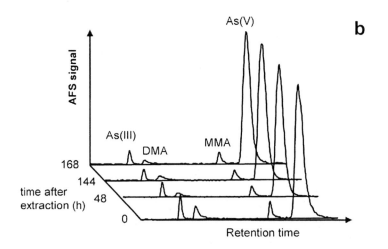

*Figure 2.* Stability arsenic species in soil extracts: (a) soil extract from sample collected after Aznalcollar (Spain) spill (adapted from Garcia-Manyes, 2002), (b) soil extract from sample collected in Massif Central (France) (adapted from Ruiz-Chancho, 2005).

## 2.2. SELENIUM IN SOILS

Regarding to selenium it is considered as an essential element but it can be toxic at high concentrations. The narrow interval of concentration between the two opposite effects requires accurate and precise knowledge of the selenium as well as each species present in the environment. The interest on selenium regarding to its toxic character related to soils is also relevant and due to this several international agreements such as that of United Nations Environmental Programme (UNEP) consider this element in the guidelines on landfill practices (Technical Guidelines, 2003). It is also stated that that the primary source of selenium in plants and humans is attributable to soils.

The selenium contents in soils varies widely according to the region, several hundreds of μg/g can be measured in seleniferous soils whereas in soils with deficiency in selenium its content can lower than be four orders of magnitude (Kabata-Pendias et al., 2001). In many countries some agricultural practices are based on the use of selenium enriched soils for correcting nutritional deficiency (Smrkolj et al., 2006) or on the addition of selenite or selenate to fertilizers (Carvalho et al., 2003). A relevant interest in this field lies on the uptake and transformations of the selenium in the crops. In this context it is stated that plants can transform inorganic selenium into seleno aminoacids, such as selenomethionine. Moreover some organic compounds present in soils can interact with selenium affecting the availability by plants. In spite of these considerations selenium speciation in soils is scarcely studied and several selenium-containing organic molecules still remain unidentified (Pezzarossa et al., 2007).

Even the determination of total selenium in soils can be a difficult task, mainly due to the complexity of the matrix composition of the soils and to the low levels that have to be determined in some soils. Even when using a standardized procedure such as the *aqua regia* digestion, the composition of the resulting solution can cause serious interferences in the final measurement where the conditions have to be rigorously established (González-Nieto et al., 2006). The literature on selenium speciation in soils is very scarce and no agreement is evidenced about the extraction systems. With regard to stability of selenium species some reports (Capelo et al., 2006) show that there is no consensus about the optimum conditions for preservation of their integrity during the analytical procedure, but in such reports soils extracts are not considered.

According to the experience of our group from a study on selenium speciation of various soils such as seleniferous natural soils and selenium contaminated soils, phosphoric acid was revealed as a suitable extracting agent, with recovery levels ranging from 60% to 70%. To our experience on the stability of inorganic species selenite and selenate in the extracts form the above mentioned soils no oxidation of selenite was observed in a short term after extraction.

## 2.3. ANTIMONY IN SOILS

With regard to antimony, some anthropogenic origins contribute to pollution of this element on soils, among them mining activities and lead alloys manufacture. Antimony is also considered as a traffic-related pollutant, since its presence in the roadways dust particles is attributed to the automotive brake pads wear of automobiles. It is stated that another source of antimony pollution is attributable to shooting activities in which bullets of PbSb alloys are used. The interest on the presence of this element in the environment was successfully presented at the first International Workshop on the Antimony in the Environment held in Heidelberg in 2005. Very few studies have been devoted to the presence and mobility of antimony in soils. Sequential extraction schemes are not suitable for determine the association of the element to the different forms of soils and no methods are established for the determination of the chemical species.

There is little literature describing the determination of antimony species in soils and the main difficulty can be attributed to the different types of binding to such matrices. Therefore, a variety of "soft extractants" were tested in the few experiments reported in the literature for extracting antimony with the aim to obtain different types of information. We describe in the present section the results from the recent studies carried out in our research group on the establishment of methodology for the determination of antimony and its species in contaminated soils (Miravet, 2007). The soils were sampled from the *Vall de Ribes*, in the eastern Pyrenees. This zone presented an active mining activity which at the beginning of the nineteenth century was producing small amounts of several metals and now is an abandoned zone colonized by vegetation. For determining the total content of antimony the digestion with *aqua regia* was applied and the higher contents (up to 3,180 mg Sb kg$^{-1}$) were measured in soil samples collected around the mine tailings. For antimony speciation the strategy followed consisted of the assay of several extractants giving different type of information and in each of the extracts analytical speciation by using HPLC-HG AFS was performed. The extractants assayed were water, acetic acid, sodium hydroxide, EDTA, citrate and calcium chloride, according to the extraction procedures established in a variety of research fields dealing on trace metal mobilisation from soils and mentioned in the first part of the present chapter.

Both chelating solutions and, particularly, the citrate buffer, were those presenting the highest extraction efficiencies for all the soil samples. The extraction efficiency was, however, below 7% with respect to the total antimony content measured in the *aqua regia* extracts. According to these observations, there is evidence that antimony is strongly bound to the soil matrix and, thus, it was solubilised only to a minor extent by the "soft extractants" employed. These results concur with previous findings reported in the literature which support that antimony is mainly bound to immobile soil fractions. However, taking into account the elevated antimony values found for some of the analysed samples, such materials could be considered as a potential source of antimony pollution. Figure 3 shows, as an example, three chromatograms obtained for some soil extracts when HPLC-HG-AFS coupling was used for the quantification of the antimony species. In general, only Sb(V) was detected in each soil extract. This study can contribute to extend the scarce information on the presence and analysis of mobile forms of antimony from soils.

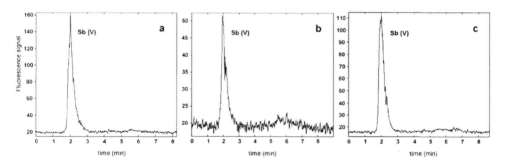

*Figure 3*. Antimony species in soil extracts: (a) deionised water, (b) 0.43 M acetic acid, and (c) 0.05 M EDTA (adapted from Miravet, 2007).

## Conclusion

From the paragraphs above it has to be stated that, in general, efforts have to be dedicated to the establishment of standardized methods for assessing metal species present in soils. These methods must include detailed protocols for assuring the integrity of the species during the critical sample pre treatment steps, their stability in the extracts as well as the knowledge of the main components of the soils as factors influencing the final measurement, even for the determination of the total metal content.

A significant point deals on the production of Certified Reference Materials (CRMs) for chemical species in soils. Great effort has been made within the scientific community for producing CRMs for extractable metals, but for analytical speciation very few projects have been developed, among them the SeAs project dealing on the feasibility studies on the production of CRMs for arsenic and selenium species in several materials, among them several soils, within the EU thematic network Speciation 21 (SeAs Project 2003). These projects should be considered as significant for the recent advances in the analytical speciation of trace elements in complex matrices of environmental interest such as soils.

Another point of interest is the development and application of new methodology that permits the identification of unknown elemental species occurring in some materials, as well as the availability of the corresponding pure standards. However it has to keep in mind that multidisciplinary studies are necessary to move forward successfully in the fields related to the knowledge of chemical species behaviour in the soil environment.

## References

Adamo P, Zampella M, Gianfreda L, Renella G, Rutigliano FA, Terribile F. 2006. Environ. Poll. 144, 308–316.

AFNOR (Association Francaise de Normalization). 1994. AFNOR Paris, p. 250.

Alvarez JM, Lopez-Valdivia LM, Novillo J, Obrador A, Rico MI. 2006. Geoderma, 132, 450–463.

Arunachalam J, Emons H, Krasnodebska B, Mohl C. 1996. Sci. Total Environ., 181, 147–159.

Aumada I, Mendoza J, Navarrete E, Ascar L. 1999. Commun. Soil Sci. Plant Anal., 30, 1507–1519.

Ayoub AS, McGaw BA, Shand CA, Midwood AJ. 2003. Plant Soil, 252, 291–300.

Barona A, Romero F. 1996. Environ. Technol., 17, 63–70.

Barona A, Aranguiz I, Elias A. 1999. Chemosphere, 39, 1911–1922.

Benitez NL, Dubois JP. 1999. Inter. J. Environ. Anal. Chem., 74, 289–303.

Bunzl K, Trautmannsheimer M, Schramel P. 1999. J. Environ. Qual., 28, 1168–1173.

Cabral AR, Lefebre G. 1998. Water, Air Soil Pollut., 102, 329–344.

Canepari S, Cardarelli E, Ghighi S, Scimonelli L. 2005. Talanta, 66, 1122–1130.

Capelo JL, Fernandez C, Pedras B, Santos P, Gonzalez P, Vazs C. 2006. Talanta 68, 1442–1447.

Carvalho KM, Gallardo-Williams TG, Benson RF, Martin DF. 2003. J. Agric. Food Chem. 51, 704–709.

Chlopecka A. 1993. Water Air Soil Pollut., 69, 127–134.

Chlopecka A, Bacon JR, Wilson MJ, Kay J. 1996a. J. Environ. Qual., 25, 69–79.

Chlopecka A. 1996b. Sci. Total Environ., 188, 253–262.

Chlopecka A. 1996c. Water, Air Soil Pollut., 87, 297–309.

Chomchoei R, Shiowatana J, Pongsakul P. 2002. Anal. Chim. Acta., 472, 147–159.

Colinet E, Gonska H, Griepink B, Muntau H. 1983. EUR Report 8833 EN, 57.

Cornelis R, Caruso J, Crews H, Heumann K, eds. 2003. Handbook of Elemental Speciation. Techniques and Methodology. Wiley. Chichester.

Davidson CM, Nordon A, Urquhart G, Ajmone-Marsan T, Biasioli M, Duarte AC, Diaz-Barrientos E, Grcman H, Hodnik A, Hossack I, Hursthouse AS, Ljung K, Madrid F, Otabbong E, Rodrigues S. 2007. Int. J. Environ. Anal. Chem., 87, 589–601.

De Gregori I, Fuentes E, Olivares D, Pinochet H. 2004. J. Environ. Monit., 6, 38–47.

Díaz-Barrientos E, Madrid L, Cardo I. 1999. Sci. Total Environ., 242, 149–165.

DIN (Deutches Institut für Normung). 1993. Bodenbeschaffenheit. Vornorm DIN V 19730. in Boden-Chemische Bodenuntersuchungsverfahren, ed DIN Berlin, p. 4.

Dove PM, Rimstidt JD. 1985. Am Mineral. 70, 838–844.

Dutta M, Chandrasekhar K, Dutta S, Das AK. 2005. Atom. Spectrosc., 26, 137–144.

Emerson RHC, Birkett JV, Scrimshaw M, Lester JN. 2000. Sci. Total Environ., 254, 75–92.

Fedotov PS, Wennrich R, Staerk HJ, Spivakov BY. 2005. J. Environ. Monit., 7, 22–28.

Flores L, Blas G, Hernández G, Alcalá R. 1997. Water, Air Soil Pollut., 1997, 98, 105–117.

Garcia-Manyes S, Jiménez G, Padró A, Rubio R, Rauret G. 2002. Talanta, 58, 97–109.

García Sánchez A, Moyano A, Nuñez C. 1999. Commun. Soil Sci. Plant Anal., 30, 1385–1402.

Geebelen W, Adriano DC, van der Lelie D, Mench M, Carleer R, Clijsters H, Vangronsveld J. 2003. Plant Soil, 249, 217–228.

González-Nieto J, López-Sánchez JF, Rubio R. 2006. Talanta, 69, 1118–1122.

Guevara-Riba A, Sahuquillo A, Rubio R, Rauret G. 2004. Sci. Total Environ., 321, 241–255.

Gupta SK, Aten C. 1993. Inter. J. Environ. Anal. Chem., 51, 25–46.

Houba VJG, Novozamski I, Lexmon TX, van der Lee JJ. 1990. Commun. Soil Sci. Plant Anal., 21, 2281–2291.

Hudson-Edwards KA, Houghton SL, Osborn A. 2004. Trends Anal. Chem. 23, 745–752.

Hughes JC, Noble AD. 1991. Commun. Soil Sci. Plant Anal., 22, 1753–1766.

Ipolyi I, Brunori C, Cremisini C, Fodor P, MacAluso L, Morabito R. 2002. J. Environ. Monit., 4, 541–548.

ISO 11466 NORM. 1995. Soil Quality-Extraction of Trace Elements Soluble in Aqua Regia.

Jeng AS, Singh BR. 1993. Soil Sci., 156, 240–250.

Juste C, Solda P. 1988. Agronomie, 8, 897–904.

Kaasalainen M, Yli-Halla M. 2003. Environ. Poll., 126, 225–233.

Kabata-Pendias A, Pendias H. 2001. Trace Elements in Soils and Plants, third ed. CRC, Boca Raton.

Kahakachchi C, Uden PC, Tyson JF. 2004. Analyst, 129, 714–718.

Kazi TG, Jamali MK, Kazi GH, Arain MB, Afridi HI, Siddiqui A. 2005. Anal. Bioanal. Chem., 383, 297–304.

Keller C, Vedy JC. 1994. J. Environ. Qual., 23, 987–999.

Kersten M, Förstner U. 1989. Chapter 8 in "Trace Element Speciation: Analytical Methods and Problems", ed. GE Batley, CRC.

Kim ND, Fergusson JE. 1991. Sci. Total Environ., 105, 191–209.

Kowalska J, Krasnodebska-Ostrega B, Golimowski J. 2002. Anal. Bioanal. Chem., 373, 116–118.

Krasnodebska-Ostrega B, Kaczorowska M, Golimowski J. 2006. Microchim. Acta., 154, 39–43.

Kubova J, Matus P, Bujdos M, Medved J. 2005. Anal. Chim. Acta., 547, 119–125.

Kubova J, Stresko V, Bujdos M, Matus P, Medved J. 2004. Anal. Bioanal. Chem., 379, 108–114.

Kuehnelt D, Goessler W. 2003. In Organometallic Compounds in the Environment. Craig PJ ed. Wiley. Chichester. Lebourg A, Sterckeman T, Cielsielki H, Proix N. 1996. Agronomie, 16, 201–215.

Li X, Coles BJ, Ramsey MH, Thornton I. 1995. Chem. Geol., 124, 109–123.

Lindsay WL, Norvell, WA. 1978. Soil Sci. Soc. Am. J., 42, 421–428.

Lo IMC, Yang XY. 1998. Waste Management, 1998, 18, 1–7.

Ma LQ, Rao GD. 1997. J. Environ. Qual., 26, 259–264.

Madrid F, Reinoso R, Florido MC, Diaz-Barrientos E, Ajmone-Marsan F, Davidson CM, Madrid L. 2007. Environ. Pollut., 147, 713–722.

MAFF (Ministry of Agriculture, Fisheries and Food) 1981. Reference Book 427 MAFF. London.

Maiz I, Arambarri I, Garcia R, Millan E. 2000. Environ. Pollut., 110, 3–9.

Maiz I. Esnaola MV, Millan E. 1997. Sci. Total Environ., 206, 107–115.

Margui E, Salvado V, Queralt I, Hidalgo M. 2004. Anal. Chim. Acta, 524, 151–159.

Melich A. 1984. Commun. Soil Sci. Plant Anal., 15, 1409–1416.

Miravet R. 2007. New Methodologies for Antimony Speciation in Environmental Matrices. Doctoral Thesis. Universitat de Barcelona. Barcelona.

Mulchi CL, Adamu CA, Bell PF, Chaney RL. 1992. Commun. Soil Sci. Plant Anal., 23, 1053–1059.

Mulligan CN, Yong RN, Gibbs BF. 1999. Environ. Progress, 18, 50–54.

Naidu R, Smith E, Owens G, Bhattacharya P, Nadebaum P, eds. 2006. Managing Arsenic in the Environment. From Soil to Human Health. CSIRO, Collingwood.

Nakazato T, Akasaka M, Tao H. 2006. Anal. Bioanal. Chem., 386, 1515–1523.

Narwal RP, Singh BR, Salbu B. 1999. Commun. Soil Sci. Plant Anal., 30, 1209–1230.

Niesiobedzka K. 2004. Inzynieria i Ochrona Srodowiska, 7, 393–399.

Novozamski I, Lexmon ThM, Houba VJG. 1993. Int. J. Environ. Anal. Chem., 51, 47–58.

Orsini L, Bermond A. 1993. Int. J. Environ. Anal. Chem., 51, 97–108.

Pazos-Capeans P, Barciela-Alonso MC, Bermejo-Barrera A, Bermejo-Barrera P. 2005. Talanta, 65, 678–685.

Perez-de-Mora A, Madejon E, Burgos P, Cabrera F. 2006a. Sci. Total Environ., 363, 28–37.

Perez-de-Mora A, Madejon E, Burgos P, Cabrera F. 2006b. Sci. Total Environ., 363, 38–45.

Pezzarossa B, Petruzzelli G, Petacco F, Malorgio F, Ferri T. 2007. Chemosphere, 67, 322–329.

Pickering WP. 1986. Ore Geol. Rev., 1, 83–146.

Planquart P, Bonin G, Prone A, Massiani C. 1999. Sci. Total Environ., 241, 161–179.

Pueyo M, Rauret G, Bacon JR, Gomez A, Muntau H, Quevauviller Ph, López-Sánchez JF. 2001. J. Environ. Monit., 3, 238–242.

Qian J, Wang Z, Shan X, Tu Q, Wen B, Chen B. 1996. Environ. Pollut., 91, 309–315.

Quaghebeur M, Rengel Z. 2005. Microchim. Acta, 151, 141–152.

Quevauviller Ph, Lachica M, Barahona E, Rauret G, Ure AM, Gomez A, Muntau H. 1996. Sci. Total Environ., 178, 127–132.

Quevauviller Ph, Rauret G, Ure AM, Bacon J, Muntau H. 1997a. EUR-report 17127. Office for official publications of the European Communities, Luxembourg (ISBN 92-827-6938-0).

Quevauviller Ph. Rauret G, Rubio R, López-Sánchez JF, Ure AM, Bacon J, Muntau H. 1997b. Fresenius J. Anal. Chem., 357, 611–618.

Quevauviller Ph, Lachica M, Barahona E, Rauret G, Ure AM, Gomez A, Muntau H. 1997c. EUR-report 17555. Office for official publications of the European Communities, Luxembourg (ISBN 92-828-0126-8).

Ramos L, Hernández LM, González MJ. 1994. J. Environ. Qual., 23, 50–57.

Rauret G. 1998. Talanta, 46, 449–455.

Rauret G, López-Sánchez JF, Sahuquillo A, Rubio R, Davidson CM, Ure AM, Quevauviller Ph. 1999. J. Environ. Monit., 1, 57–61.

Rauret G, López-Sánchez JF, Sahuquillo A, Barahona E, Lachica M, Ure AM, Muntau H, Quevauviller Ph. 2000a. EUR-report 19503. Office for official publications of the European Communities, Luxembourg (ISBN 92-828-3010-1).

Rauret G, López-Sánchez JF, Sahuquillo A, Barahona E, Lachica M, Ure AM, Davidson CM, Gomez A, Luck D, Bacon J, Yli-Halla M, Muntau H, Quevauviller Ph. 2000b. J. Environ Monit., 2, 228–233.

Rauret G, López-Sánchez JF, Bacon J, Gomez A, Muntau H, Quevauviller Ph. 2001. EUR-report 19774. Office for official publications of the European Communities, Luxembourg (ISBN 92-894-0566-X).

Rosenberg E, Ariese F in Trace Element Speciation for Environment, Food and Health. Ebdon L, Pitts L, Cornelis R, Crews H, Donard OFX Quevauviller Ph, eds. RSC Cambridge 2001.

Ruiz-Chancho MJ, Sabé R, López-Sánchez JF, Rubio R, Thomas P. 2005. Microchim. Acta, 151, 241–248.

Ruiz-Chancho MJ, López-Sánchez JF, Rubio R. 2007. Anal. Bioanal. Chem., 378, 627–635.

Sahuquillo A, López-Sánchez JF, Rubio R, Rauret G, Thomas RP, Davidson CM, Ure AM, 1999. Anal. Chim. Acta, 382, 317–327.

Schalscha EB, Escudero P, Salgado P, Ahumada I. 1999. Commun. Soil Sci. Plant Anal., 30, 497–507.

Schramel O, Michalke B, Kettrup A. 2000. Sci. Total Environ., 263, 11–22.

SeAs project. Feasibility studies for Speciated CRMs for arsenic in chicken, Rice, Fish and Soil and Selenium in Yeast and Cereal. Standards Measurements and Testing Programme EU. Nr G6RDCT200100473. L. Ebdon coordinator.

Simeonova B, Simeonov L. 2006. An application of a phytoremediation technology in Bulgaria. The Kremikovtzi Steel Works experiment Remediation Journal, Spring edition 2006, Wiley, New York, pp. 113–123.

Smrkolj P, Stibilj V, Kreft I, Germ M. 2005. Food Chem., 96, 675–681.

Sommers LE, Harris RF, Williams JDH, Armstrong DE, Syers JK. 1972. Soil. Sci. Soc. Am. Proc., 36, 51–54.

Song QJ, Greenway GM. 2006. Int. J. Environ. Anal. Chem., 86, 359–366.

Sutherland RA, Tack FMG. 2000. Sci. Total Environ., 256, 103–113.

Sutherland RA, Tack FMG. 2002. Anal. Chim. Acta., 454, 249–257.

Száková J, Tlustos P, Balik J, Pavlikova D, Vanek V. 1999. Fresenius J. Anal. Chem., 363, 594–595.

Száková J, Tlustos P, Goessler W, Pavlíková D, Balík J, Schlagenhaufen C. 2005. Anal. Bioanal. Chem., 382, 142–148.

Technical Guidelines on Specially Engineered Landfill (D5) 2003. Basel Convention Series no 3, UNEP, Geneva. Tessier A, Campbell PGC, Bisson M. 1979. Anal. Chem., 51, 844–851.

Thöming J, Calmano W. 1998. Acta Hydrochim. Hydrobiol., 26, 338–343.

Tokalioglu S, Kartal S, Guenes AA. 2004. Int. J. Environ. Anal. Chem., 84, 691–705.

Tokalioglu S, Kartal S, Birol G. 2003. Turkish J. Chem., 27, 333–346.

Tongtavee N, Shiowatana J, Mclaren, RG. 2005. Int. J. Environ. Anal. Chem., 85, 567–583.

UNICHIM (Ente Nazionale Italiano di Unificazione) 1991. UNICHIM, Milan.

Ure AM, Quevauviller Ph, Muntau H, Griepink B. 1992 EUR report EN 14472 European Commission, Brussels.

Ure AM, Quevauviller Ph, Muntau H, Griepink B. 1993a. Int. J. Environ. Anal. Chem., 51, 135–151.

Ure AM, Thomas R, Litlejohn D. 1993b. Int. J. Environ. Anal. Chem., 51, 65–84.

Van der Sloot HA, Heasman L, Quevauviller Ph. eds. 1997. Chapter 3 in "Harmonization of Leaching/extraction Tests". Studies in Environmental Science 70 series, Elsevier.

VSBo (Veordnung über Schadstoffgehalt im Boden) 1986 Nr. 814.12, Publ. eidg. Drucksachen und Materialzentrale, Bern, 1–4.

Williams JDH, Syers JK, Harris RF, Armstrong DE.1971a. Soil. Sci. Soc. Am. Proc., 35, 250–255.

Williams JDH, Syers JK, Armstrong DE, Harris RF. 1971b. Soil. Sci. Soc. Am. Proc., 35, 556–561.

Yehl PM, Gurleyuk H, Tyson JF, Uden PC. 2001. Analyst, 126, 1511–1518.

Yu Y, Zhu Y, Williams-Jones AE, Zhenmin G, Gao Z, Li D. 2004. Appl. Geochem., 19, 435–444.

Zemberyova M, Bartekova J, Zavadska M, Sisolakova M. 2007. Talanta, 71, 1661–1668.

# PCBS CONTENT IN TOPSOILS AROUND THE METALLURGICAL PLANT KREMIKOVTSI

ANNA DIMITROVA[*]
*National Center of Public Health Protection*
*15 Acad. Ivan Geshov Blvd., 1431 Sofia, Bulgaria*

**Abstract.** Polychlorinated biphenyls (PCBs) are ubiquitous environmental contaminants. They are highly persistent and tend to accumulate in many environmental compartments including soils. The distribution of PCBs in the environment has not been well studied in Bulgaria in spite of their negative effect on the environment and human health. In this study, soil samples from a metallurgical plant near Sofia were analyzed for indicator and some dioxin-like PCBs. The sums of concentrations of indicator-PCBs were in the range 7.3–19.2 µg kg$^{-1}$. Small amounts of dioxin-like PCBs also were found (2.3–5.1 µg kg$^{-1}$). High chlorinated indicator-PCBs: 138, 153 and 180 were the most abundant in the soil samples. Dioxin-like PCB 77 (non ortho-PCB) and PCB 105 (mono ortho-PCB) were in minor concentrations (0.3 µg kg$^{-1}$) but they are more toxic contaminants than indicator PCBs. The PCBs concentrations found in this study are below the maximum admissible concentrations in soils according to the Bulgarian Legislation.

Keywords: polychlorinated biphenyls (PCBs), soil, metallurgical plant, total toxic endorsed equivalents (TEQ)

## 1. Introduction

Polychlorinated biphenyls (PCBs) are a class of synthetic chlorinated organic chemicals made up 209 species (called congeners). The PCB congeners consist of a biphenyl with 1 to 10 chlorine atoms distributed around the biphenyl rings. They have similar chemical and physical properties with no taste and odour (WHO, 1993). The general formula of the PCBs is shown in Figure 1.

---

[*] To whom correspondence should be addressed: a.dimitrova@ncphp.government.bg

L. Simeonov and V. Sargsyan (eds.),
*Soil Chemical Pollution, Risk Assessment, Remediation and Security.*
© Springer Science+Business Media B.V. 2008

Commercial PCBs products contain 60–90 congeners, and they were widely used as additives to oils in electrical equipment, as dielectric fluids – transformers and capacitors. They used in the electronics industry and as components of adhesives and plastic materials, too. PCBs are chemically and thermally stable, so, although their use has been reduced, they still are found as pollutants throughout the ecosystem and persist in soils and sediments (Zhao, 2006). They are enlisted in UNEP list of persistent organic pollutions (POPs) which have to be globally banned.

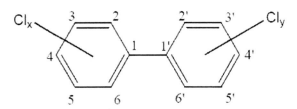

*Figure 1.* General formula of PCBs.

It is well known that PCBs are bioaccumulated in the food chain, due to their lipophilic behavior, and constitute a serious health problems to human and animals, such as cancer, damage to the nervous system, reproductive disorders and disruption of the immune system (Sedlak et al., 1991). The International Agency for Research on Cancer (IARC) classifies PCBs as probably carcinogenic to humans – 2A group (WHO, 1987).

In the past environmental contamination with PCBs has often resulted from the manufacture, handling, use and disposal of these chemicals. Now their production is stopped and their application – banned.

Nowadays the major source for PCBs into the environment is incineration activities (Sog et al., 1992), chemical, petrochemical, metallurgical industries (Yu et al., 2005; Nadal et al., 2007) and uncontrolled natural combustion processes.

Dry and wet atmospheric deposition constitutes the main input of PCBs to soil (Tremolada et al., 1996). They are transported in the atmosphere at over short and long distances in both gaseous and particulate forms. Because of their persistence and hydrophobicity, these omnipresent compounds accumulate in soils where they are likely to be retained for many years. Consequently, soils are an important reservoir for these compounds (Harrad et al., 1994; Ockenden et al., 2003).

The distribution of PCBs in the environment has not been well studied in Bulgaria in spite of their negative effect on the environment and human health (Rizov et al., 2003, 2005).

The aim of this work was to determine the levels of indicator PCBs (28, 52, 101, 138, 153, 170 and 180) and some dioxin-like PCBs (77, 105, 118, 126, 156 and 169) in soil samples around coke-chemical and agglomerate production shop at the metallurgical plant "Kremikovtsi" and to determine TEQ contributions of dioxin-like PCBs to the total toxic endorsed equivalents (TEQ).

## 2. Materials and Methods

### 2.1. THE METALLURGICAL PLANT KREMIKOVTSI

The plant is situated in the north-eastern suburbs of Sofia. It occupies 2,357 sq. km on the territory of Sofia hollow. The average altitude in the region is 550 m.

The plant covers the full metallurgical cycle – from output of iron ore to ready products – cast iron, steel and ferrous alloys. It also produces tin-plated sheet iron and metalloplasts.

The industrial activities began during the 1960s. According to the amounts of production, it is the most important metallurgical complex of Bulgaria. The plant was identified as important source of inorganic contaminants. But it is likely source of toxic persistent chemicals (PCBs, PCDDs/Fs and PAHs).

### 2.2. MATERIALS

Acetone and n-hexane were of analytical grade quality (Merck, Darmstadt, Germany). Anhydrous sodium sulfate and silica gel (70–230 mesh ASTM) for analysis were also from Merck. Cellulose extraction thimbles were from Whatman Ltd (Maidstone, England). Cellulose thimbles were cleaned by Soxhlet extraction with dichloromethane before analysis. Standard solutions of PCB 30, PCB 204 and MIX 20 (mixture of 15 PCB congeners) were from Dr Ehrenstorfer (Augsburg, Germany).

### 2.3. SAMPLE COLLECTION

Soil samples were collected using the "envelope" method: five individual subsamples – four along the boundaries and one in the center. They were taken from the topsoil (10 cm) over a 50 $m^2$ area typical for the site. Subsamples were combined and carefully homogenized and then an average sample weight of approximately 0.3 kg was taken.

Fifteen average soil samples were prepared. Eight soil samples were collected at 1 km distance (Soils 1–8) and another seven at 2 km distance (Soils 9–15) around a coke-chemical production and agglomerate shop.

## 2.4.   SAMPLE PREPARATION

The soil samples were air dried ground and sieved through a <2 mm sieve before extraction.

## 2.5.   EXTRACTION

Soil samples were weighed into Whatman Soxhlet cellulose thimbles, spiked with PCB 30 and PCB 204, covered with anhydrous sodium sulfate and extracted by Soxhlet technique for 18 h with acetone: hexane (1:1). All extracts were concentrated by rotary vacuum evaporator to 1 mL. The concentrated organic extracts were cleaned with an acid-base-silica column. PCBs were eluted with 50 mL hexane. Elution solvents were concentrated to 1 mL in hexane after rotary vacuum and nitrogen stream evaporation.

## 2.6.   INSTRUMENTAL ANALYSIS

The analysis was performed with Hewlett Packard Model 5890 Series II PLUS gas chromatograph equipped with Hewlett Packard 5972 Mass Selective Detector (GC/MS). A HP-5 fused silica capillary column (30 m length, 0.25 mm i.d., 0.25 μm film thickness) coated with 5% phenyl – 95% methyl-polysiloxane was used for the analysis.

   The column oven temperature program started at 120°C (holding time 1 min), increased to 190°C at 20°C min$^{-1}$, increased to 230°C at 5°C min$^{-1}$ and finally to 300°C at 25°C min$^{-1}$ (holding time 10 min). Injector and detector temperatures were 280°C and 300°C, respectively. Helium was used as carrier gas at a constant flow rate of 0.8 mL min$^{-1}$. Injection was performed in splitless mode.

   The mass spectrometer was operated in electron impact ionization (EI) positive-mode using automatic gain control. The storage window was set between $m/z$ 200 and $m/z$ 500 and selected ion monitoring (SIM) was used. The scan time during data acquisition was set at 1.0 s with four microscans per second. Quantification of the target compounds was performed by monitoring of the characteristic ions $m/z$ 258, 292, 326, 362, 396 and 430. These ions were selected considering the parent ions of the group of PCBs present in MIX 20. Calibration was performed by injections of standard solution of MIX 20 at 7 calibration levels (0.01, 0.02, 0.05, 0.1, 0.2, 0.5 and 1 μg mL$^{-1}$). Within this range the method was linear and the $R^2$ values were found to be between 0.992 and 0.999. For all the investigated PCBs no peaks areas in the blanks were found and the limit of detection was 0.1 μg kg$^{-1}$ defined as the concentration giving a signal-to-noise ratio of 3.

## 3. Results and Discussion

This paper reports on the results obtained from an initial study in the metallurgical plant Kremikovtsi near Sofia, for levels of soil contamination with PCBs.

The results of the analytical chemical analysis are summarized in Table 1 for each of the PCBs categories – indicator-PCBs (I-PCBs) and dioxin-like PCBs (dl-PCBs).

Analyzed PCBs were detected in all samples. It was found that the concentration of total PCBs (sum of indicator-PCBs and dioxin-like PCBs) in all samples were similar. The levels ranged between 9.6 μg kg$^{-1}$ and 21.9 μg kg$^{1}$ and did not depend on the distance from the production shop (Table 1). A possible reason for this is that all samples were taken relatively close to the shops.

TABLE 1. Concentrations of PCBs in topsoil samples (μg kg$^{-1}$ dry weight; each value is mean of two measurements)

| Samples | PCB 28 | PCB 52 | PCB 101 | PCB 128 | PCB 138 | PCB 153 | PCB 170 | PCB 180 | total I-PCBs |
|---------|--------|--------|---------|---------|---------|---------|---------|---------|--------------|
| Soil 1  | 1.4 | 1.5 | 1.6 | 0.5 | 2.6 | 2   | 1.1 | 1.7 | 12.3 |
| Soil 2  | 1.7 | 1.4 | 1.5 | 0.3 | 1.9 | 2.3 | 1.6 | 1.6 | 19.2 |
| Soil 3  | 1.3 | 1.2 | 2.4 | 0.5 | 5   | 3.6 | 0.5 | 4.7 | 12.1 |
| Soil 4  | 0.2 | 1.9 | 2   | 0.5 | 2.9 | 2   | 0.8 | 1.8 | 11.7 |
| Soil 5  | 1.3 | 1.2 | 2.1 | 0.5 | 2.3 | 2.1 | 0.4 | 1.8 | 16.8 |
| Soil 6  | 2.1 | 1.8 | 2.6 | 0.9 | 3.3 | 2.7 | 1.3 | 2.1 | 8.8 |
| Soil 7  | 0.2 | 0.3 | 0.2 | 0.5 | 3.4 | 1.6 | 1.1 | 1.5 | 7.3 |
| Soil 8  | 0.2 | 0.5 | 0.2 | 0.5 | 1.4 | 1.1 | 1.2 | 2.2 | 8.6 |
| Soil 9  | 0.2 | 0.6 | 0.2 | 0.5 | 1.8 | 1.3 | 1   | 3   | 14.1 |
| Soil 10 | 0.2 | 0.6 | 0.2 | 0.5 | 2.3 | 1.6 | 3.7 | 5   | 11.1 |
| Soil 11 | 0.2 | 0.5 | 0.8 | 0.5 | 2.3 | 1.7 | 1.5 | 3.6 | 11.7 |
| Soil 12 | 0.2 | 1.4 | 0.2 | 0.5 | 2.4 | 1.5 | 1.3 | 4.2 | 11.6 |
| Soil 13 | 0.2 | 1.2 | 1.4 | 1.5 | 1.5 | 1.8 | 1.9 | 2.1 | 11.1 |
| Soil 14 | 1.1 | 1.4 | 1.3 | 0.6 | 2.1 | 1.9 | 1.2 | 1.5 | 13.1 |
| Soil 15 | 1.2 | 1.6 | 1.8 | 0.9 | 2.3 | 2.1 | 1.5 | 1.7 | 13.1 |
| Mean    | 0.8 | 1.1 | 1.2 | 0.6 | 2.5 | 2.0 | 1.3 | 2.6 | 12.2 |
| SD      | 0.66 | 0.50 | 0.84 | 0.28 | 0.87 | 0.59 | 0.74 | 1.18 | 2.91 |

TABLE 1. (continued)

| Samples | dl-PCB 77 | dl-PCB 105 | dl-PCB 118 | dl-PCB 126 | dl-PCB 156 | dl-PCB 169 | total dl-PCBs | Total PCBs |
|---------|-----------|------------|------------|------------|------------|------------|---------------|------------|
| Soil 1  | 0.3 | 0.3 | 0.6 | 0.5 | 0.5 | 0.4 | 2.6 | 15   |
| Soil 2  | 0.4 | 0.3 | 0.5 | 0.5 | 0.4 | 0.4 | 2.5 | 14.8 |
| Soil 3  | 0.3 | 0.3 | 0.3 | 0.5 | 0.5 | 0.4 | 2.3 | 21.5 |
| Soil 4  | 0.3 | 0.3 | 0.9 | 0.5 | 0.5 | 0.4 | 2.9 | 15   |
| Soil 5  | 0.3 | 0.3 | 0.4 | 0.5 | 0.5 | 0.4 | 2.4 | 14.1 |
| Soil 6  | 0.3 | 0.7 | 1.1 | 0.8 | 1.1 | 1.1 | 5.1 | 21.9 |
| Soil 7  | 0.3 | 0.3 | 0.3 | 0.5 | 0.5 | 0.4 | 2.3 | 11.1 |
| Soil 8  | 0.3 | 0.3 | 0.3 | 0.5 | 0.5 | 0.4 | 2.3 | 9.6  |
| Soil 9  | 0.3 | 0.3 | 0.3 | 0.5 | 1.3 | 1.5 | 4.2 | 12.8 |
| Soil 10 | 0.3 | 0.3 | 0.3 | 0.5 | 0.5 | 0.4 | 2.3 | 16.4 |
| Soil 11 | 0.3 | 0.3 | 0.3 | 0.5 | 0.5 | 0.4 | 2.3 | 13.4 |
| Soil 12 | 0.3 | 0.3 | 0.3 | 0.5 | 0.5 | 0.4 | 2.3 | 14   |
| Soil 13 | 0.3 | 0.3 | 0.5 | 0.5 | 0.4 | 0.4 | 2.4 | 14   |
| Soil 14 | 0.4 | 0.4 | 0.3 | 0.4 | 0.5 | 0.5 | 2.5 | 13.6 |
| Soil 15 | 0.3 | 0.3 | 0.4 | 0.4 | 0.4 | 0.5 | 2.3 | 15.4 |
| Mean    | 0.3 | 0.3 | 0.5 | 0.5 | 0.6 | 0.5 | 2.7 | 14.8 |
| SD      | 0.03 | 0.10 | 0.24 | 0.09 | 0.25 | 0.31 | 0.79 | 3.14 |

Indicator-PCBs dominated in soil samples (Figure 2). The sum of indicator PCB congeners was about 80% of the total PCBs. Their concentrations ($\Sigma$ I-PCBs) in samples ranged from 7.3 µg kg$^{-1}$ to 19.2 µg kg$^{-1}$ (Table 1). From those indicator PCBs, the heptachlorinated PCB 180 and hexachlorinated PCB 138 and PCB 153 were most abundant in soil samples (Figure 3). This may be due to their differences in physicochemical properties. Low chlorinated congeners are more volatile and hydrophilic while high chlorinated biphenyls are more hydrophobic and strongly adsorbed to organic carbon of the soil.

The results from Wilcke et al. (1998) and Nadal et al. (2007) have demonstrated that hexachlorinated PCBs are also most abundant in industrial soils). In addition, PCB congeners 138 and 153 have been generally characterized as the most abundant and frequently detected in soils of metropolitan centers. Lee et al. reported that hexa-CBs were the predominant congeners in the ambient air of petrochemical and chemical areas (Lee et al., 1996).

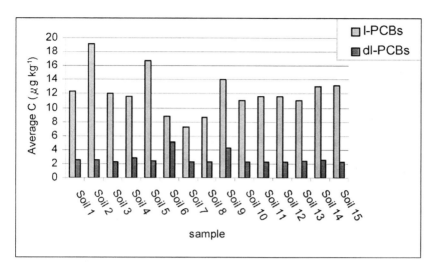

*Figure 2.* Comparison between concentration of indicator PCBs and dioxin-like PCBs.

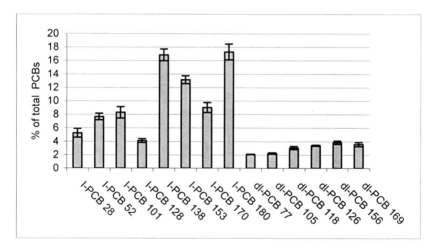

*Figure 3.* Average PCB congener profiles of soils 1–15 (error bars indicating standard deviations).

The concentrations of indicator PCBs found in this study are below the maximum admissible concentrations (MAC) of 0.2 mg kg$^{-1}$ in soils according to the Bulgarian Legislation (Regulation Number 3 Concerning Limit Values of Hazardous Substances in Soil; State Gazette, Sofia, No. 39/2002). They are also below the precautionary target values, but exceed the referent background values (0.005 mg kg$^{-1}$) as set by the ordinance. Accordingly, it can assume that the Kremikovtsi soils were not severely contaminated by PCBs.

The levels of I-PCBs in soils from Kremikovtsi region were much lower than those in soils adjacent to industrial areas in the Seine River basin in France, where the average value of Σ I-PCBs reached 73.9 µg kg$^{-1}$ (Motelay-Massei et al., 2004). This values were also lower than the concentrations in urban soils of Taiwan (94.9 µg kg$^{-1}$), where an increase in PCB usage and disposal occurred due to the development of industrial technology in the 1990s (Thao et al., 1993).

The range of PCB concentrations found in this study was lower also than that reported for soils from industrial areas of Austria (6.4–95 µg kg$^{-1}$) (Weiss et al., 1994) and Poland (4.6–3400 µg kg$^{-1}$) (Falanysz et al., 2001). In Romania, Covaci et al. (2001) found average contents of 57.3 and 722 µg kg$^{-1}$ (sum of 9 PCBs) for urban and industrial sites, respectively. Notarianni et al. (1998) reported an average value of 25 µg kg$^{-1}$ for urban sites in northern Italy.

But the concentrations in soil samples from this metallurgical area are higher than those found in woodland regions of Germany (0.2–4.8 µg kg$^{-1}$) (Krauss et al., 2000), Austria (0.2–7.5 µg kg$^{-1}$) (Weiss et al., 1998), Brazil (0.61 µg kg$^{-1}$) (Rissato, 2006) and the Tatra Mountains (Poland) (0.87–1.5 µg kg$^{-1}$) (Grimalt et al., 2004). Borgini et al. determined PCB content in soils from Victoria Land (Antarctica) in the range: 0.36–0.59 µg kg$^{-1}$ (Borghini et al., 2005).

Dioxin – like PCB congeners were detected in all soil samples, presumably sourced from local manufacture activity of the metallurgical plant and deposition of contaminated ashes and wastes.

Concentration of dioxin-like PCBs in samples ranged from 2.3 to 5.1 µg kg$^{-1}$ (Table 1). The dl-PCB 156 (mono ortho-PCB) and dl-PCB 169 (non ortho-PCB) were in major concentration and non ortho-PCB 77 and mono ortho-PCB 105 were in minor concentrations (Figure 3).

Yet in Bulgarian legislation there are no maximum limits for dl-PCBs in soils in spite of their higher toxicity compared to the I-PCBs.

Dl-PCB concentrations were generally low in comparison with those of other countries (as shown in Table 2), indicating that the Northeastern industrial part of Sofia was not grossly polluted by these persistent toxic substances.

TABLE 2. Comparison of dl-PCBs in soil samples (µg kg$^{-1}$) from other locations of the world

| Location | dl-PCBs | References |
|----------|---------|------------|
| Korea | 0.23–13.78 | Shin et al., 1998 |
| Germany | 2.3–70.2 | Krauss et al., 2003 |
| Spain | 32 | Garcia-Alonso et al., 2003 |
| Italy | 0.7–30.1 | Capuano et al., 2005 |
| Japan | 0.024–16.7 | Nakao et al., 2006 |

Toxicity of dioxin-like PCBs is expressed in toxic equivalents (TEQ). The levels of dioxin-like PCBs were converted to 2, 3, 7, 8-tetrachlorodibenzo-*p*-dioxin equivalents – TEQ (Table 3). This dioxin is the most toxic and has TEF = 1 (Van der Berg et al., 2006). TEQ is the product of the concentration of individual dioxin-like PCB in a sample and toxicity equivalence factor (TEF) for that compound, as defined by the World Health Organization (WHO).

$$TEQ = TEF_{congener} \times concentration_{congener};$$
$$\Sigma\, TEQs = \Sigma\, (TEF_1 \times concentration_1 + \dots TEF_n \times concentration_n)$$

TABLE 3. TEQ values of dioxin-like PCBs in soil samples (x $10^{-3}$ µg $kg^{-1}$)

| dl-PCBs | dl-PCB 77 | dl-PCB 105 | dl-PCB 118 | dl-PCB 126 | dl-PCB 156 | dl-PCB 169 | Σ TEQ |
|---------|-----------|------------|------------|------------|------------|------------|-------|
| TEQ Soil 1 | 0.03 | 0.009 | 0.02 | 50 | 0.015 | 12 | 62.1 |
| TEQ Soil 2 | 0.04 | 0.009 | 0.02 | 50 | 0.012 | 12 | 62.1 |
| TEQ Soil 3 | 0.03 | 0.009 | 0.01 | 50 | 0.015 | 12 | 62.1 |
| TEQ Soil 4 | 0.03 | 0.009 | 0.03 | 50 | 0.015 | 12 | 62.1 |
| TEQ Soil 5 | 0.03 | 0.009 | 0.01 | 50 | 0.015 | 12 | 62.1 |
| TEQ Soil 6 | 0.03 | 0.021 | 0.03 | 80 | 0.033 | 33 | 113 |
| TEQ Soil 7 | 0.03 | 0.009 | 0.01 | 50 | 0.015 | 12 | 62.1 |
| TEQ Soil 8 | 0.03 | 0.009 | 0.01 | 50 | 0.015 | 12 | 62.1 |
| TEQ Soil 9 | 0.03 | 0.009 | 0.01 | 50 | 0.039 | 45 | 95.1 |
| TEQ Soil 10 | 0.03 | 0.009 | 0.01 | 50 | 0.015 | 12 | 62.1 |
| TEQ Soil 11 | 0.03 | 0.009 | 0.01 | 50 | 0.015 | 12 | 62.1 |
| TEQ Soil 12 | 0.03 | 0.009 | 0.01 | 50 | 0.015 | 12 | 62.1 |
| TEQ Soil 13 | 0.03 | 0.009 | 0.02 | 50 | 0.012 | 12 | 62.1 |
| TEQ Soil 14 | 0.04 | 0.012 | 0.01 | 40 | 0.015 | 15 | 55.1 |
| TEQ Soil 15 | 0.03 | 0.009 | 0.01 | 40 | 0.012 | 15 | 55.1 |

In environmental and living samples including food TEQ of dl-PCBs is higher than TEQ of PCDD/Fs (Eljarrat, 2003). Possible reasons for that are because PCBs are ubiquitous pollutants. They released in one part of the world and can travel to regions far from their source of origin. They circulate globally via atmosphere, oceans, soils and other pathways. They are lipophilic compounds and have high bioaccumulation in the environmental samples and subsequent

magnification in organisms. While PCDD/Fs formed mainly by various incineration processes, where dl-PCBs are byproducts. For incineration emissions
TEQ (PCDD/Fs) > TEQ dl-PCBs.

Samples 6 and 9 have the highest TEQ – 0.11 TEQ μg kg$^{-1}$ and 0.09 TEQ
μg kg$^{-1}$ respectively. The other samples have similar TEQ.

PCB 126 and PCBs 169 have the highest TEF (0.1 and 0.3, respectively)
and contribute most to the total TEQ (Figure 4).

The TEQ of dioxin-like PCBs in soil samples from this industrial area are in
the range 0.06–0.11 μg kg$^{-1}$.

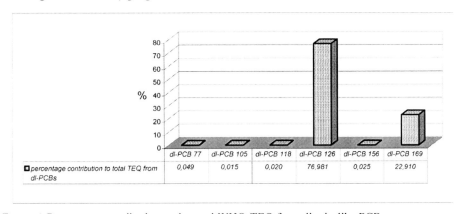

*Figure 4.* Percentage contribution to the total WHO-TEQ from dioxin-like PCBs.

## Conclusion

The metallurgical plant "Kremikovtsi" produces cast iron and steel. The plant is
a probable source of PCBs because in thermal metallurgical processes such
as ferrous foundry and metal smelting operations, large flows of off-gases are
generated. And chlorinated compounds are always present in the feed. Nevertheless up to that moment PCBs contents in soils and emissions at the plant
territory and around it have not been monitored.

The present initial study has provided the first data on the levels of indicator
and some dioxin-like PCBs in soils near coke-chemical and agglomerate shop
at the metallurgical plant.

The results indicated that PCBs present in all soil samples. The high
chlorinated biphenyls (hexa- and hepta-CBs) were predominated in the samples.
The probable reason is that low chlorinated biphenyls are more volatile while
the high chlorinated congeners are more adsorbed to soil particles.

This study revealed that the concentrations of dioxin-like PCBs were lower
than those of indicator PCBs. However, the dioxin-like PCBs should be closely
monitored in the environmental compartments because of their higher toxicity.

Determined PCBs levels are below the maximum admissible concentrations of the Bulgarian Legislation but higher compared to those measured in pristine environments in the world. The levels are slightly lower than those described in the literature for PCBs content in soils of industrial areas.

To answer the question if the metallurgical plant is a source of PCBs to the nearby soils more data are need. The PCB content in and around a metallurgical plant "Kremikovtsi" should be studied in details in relation to emission diffusion to different longer distances from the plant. Then the data would be compared with pristine soils and other urban soils from Sofia.

But regular monitoring would be advisable, particularly for industrial and waste incinerator areas, which are main sources of PCBs in the environment.

## References

Borghini, F., J. Grimalt, R. Sanchez-Hernandez, 2005. Chemosphere, 58, 271–278.

Capuano, F., B. Cavalchi, G. Martinally, G. Pecchini, E. Renna, I. Scaroni, M. Bertacchi, G. Bigliardi, 2005. Chemosphere, 58, 1563–1569.

Covaci A., C. Hura, P. Schepens, Sci. Total Environ., 2001. 280, 143–152.

Eljarrat, E., D. Barcelo, 2003. Trends in Anal. Chem., 22 (10), 655–665.

Falanysz, J., B. Brudnowska, M. Kawano, T. Wakimoto, 2001. Environ. Contam. Toxicol., 40, 173–178.

Garcia-Alonso, S., R.M. Peretz-Pastor, 2003. Water Air Soil Pollut., 146, 283–295.

Grimalt, J., B. Van Drooge, A. Ribes, R. Vilanova, P. Fernandez, P. Appleby, 2004. Chemosphere, 54, 1549–1561.

Harrad, S., A. Sewart, R. Alcock, R. Boumphrey, V. Burnett, R. Duarte-Davidson, C. Halsall, G. Sanders, K. Waterhouse, S. Wild, K. Jones, 1994. Environ. Pollut., 85, 131–146.

Krauss, M., W. Wilcke, W. Zech, 2000. Environ. Pollut., 110, 79–88.

Krauss, M., W. Wilcke, 2003. Environ. Pollut., 122, 75–89.

Lee, W.J., S.J. Lin Lewis, Y.Y. Chen, Y.F., Wang, H.L., Sheu, C.C. Su, Y.C. Fan, 1996. Atmos. Environ., 30, 2371–2378.

Motelay-Massei, A., D. Ollivon, B. Garban, M.J. Teil, M. Blanchard, M. Chevreuil, 2004. Chemosphere, 55, 555–565.

Nadal, M., M. Schuhmacher, J. Domingo, 2007. Chemosphere, 66, 267–276.

Nakao, T., O. Aozasa, S. Ohta, H. Miyata, 2006. Chemosphere, 62, 459–468.

Notarianni, V., M. Calliera, P. Tremolada, A. Finizio, M. Vighi, 1998. Chemosphere, 37, 2839–2845.

Ockenden, W.A., K. Breivik, S. Meijer, E. Steinnes, A. Sweetman, K. Jones, 2003. Environ. Pollut., 121, 75–80.

Regulation Number 3 Concerning Limit Values of Hazardous Substances in Soil; State Gazette, Sofia, No. 39/2002.

Rizov, N., F. Kaloyanova, Y. Simeonov, I. Benchev., 2003. Priority toxic substances in Bulgaria. In Proceedings of the Workshop "Bulgarian Priorities in Chemical Risk Assessment and Management", September 2003, Sofia, Bulgaria, 30–37.

Rizov, T., V. Kambourova, Y. Simeonov, I. Benchev, 2005. Is environmental pollution with polychlorinated biphenyls a problem for Bulgaria. In: Management of Intentional and Accidental Water Pollution, 137–150.

Sedlak, D., A. Andrean, 1991. Environ. Sci. Technol., 25, 1419–1427.

Shin, K.J., Y.S. Chang, 1999. Chemosphere, 38 (11), 2655–2666.

Sog, X., A. Hase, A. Laukkarinen, S. Salonen, E. Hakala, 1992. Chemosphere, 24, 249–259.

Thao, V.D., M. Kawano, R. Tatsukawa, 1993. Environ. Pollut., 81, 61–71.

Tremolada, P., V. Burnett, D. Calamari, K. Jones, 1996. Environ. Sci. Technol., 30, 3570–3577.

Van der Berg, M., L. Birnbaum, A. Bosveld, B. Brunstrom, Ph. Feeley, J. Giesy, A. Hanberg, R. Hasegawa, S. Kennedy, T. Kubiak, J. Larsen, F. Van Leeuwen, A. Liem, C. Nolt, R. Peterson, L. Poellinger, S. Safe, D. Schrenk, D. Tillit, M. Tysklind, M. Younes, F. Waern, T. Zacharewski, 2006. Environ. Health Perspect., 106, 775–792.

Weiss, P., A. Riss, E. Gschmeidler, H. Schentz, 1994. Chemosphere, 29, 2223–2236.

Weiss, P., G. Lorbeer, S. Scharf, 1998. Organohalogen Compd., 39, 381–384.

Wilcke, W., W. Zech. Z. Pflanzenernahr. Bodenk., 1998. 161, 289–295.

World Health Organization, 1993. Environmental Health Criteria 140: Polychlorinated Biphenyls and Terphenyls (Second ed.), Geneva, 21–24.

World Health Organization, 1987. International Agency for Research on Cancer, Lyon, Evaluation of the Carcinogenic Risk to Humans. Suppl. 7, 1–42.

Yu, B.W., G.J. Jin, Y.H. Moon, M.K. Kim, J.D. Kyoung, Y.S. Chang , 2006. Chemosphere, 62, 494–501.

Zhao, X., M. Zheng, B. Zhang, Q. Zhang, W. Liu, 2006. Sci. Total Environ., 386, 744–752.

# NEW BULGARIAN SOIL POLLUTION STANDARDS

IVAN ATANASSOV
*Institute for Sustainable Development*
*Yaroslav Veshin Str. Bl. 10, 1408 Sofia, Bulgaria*
*E-mail: i_sd@abv.bg*

**Abstract.** Recently in Bulgaria have been adopted new quality standards for soils polluted by arsenic, cadmium, copper, chromium, nickel, lead, mercury, zinc and cobalt, set on the base of three level system according to precautionary, maximum permissible (trigger) and intervention concentrations. The precautionary values are derived on the base of the background concentrations of the above mentioned inorganic substances into the soils. Maximum permissible (trigger) values are derived on the base of (eco) toxicity considerations for three pathways: soil-humans (direct intake); soil-food and fodder plants and soil-groundwater. The maximum permissible values are set for four land use types: arable lands, permanent grasslands, residential areas and for lands occupied by industrial/commercial enterprises. These values are differentiated for three soil pH ($H_2O$) and for two soil textural classes: soil with coarse and medium texture and fine textured soils. The intervention values are defined as 3–5 times of maximum permissible concentrations for three land-use types: agricultural lands (arable lands + grasslands), residential and industrial/commercial areas. The purpose of the presentation is to describe the methods used for derivation the soil quality standards and to show the ways for their implementation.

Keywords: soil quality standards, precautionary, maximum permissible (trigger) and intervention values, inorganic soil contaminants

## 1. Introduction

Soils, polluted by heavy metals and arsenic in Bulgaria are more or less well investigated and mapped. According to some estimations (Todorova, 2002) the total area of soils, polluted by different chemical substances is approx. 43, 600 ha. They are located in vicinity of ferous and non-ferrous metallurgical industrial enterprises, near highways and areas with chemical industry. In most cases the polluted soils contain a mixture of heavy metals. Between more widespread inorganic pollutants can be mentioned lead, copper, zinc, cadmium. Some soils are contaminated by arsenic.

Chemically polluted soils occupy comparatively small area of the country (only 0.2% from the total territory), but the risks created by them on humans, animals, plants and ground waters is too high. Because of that it is important to develop criteria and set up standards which can be used for assessment the degree of the risk coming from contaminated soils.

The first standards for assessment the lead, copper and zinc polluted soils were set up in 1979. In 1994 the Ministry of Agriculture and Forestry (MAF) of Bulgaria has adopted new quality standards (still in force) for assessment the degree of pollution of the agricultural soils caused by heavy metals, oil products, radionuclides, high salts content and affected by alkaline substances. For inorganic substances including arsenic, cadmium, copper, chromium, nickel, lead, mercury, zinc and selenium have been set two categories of quality standards: maximum permissible and intervention values, measured as concentration of each individual pollutant (mg/kg, total amount) in the upper soil horizon. The both were designed as generic numerical values, and the maximum permissible concentration values being differentiated for 6 pH ($H_2O$) of the soil. According to the concentration of each individual pollutant the soils are divided into three main categories: non-polluted (level A), weakly, moderately and strongly polluted (level B) and extremely polluted (level C).

The 1994 soil quality standards are not effect-based and do not reflect other land use types besides agriculture. Some numerical values are not corrected on the base of the current research data. These and other disadvantages, analysed by Sauerbeck et al. (1996) made necessary development of new soil quality standards for polluted soils, now adopted officially by the Ministry of Environment and Water of Bulgaria.

This article present the methods used for derivation the new soil quality standards and their implementation for risk assessment the metal polluted soils.

## 2.  Basic Concept

The Bulgarian Law of Environment Protection (2002) considers the soil as a limited, indispensable and non-renewable natural resource which has to be protected, used in a sustainable way and restored in order to protect human health and soil multifunctionality. One of purposes of the soil protection is prevention or elimination of adverse changes of soil quality due to the processes of soil contamination. Like other European countries, the Bulgarian policy to contaminated lands is based on the recognition of the pollution prevention, polluter pays principle, the precautionary principle and to use a procedure for risk assessment in order to identify the needs for remedial action. According to the Law, a special soil quality standards has to be developed as an instrument for appraisal the degree of pollution and risk assessment.

The updated soil quality standards are set on the base of the concept of three level system including precautionary, maximum permissible (trigger) and intervention concentrations. The three level system is widespread and is implemented in many European countries (Anonym, 1998, 1999; Fergusson, 1999). This system is accepted in the new Bulgarian Soil Protection Act – not yet adopted by the Parliament – where definitions for "precautionary", "maximum permissible" and "intervention" values are presented.

The precautionary level of concentrations are defined as values which indicate that the soil is not contaminated but contains contaminants higher than background concentrations. This shows possible unfavourable soil changes in future which have to be avoided.

The maximum permissible (trigger) concentrations are perceived as values, which indicate that soil is contaminated but in the frame of acceptable risk. Investigations are needed in order to define the real risk related to current or intended lands use.

Intervention values are defined as a concentration of a contaminant in the upper soil horizon, which if exceeds, should be considered as a harmful and create an unacceptable risk for human under the specific soil use. Measures are required (interventions) to clean up affected soils or to change the land use type. According to this concept, precautionary, maximum permissible (trigger) and intervention values have been derived.

## 3. Derivation of Soil Quality Standards

The precautionary values are derived on the base of reference background values of the inorganic substances in the soil. During 1999–2002 a special investigation has been carried out (Atanassov et al., 1999, 2002) in order to summarize existing data on the concentration of As, Cd, Cu, Cr, Ni, Pb, Hg and Co in the soils, located in non contaminated or background territories of the country. The obtained background concentrations didn't vary significantly between different soils and because of that it was possible to establish generalized background values equal to all soils. The 90 percentiles values of the observed background concentrations have been accepted as *reference* background values for above mentioned substances. The soil reference background values are related to the top (A) horizons of the mineral soils and can be defined according LABO (1995) as values, which indicate the geogenic basic contents as well as the general antropogenic additional pollution of the soils.

For setting precautionary standards the calculated soil reference background values have been doubled (Table 1). The precautionary values have been designed for standard soil (soil with pH 6.0) and for three soil textural classes: coarse, medium and fine textured soils. The values may be applied all over the country with exception the soils with extremely high natural (geogenic) metal content or metal polluted soils.

TABLE 1. Reference background concentrations and Precaution values for As, Cd, Cu, Cr, Ni, Pb, Zn, Hg and Co in the soils of Bulgaria (in mg/kg dry soil, aqua regia extraction)

| Soil | As | Cd | Cu | Cr | Ni | Pb | Zn | Hg | Co |
|------|-----|-----|-----|-----|-----|-----|-----|------|-----|
| Reference background concentration | | | | | | | | | |
| Standard soil | 10 | 0.4 | 34 | 65 | 46 | 26 | 88 | 0.03 | 20 |
| Precaution values | | | | | | | | | |
| Coarse textured soils | 15 | 0.6 | 50 | 90 | 60 | 40 | 110 | 0.05 | 30 |
| Medium textured soils | 15 | 0.6 | 60 | 110 | 70 | 45 | 160 | 0.07 | 35 |
| Fine textured soils | 20 | 1.0 | 70 | 130 | 70 | 50 | 180 | 0.08 | 40 |
| Soils with increased natural content | To be established on the base of local background concentrations | | | | | | | | |

Explanations: (1) Standard soils – soil with pH (H$_2$O) = 6.0; (2) Coarse, medium and fine textured soils; with content of particles < 0.01 mm respectively up to <20%, 20–60 and >60%.

The maximum permissible (trigger) as well as the intervention values have been derived on the base of (eco) toxicity considerations, taking into account the impact of metal polluted soils on humans (direct intake), food and fodder plants and underground water. To derive maximum permissible (trigger) values for pathway: soil-human beings (direct intake) a calculation can be used proposed by Hammann and Gupta (1998):

$$TVsoil, mg/kg = bw\ ADI.e\ /\ SI \qquad (1)$$

were:

TV – trigger values
bw – body weight of a sensitive group of children = 20 kg
ADI – acceptable daily intake of a pollutant from soil
e – % of daily direct intake of a pollutant from soil
SI – accepted daily direct intake of soil (oral or through inhalation) from a member of a sensitive group of the population (g/daily)

The maximum permissible (trigger) values for the pathway: soil – food and fodder plants have been derived on the base of five step procedure used by W. Kördel et al. (2003), and K. Terytze et al. (2006) for derivation trigger values for soils polluted by persistent organic pollutants (POPs). The procedure includes:

- Consideration of maximum residue (tolerable) level (MRL) in four groups of plants: cereals, vegetables, fruits and fodder plants
- Quantitative description of soil-plant uptake including calculation of transfer coefficients (Tc) and derivation of maximum permissible soil concentration (MPC) for each individual inorganic substance
- Plausibility check, including expert judgement
- Final stipulation of trigger values on the basis of Realistic Worst Case (RWC) scenario
- Comparison with trigger values adopted in other countries for inorganic substances

A key point of the procedure is obtaining data for maximum residue (tolerable) level. To calculate MRL values Atanassov et al. (2004) have been carried out an investigation to summarize information from three sources: (1) data from field, greenhouse and laboratory experiments with agricultural plants carried out on metal-polluted soils, on soils with added or known concentration of a pollutant in the soil and plant; (2) data from field investigations of metal polluted soils and plants growing on these soils and (3) calculated MRL according the method proposed by W. Kördel et al. (2003) in case of absence of reliable data.

The major disadvantages of this approach is lacking of sufficient data and knowledge concerning the thresholds and (eco)toxicity data for pathways soil-humans, soil-plants and soil-ground water. This is because the field and/or greenhouse experiments in many cases have been carried out for other purposes but not directly for establishement the relation between the concentration of the pollutants in the soil and impact on plant-water-other organisms.

To obtain numerical values from maximum permissible (trigger) concentrations in soil for inorganic substances an equation has been used, proposed by W. Kördel et al. (2003):

$$\text{MPC soil} = \text{HF.MRLi (mg/kg)} / \text{SF . Tc} \qquad (2)$$

were:

MPC soil – maximum permissible concentration of metal in soil, mg/kg
MRLi – maximum residual level for plant $i$
Tc – transfer coefficient

HF – hazard factor in order to relate the value to the precautionary aspect to hazard

SF – safety factor, used in case where not an optimal soil data are available

The obtained maximum permissible (trigger) concentrations (Table 2) are designed for four land use type: (1) Arable lands; (2) Permanent grasslands; (3) Residential areas, parks, and recreation site/facilities; and (4) Lands occupied by industrial/commercial enterprises. These values in addition are designed for three soil pH (H₂O): (a) for soils with pH < 6.0; (b) soils with pH between 6.0 and 7.4; and (c) soils with pH > 7.4. The maximum permissible (trigger) values are related to the soil texture. Approximately 80% of the Bulgarian soils are

TABLE 2. Maximum permissible (trigger) and intervention values for As, Cd, Cu, Cr, Ni, Pb, Hg and Zn in agricultural lands and Grasslands according to texture and pH of the soil (mg/kg dry soil, aqua regia extract)

| Substances | pH (H₂O) | Trigger values | | Correction coefficient Cc | Intervention values for land used types in columns 3 and 4 |
|---|---|---|---|---|---|
| | | Agricultural lands | Grasslands | | |
| 1 | 2 | 3 | 4 | 5 | 6 |
| Arsenic – As | – | 25 | 30 | 1.2 | 90 |
| Cadmium – Cd | <6.0 | 1.5 | 2.0 | | |
| | 6.0–7.4 | 2.0 | 2.5 | 1.3 | 12 |
| | >7.4 | 3.0 | 3.5 | | |
| Copper – Cu | <6.0 | 80.0 | 80.0 | | |
| | 6.0–7.4 | 170.0 | 170.0 | 1.2 | 500 |
| | >7.4 | 200.0 | 200.0 | | |
| Chromium – Cr | – | 200.0 | 250.0 | 1.2 | 550 |
| Nickel – Ni | <6.0 | 77.0 | 70.0 | | |
| | 6.0–7.4 | 75.0 | 80.0 | 1.2 | 300 |
| | >7.4 | 90.0 | 110.0 | | |
| Lead – Pb | <6.0 | 60.0 | 90.0 | | |
| | 6.0–7.4 | 100.0 | 130.0 | 1.3 | 500 |
| | >7.4 | 120.0 | 150.0 | | |
| Mercury – Hg | | 1.0 | 1.0 | 1.2 | 10 |
| Zinc – Zn | <6.0 | 150.0 | 110.0 | | |
| | 6.0–7.4 | 250.0 | 270.0 | 1.3 | 900 |
| | >7.4 | 300.0 | 320.0 | | |

medium, 9% – coarse and 11% – fine textured soils. The derived maximum permissible values are related to the medium and coarse textured soils. For fine textured soils (soils, containing in the upper soil layer particles smaller than 0.01 mm more than 60%) the numerical values set for medium/coarse textured soils have to be multiplied by the empirically derived correction factors which vary between 1.0 up to 1.3 (Table 3). Other soil properties like organic matter content are not taken into account. The reason is that more soils of the country (more than 90%) contain organic matter in the upper soil horizons between 1% and 3.5% (average 2.2%). The soils with humus content more than 5% occupy only 0.4% of the territory.

Explanations: (1) pH ($H_2O$); (2) Cc is applied when the content of particles smaller then 0.01 mm in upper soil layer is more 60%.

The intervention values have been designed for three land use types: for agricultural lands (arable lands + grasslands), for residential areas/parks and for industrial areas. For pathway: soil-food and fodder plants, the intervention values are defined as 3–5 times of maximum permissible concentrations, found according equation (2) and applying Best Case (BS) scenario. For the areas occupied by industrial/commercial enterprises the intervention values are based mainly on expert judgement and comparision with intervention values set in other countries. The established numerical intervention values are not corrected according soil properties. These values are equal for all soil under given land use type and include all relevant exposure partways, e.g. soil ingestion, crops consumption, inhalation of soil contaminants, consumption of groundwater or dermal contact with contaminated soils.

TABLE 3. Trigger and intervention values for As, Cd, Cu, Cr, Ni, Pb, Hg and Zn in soils for urban and industrial areas (in mg/kg dry soil, aqua regia extraction)

| Substances | Residential areas, parks, sport facilities | | Industrial/commercial areas | |
|---|---|---|---|---|
| | Trigger values | Intervention values | Trigger values | Intervention values |
| Arsenic – As | 25 | 50 | 40 | 120 |
| Cadmium – Cd | 8 | 12 | 10 | 40 |
| Copper – Cu | 300 | 500 | 500 | 1,000 |
| Chromium – Cr | 200 | 550 | 300 | 600 |
| Nickel – Ni | 100 | 300 | 250 | 700 |
| Lead – Pb | 200 | 500 | 500 | 1,000 |
| Mercury – Hg | 8 | 10 | 10 | 40 |
| Zinc – Zn | 400 | 300 | 600 | 1,500 |

## 4. Discussion

Soil quality standards are impotant instrument for implementation the soil protection policy. The quality standards can be applied as a decision-support tool in risk assessment of polluted soils and their impact on human health, water resources and other environmental compartements.

The assessment procedure is based on the measurment of the content of a particular substance in the soil and comparision the observed concentration with soil quality standards, expressed in Tables 1, 2 and 3. Generally, three levels of concentrations can be defined:

- The first, A-level, when the observed concentration of a particular substance (inorganic compaund) in the soil is less than maximum permissible (trigger) value set for this substance or is closer to the refference background or precautionary values. The soil can be considered as multifunctional, without suspicion to be hazardous for public health or the environment.
- The second B- level, when the observed concentration of one or more substances in the soil is between maximum permissible and intervention values: the soil can be considered as contaminated, but the level of concentration does not create immidiete risk for humans, environment or particular land use. Further investigations are needed to ascertain whether the found level of concentration implies a danger.
- The third C-level, when the concentration of a particular substance in the soil is higher than the intervention value set for this pollutant: the soil can be considered as unacceptably polluted. The level of concentration of the particular substance implies a danger which has to be avoided through provididing of remedial action (intervention) or changes of actual use of the land.

The application of soil quality standards for assessment the level of contamination needs obtaining reliable data for concentration of the pollutants into the soil. This can be made through field investigations, laboratory analyses and appropriate statistical treatment of the data obtained.

The procedure for investigation the contaminated soils includes the following concecutive steps:

- Preliminary investigation of the suspected contaminated soil/sites and the polluted areas
- Detailed investigation of the contaminated soil/sites and mapping
- Relative risk rating to prioritise contaminated soils within each locality
- Development of proposals for safety land use and/or projects for remediation of the damaged soils
- Monitoring and management the area where remediation/reclamation activities have been carried out

Soil sampling strategy includes preparing of mixed soil samples, each composed of nine individual samples equal in volume, taken in hexagonal grid of sampling pattern. The density of sampling points depends on the purposes of the investigation. The sampling depts is fixed to upper soil horizons: for arable lands – 0–20 cm and for non-cultivated soil – 0–10 cm. The laboratory methods used for analysing the metal content in the soil include aqua regia extraction according ISO/DIN 11466 (1997). These methods permit the determination of the fraction closely related to the total content of the metal in the soil which potentially can be released in the course of weathering processes and expresses the maximum hazard potential (Terytze, 2002).

Nevertheless, the procedure for investigation the contaminated soils/sites is still not yet thoroughly developed. The contaminated soil risk assessment is not adequately integrated with risk assessment of surface and underground water and biodiversity. In relation to contaminated land remediation practices there is no defined rules how to implement the intervention values as remediation targets. These and other problems have to be solved in the near future.

## Conclusions

New soil quality standards have been set in Bugaria according to precautionary, maximum permissible (trigger) and intervention values in order the procedure for assessment of the soil contamination by inorganic substances to be more appropriate and realistic.

The precautionary values have been derived through doubling the calculated soil reference background concentrations for arsenic, cadmium, copper, chromium, nickel, lead, mercury, zinc and cobalt and subsequent differentiation for coarse, medium and fine textured soils.

For derivation the maximum permissible (trigger) and intervention values have been used (eco) toxicity data taking into account the impact of metal polluted soils on humans, food and fodder plants and underground water.

The obtained maximum permissible (trigger) concentrations are designed for four land use types, namely, arable lands, permanent grasslands, residential areas and territories occupied by industrial/commercial enterprises. The values are differentiated according site specific characteristics like soil pH ($H_2O$) and soil textural classes.

The new soil quality standards can serve as an instrument for implementation the soil protection policy in the country. The intervention values can be used as soil clean-up objectives. Further improvements of soil risk assessment is needed through integration with risk assessment of water, air and biodiversity.

# References

Anonym, 2002. Law of Environment Protection – State Gazette No. 91/2002, Sofia.

Anonym, 1998. Federal Soil Protection Act, Germany – Federal Law Gazette, I, p. 502.

Anonym, 1999a. B BodSchV – Federal Soil Protection and Contaminated Sites Ordinance Germany BGB I, 1554.

Atanassov, I, K. Terytze and A. Atanassov, 1999. Determination of standard values for POP's polluted soils in Bulgaria. Report No. 535, part I and II, RPSL – CM, Sofia, Bulgaria.

Atanassov, I, K. Terytze, A. Atanassov, U. Christova, and D. Christov, 2000. Development of precautionary values for heavy metals in soils. Report No. 874-2324, pp. 1–241, MEW, Sofia, Bulgaria.

Atanassov, I., K. Terytze and A. Atanassov, 2002. Background values for heavy metals, PAHs and PSBs in the soils of Bulgaria. In: "Assessment of the Quality of Soils and Sites in Central and Eastern European Countries (CEEC) and New Independent States (NIS). Proceedings (K. Terytze and I. Atanassov, eds), pp. 83–103, GorexPress, Sofia.

Atanasov, I. et al., 2004. Effect-based quality standards for heavy metals in soils of Bulgaria depending on the type of land use. MEW, Report No. 3327/370, pp. 1–218.

Biber, A. 1998. Risk assessment of Contaminated Sites according to the Federal Soil Protection Act. In Proceedings of CARACAS workshop, pp. 73–87, UBA, Berlin.

Ferguson, C.C., 1999. Assessing Risks from Contamenated Sites: Policy and Practice in 16 European Countries, Land Contamination and Reclamation, 7 (2), pp. 33–54, EPP Publication.

Hammann, M. and Gupta, 1998. Derivation of trigger and clean-up values for inorganic pollutanys in the soils. Env. Documentation No. 83: Soil, SAEFL, Bern, pp. 1–105.

Kördel, W., M. Herrchen and K. Terytze, 2003. Concept for the derivation of trigger values for the soil-plant uptake. In: "7th International HCH and Pesticides Forum", Proceedings, pp. 118–122, Kiev, Ukraine.

LABO, 1995. Soil Background and Refference Values in Germany. Rep. LABO, FEA, pp. 1–106.

MEW, 1979. Regulation No. 3. On the admissible content of harmful substances in soils. State gazette No. 36/8.05.1979.

MAF, 1994. Instruction RD 0011 on the kind and degree of contamination of agricultural lands and ways of their safety use. Bul. No. 27/MAF, Sofia.

Sauerbeck, D.R. et al., 1996. A study of agricultural soil pollution by the imition of denger substances and assessment of damages for soils in Bulgaria. UBA, Berlin, pp. 1–96.

Terytze, K. 2002. Results obtained in harmonizing soil investigation methods within the framework of cooperation with countries in Central and Eastern Europe. In: "Assessment of the Quality of Soils and Sites in Central and Eastern European Countries (CEEC) and New Independent States (NIS). Proceedings (K. Terytze and I. Atanassov, eds), pp. 149–159, GorexPress, Sofia.

Terytze, K., W. Kördel, J. Müller, M. Herrchen, A. Nester, 2006. Soil-plant transfer of organic chemicals and derivation of trigger values. In "Environmentally sound management (ESM). Practices on cleaning up obsolete stokpiles of Pesticides for Central European and EECCA countries". Proceedings of 8th International HCH and Pesticides Forum (N. Schulz, I. Atanassov and J. Vijgen, eds), pp. 171–174, GorexPress, Sofia.

Todorova, I., 2002. Soil protection in environmental aspects status and outlook in Bulgaria. In: "Assessment of the Quality of Soils and Sites in Central and Eastern European Countries (CEEC) and New Independent States (NIS). Proceedings (K. Terytze and I. Atanassov, eds), pp. 17–20, GorexPress, Sofia.

# STUDIES ABOUT REMEDIATION OF CONTAMINATED SOILS WITH TOXIC METALS

ELISABETA CHIRILA* AND IONELA POPOVICI CARAZEANU
*Chemistry Department, Ovidius University*
*124 Mamaia Blvd., 900527 Constanta, Romania*

**Abstract.** The heavy metals pollution of soils represents sometimes a serious concern regarding the earth's health, and therefore studies about heavy metals removal from contaminated soils are of increased interest. The aim of this work was to study in situ Cd and Zn immobilization in three types of contaminated soils with synthetic calcium phosphate and also to present new results concerning the influence of EDTA on the extractability of Cd and Zn in both natural and metal-amended soils. The proposed treatment stabilizes efficiently cadmium and zinc, 97.6–98% for cadmium and 100% for zinc on all types of studied soils. The obtained extraction efficiencies were for naturally occurring metals 8.9–27.7.1% for cadmium and 20.4–37.2% for zinc; in the case of metal-amended soils the extraction efficiencies were higher: 20.4–56.1% for cadmium and 62.3–80.9% for zinc.

**Keywords:** metals immobilization, polluted soils, EDTA extraction

## 1. Introduction

Soils consist of a mixture of weathered minerals and varying amounts of organic matter and can be contaminated as a result of spills or direct contact with contaminated waste streams such as airborne emissions, process solid waste, sludge or leachate from waste materials.

---

* To whom correspondence should be addressed: echirila@univ-ovidius.ro

L. Simeonov and V. Sargsyan (eds.),
*Soil Chemical Pollution, Risk Assessment, Remediation and Security.*
© Springer Science+Business Media B.V. 2008

There are thousands of contaminant sources and pollutant types, but the following list is illustrative (pollutants indicated by parentheses):

1. Petroleum hydrocarbons from rupture of underground storage tanks (benzene, ethylbenzene, toluene, xylene, alkanes, alkenes, MTBE)
2. Spillage or leakage of solvents and dry cleaning agents (acetone, trichloro-ethylene, formaldehyde) and perchloroethylene
3. Leaching of contaminants from solid waste disposal sites (lead, mercury, chromium, cadmium, bacteria, hydrocarbons)
4. Water runoff which carries pollutants and may deposit them at a point of percolation
5. Percolation into soils from pesticides and herbicides uses (wide variety of chemicals including DDT, lindane, organochlorines, organophosphates, car-bamates, cyclodienes)
6. Deposition of dust from smelting operations and coal burning power plants (zinc, cadmium, lead, mercury)
7. Lead deposition from lead abatement or construction demolition (lead)
8. Leakage of transformers (Polychlorinated Biphenyls – PCBs)
9. Radioactive materials, these include natural radioisotopes which have been moved (for instance radium bearing waste from either the production of uranium or phosphoric acid) and man made radioisotopes such as those present within Chernobyl fallout.

The soil contaminants can be classified in two categories: organic (oil products, pesticides, dioxins, PCBs) and inorganic (heavy metals or fertilizers).

Cleanup or remediation soil polluted with organic compounds is analyzed by environmental scientists who utilize field measurement of soil chemicals and also apply computer models for analyzing transport and fate of soil chemicals.

There are several principal strategies for remediation:

- Excavate soil and remove it to a disposal site away from ready pathways for human or sensitive ecosystem contact. This technique also applies to dredging of bay mud containing toxins
- Aeration of soils at the contaminated site (with attendant risk of creating air pollution)
- Bioremediation, involving microbial digestion of certain organic chemicals. Techniques used in bioremediation include landfarming, biostimulation and bioaugmentation soil biota with commercially available microflora
- Extraction of groundwater or soil vapor with an active electromechanical system, with subsequent stripping of the contaminants from the extract
- Containment of the soil contaminants (such as by capping or paving over in place).

The solubility of metals in soils is influenced by the chemistry of the soil and groundwater (Evans, 1989).

Increasing awareness of the hazard that toxic elements can cause to the environment and to humans makes it necessary to rehabilitate metal contaminated sites (Melamed et al., 2003; Simeonov et al., 1987).

Several technologies exist for the remediation of metals-contaminated soil and water.

These technologies are contained within five categories of general approaches to remediation: isolation, immobilization, toxicity reduction, physical separation and extraction.

Among the remediation technologies available for contaminated sites, in situ immobilization techniques are of particular interest because they are relatively more cost-effective compared to conventional techniques, e.g. excavation and off-site disposal (Yang et al., 2001). Immobilization technologies are designed to reduce the mobility of contaminants by changing the physical or leaching characteristics of the contaminated matrix.

Application of phosphate amendments to soils has been identified as a potentially effective in situ remediation technology (Hettiarachchi et al., 2000). The mechanism involved in the immobilization process of lead has been studied and discussed (Melamed et al., 2000). Other authors studied the efficiency of As, Cd, Pb and Zn immobilization process with monobasic calcium phosphate (Ying et al., 1995; Zenghua et al., 2001; Teodoratos et al., 2002), with zeolites (Zorpas et al., 2002), or with waste materials as agricultural (Gardea-Torresdey et al., 2003) or algal (Jalali et al., 2002).

Some techniques for removing metals from soils generally involve bringing the soil into contact with an aqueous solution (Jabali et al., 2003). Methods such as flotation and water classification are solid/liquid separation processes. Metal contamination is usually found on the finer soil particles. Because metals are often preferentially bound to clays and humic materials (Itabashi et al., 2002), separating the finely divided material may substantially reduce the heavy metal content of the bulk soil. Solid-liquid separation processes represent a preliminary step in the remediation of contaminated soils. Formation of complexes using strong complexing agent like EDTA is one of the methods to mobilize metals from soils.

The ability of chelating agents to form stable metal complexes makes materials such as ethylenediaminetetraacetic acid promising extractive agent for the treatment of soil polluted with heavy metals (Fingueiro et al., 2002). EDTA has been used to extract metals freshly adsorbed onto organic matter (Itabashi et al., 2001) to extract trace metals from the Fe oxide or Fe hydroxide species and to release metals by dissolution of carbonates.

Although complexation is the mechanism responsible for the metal solubilization, the overall release process depends on the hydrogen ion concentration and the system ionic strength. Because hydrous oxides of iron and manganese can coprecipitate and adsorb heavy metals, they are believed to play an important role in the fixation of heavy metals in polluted soils. Their dissolution under reducing conditions may weaken the solid heavy-metal bond and thereby promote solubilization of the metal ion. The presence of EDTA in soil may alter the mobility and transport of Cd and Zn in soils because of the formation of water-soluble chelates, thus increasing the potential for metal pollution of natural waters. Mobility of metals is related to their extractability.

The aim of this work was to study in situ Cd and Zn immobilization in contaminated soils with synthetic calcium phosphate and also to present new results concerning the influence of EDTA on the extractability of Cd and Zn in both natural and metal-amended soils.

## 2. Experimental

Three composite soils samples collected from different Constanta district's sites were used in our studies: A (limestone chernozem earth) – from Baneasa, B (argillaceous limestone) and C (alluvium soil) – from Harsova. Samples were taken from the surface layer (0–0.20 m), air-dried, and passed through a 2-mm sieve. For the determination of naturally occurring metals, samples were digested with nitric acid – hydrogen peroxide at 170°C in a Digesdahl device provided by Hach Company and then analyzed using flame atomic absorption spectrometry (Chirila et al., 2004).

For the FAAS analysis a Shimadzu AA 6200 spectrometer was used. The calibration of spectrometer was done using standard solutions of each metal obtained from spectral pure substances. Measurements were made in triplicate and the mean value was reported.

### 2.1. STUDIES ABOUT CD AND ZN IMMOBILIZATION IN CONTAMINATED SOILS WITH SYNTHETIC CALCIUM PHOSPHATE

10 grams of soil samples were contaminated with 5 mg/kg Cd and 600 mg/kg Zn (corresponding to the intervention limit), by the addition of appropriate volume of cadmium or zinc acetate solution in plastic beakers (Chirila et al., 2004). After 24 hours 0.5 g of synthetic calcium phosphate and 25 mL distilled water was added.

The efficiency of immobilization, evaluated based on the reduction of metals mobility was calculated with the relationship:

$$Ef(\%) = \frac{c_i - c_f}{c_i} \times 100$$

where $c_i$ and $c_f$ are metal concentrations in soil solution before and after the treatment.

The schema of experimental procedure is presented in Figure 1.

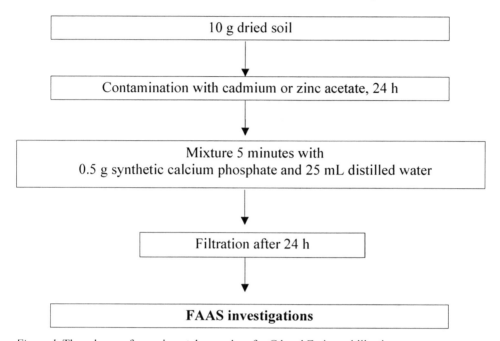

*Figure 1.* The schema of experimental procedure for Cd and Zn immobilization.

## 2.2. STUDIES ABOUT CD AND ZN EXTRACTION FROM CONTAMINATED SOILS WITH EDTA

10 grams of soil samples were contaminated with 5 mg/kg Cd and 600 mg/kg Zn (corresponding to the intervention limit), by the addition of appropriate volume of cadmium or zinc acetate solution in plastic beakers. After 24 hours, 25 mL solution of EDTA 0.01N was added (Chirila and Popovici, 2005). The schema of experimental procedure is presented in Figure 2.

In parallel there were investigated the extraction with EDTA of naturally occurring Cd and Zn using the same procedure except the contamination step.

The efficiency of extraction, evaluated based on the reduction of metals concentration in soils was calculated with the relationship:

$$Ef(\%) = \frac{c_i - c_f}{c_i} \times 100$$

where $c_i$ and $c_f$ are metal concentrations in soil before and after the treatment.

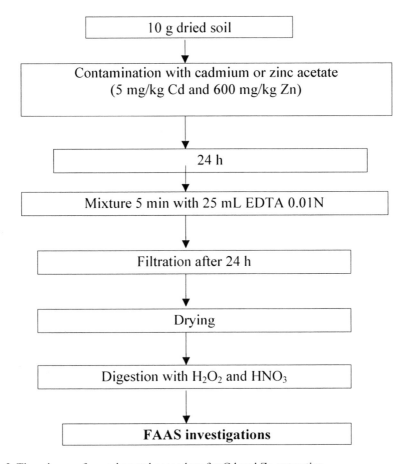

*Figure 2.* The schema of experimental procedure for Cd and Zn extraction.

## 3. Results and Discussions

The goal of this work was to study the immobilization process with calcium phosphate of cadmium and zinc from contaminated soils and also the extraction process of cadmium and zinc from contaminated soils using EDTA using flame atomic absorption spectrometry.

Chemical and physical properties of the contaminated matrix influence the mobility of metals in soils and groundwater. Contamination exists in three forms in the soil matrix: solubilized contaminants in the soil moisture, adsorbed contaminants on soil surface and contaminants fixed chemically as solid compounds. The chemical and physical properties of the soil will influence the form of the metal contaminant, its mobility and the technology selected for remediation.

Cadmium occurs naturally in the form of CdS or $CdCO_3$. Cadmium is recovered as a by-product from the mining of sulfide ores of lead, zinc and copper. Sources of cadmium contamination include plating operations and the disposal of cadmium-containing wastes. The form of cadmium encountered depends on the solution and soil chemistry as well as on the treatment of the waste prior to disposal. The most common forms of cadmium include $Cd^{2+}$, cadmium-cyanide complexes or $Cd(OH)_2$ solid sludge. Hydroxide and carbonate solids dominate at high pH whereas Cd ions and aqueous sulfate species are the dominant forms of cadmium at lower pH (<8). Under reducing conditions when sulfur is present, the stable solid CdS is formed. Cadmium will also precipitate in the presence of phosphate, arsenate, chromate and other anions, although solubility will vary with pH and other chemical factors. Cadmium is relatively mobile in surface water and ground water systems and exists primarily as hydrated ions or as complexes with humic acids and other organic ligands (Theodoratos et al., 2002). Under acidic conditions, cadmium may also form complexes with chloride and sulfate.

Zinc does not occur naturally in elemental form. It is usually extracted from mineral ores to form zinc oxide. The primary industrial use for zinc is as a corrosion resistant coating for iron or steel. Zinc may precipitate as hydroxide, carbonate, sulfide or cyanide. Zinc is one of the most mobile heavy metals in surface waters and groundwater because it is present as soluble compounds at neutral and acidic pH values. At higher pH values, zinc can form carbonate and hydroxide complexes, which control zinc solubility. Zinc readily precipitates under reducing conditions and in highly polluted systems when is present at very high concentrations, and may coprecipitate with hydrous oxides of iron or manganese.

The mean metal concentrations from the analyzed soils taken from different sampling sites are listed in Table 1. Errors were calculated using standard deviation of absorbance measurements.

TABLE 1. The naturally occurring metal concentrations from studied soil samples

| Soil sample | Metal concentration, mg/kg dry weight ± SD | |
| --- | --- | --- |
| | Cd | Zn |
| A | 1.59 ± 0.16 | 15.32 ± 0.30 |
| B | 1.24 ± 0.08 | 19.86 ± 0.27 |
| C | 1.23 ± 0.04 | 7.11 ± 0.22 |

The obtained levels of Cd and Zn concentrations naturally occurring in analyzed soils varied depending on soil characteristics but are in good accordance with literature data (Chen et al., 1999)

## 3.1. CD AND ZN IMMOBILIZATION

Figure 3 presents the efficiency of metal immobilization in studied soils using synthetic calcium phosphate. From the figure it can be observed that zinc was total immobilized in all types of studied soils. The obtained immobilization efficiencies for cadmium were between 97.6% and 98% depending on the soil characteristics.

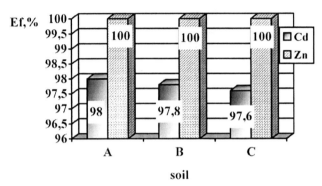

*Figure 3.* The cadmium and zinc immobilization efficiency using synthetic calcium phosphate.

Mobility of studied toxic metals decreased by physically restricting contact between the contaminant and the surrounding groundwater or by chemically altering the metal to make it more stable with respect to dissolution in water.

## 3.2. CD AND ZN EXTRACTION

Figures 4 and 5 present the extraction efficiencies of cadmium and zinc from the studied three different types of soils using EDTA having in view the natural occurring metals (Figure 4) or artificial contaminated soils (Figure 5).

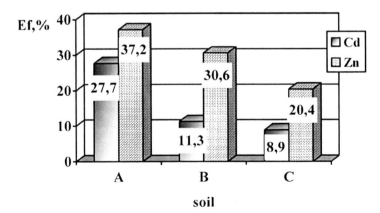

*Figure 4.* The naturally occurring cadmium and zinc extraction efficiencies from soils with EDTA.

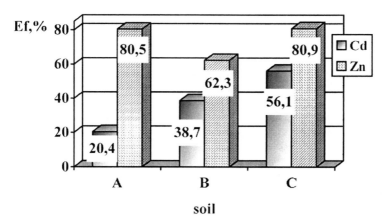

*Figure 5.* The cadmium and zinc extraction efficiencies from contaminated soils with EDTA.

The extractable Zn was higher than cadmium in all studied situations.

So the naturally occurring Cd was extracted in percents between 8.9% (soil C) and 27.7% (soil A) and Zn between 20.4% (soil C) and 37.2% (soil A).

The extraction efficiencies of cadmium and zinc from metal-amended soils were different, so cadmium was extracted in percents between 20.4 (soil A) and 56.1 (soil C) and zinc was removed in percents between 60.5 (soil B) and 80 soils A and C.

**Conclusions**

In this experimental work, the phosphate treatment was tested as a potential cost effective remediation option for the rehabilitation of contaminated sites with cadmium and zinc. As a conclusion, the proposed treatment stabilizes efficiently cadmium and zinc.

The obtained immobilization efficiencies were 100% for zinc on all types of studied soils and 97.6–98% for cadmium.

In this experimental work, the EDTA treatment was tested as a potential cost effective remediation option for the rehabilitation of contaminated sites with cadmium and zinc. The complexation of metals with EDTA reduced the activity of the free metallic ions in the soil solution, and then decreased toxicity of metals to plants. The obtained extraction efficiencies were for naturally occurring metals 8.9–27%, 7.1% for cadmium and 20.4–37.2% for zinc; in the case of metal-amended soils the extraction efficiencies were higher: 20.4–56.1% for cadmium and 62.3–80.9% for zinc.

# References

Chen M., Ma L.Q. and Harris W., 1999, Baseline concentrations of 15 trace elements in Florida surface soils. *Journal of EnvironmentalQuality*, 28, 1173–1181.

Chirila E., Carazeanu I., Belc M., 2004, Heavy metal determination in different types of sewage sludge by FAAS, *Environmental Engineering and Management Journal*, 3, 713–720.

Chirila E., Popovici I., 2005, Studies about cadmium and zinc immobilization from contaminated soils with calcium phosphate using FAAS, *Proceedings "Bramat'2005"*, 6, 60–63, ISBN 973-635-474-7, Ed. Univ. Transilvania Braşov.

Chirila E., Popovici I. and Ursu M., 2004, About cadmium and zinc extraction from contaminated soils using EDTA, *Ovidius University Annals of Chemistry*, 15, 16–18.

Evans L.J., 1989, Chemistry of metal retention by soils, *Environmental Science and Technology*, 23, 1046–1056.

Fingueiro D., Bermond A., Santos E., Carapuca H. and Duarte A., 2002, Heavy metal mobility assessment in sediments based on a kinetic approach of the EDTA extraction: search for optimal experimental conditions, *Analytica Chimica Acta*, 459, 245–256.

Gardea-Torresdey J., Hejazi M., Tiemann K., Parsons J.G., Duarte-Gardea M. and Henning J., 2002, Use of hop (*Humulus lupulus*) agricultural by-products for the reduction of aqueous lead (II) environmental health hazards, *Journal of Hazardous Materials*, B91, 95–112.

Hettiarachchi G.M., Pierzynski G.M. and Ransom M.D., 2000, In situ stabilization of soil lead using phosphorus and manganese oxide, *Environmental Science and Technology*, 34, 4614–4619.

Itabashi H., Yamazaki D., Kawamoto H. and Akaiwa H., 2001, Evaluation of the complexing ability of humic acids using solvent extraction technique, *Analytical Sciences*, 17, 785–787.

Jalali R., Ghafourian H., Asef Y., Davarpanah S.J. and Sepehr S., 2002, Removal and recovery of lead using nonliving biomass of marine algae, *Journal of Hazardous Materials*, B92, 253–262.

Melamed R., Cao X., Chen M. and Ma L.Q., 2003, Field assessment of lead immobilization in a contaminated soil after phosphate application, *The Science of the Total Environment* 305, 117–127.

Melamed R., Self P. and Smart R., 2000, Kinetics and spectroscopy of Pb immobilization on rock phosphate at constant pH in *Environmental Issues and Management of Waste in Energy and Mineral Production*, Singh-al R.K., Mehrotra A.K. (Eds), A.A. Balkema/Rotterdam/Brookfield, 301–306.

Obiajunwa E.I., Pelemo D.A., Owolabi S.A, Fasasi M.K. and Johnson-Fatokun F.O., 2002, Characterisation of heavy metals pollutants of soils and sediments around a crude-oil production terminal using EDXRF, *Nuclear Instruments and Methods in Physics Research B*, 194, 61–64.

Simeonov L.I., Managadze G.G., Schmitt C. and Scheuermann K., 1997, Ecology screening of heavy metal pollution of the soil with laser mass spectrometry, *Comptes rendus de l'Académie bulgare des Sciences*, 51, 5–6, 29–32.

Theodoratos P., Papassiopi N. and Xenidis A., 2002, Evaluation of monobasic calcium phosphate for the immobilization of heavy metals in contaminated soils from Lavrion, *Journal of Hazardous Materials*, B94, 13–146.

Yang J., Mosby D.E., Casteel S.W., Blancher R.W., 2001, Lead immobilization using phosphoric acid in a smelter-contaminated urban soil, *Environmental Science and Technology*, 35, 3553–3559.

Ying Ma Qi, Logan T.J., Traina S.J., 1995, Lead immobilization from aqueous solutions and contaminated soils using phosphate rocks, *Environmental Science and Technology*, 29, 1118–1126.

Zenghua W., Xiaorong W., Yufeng Z., Lemei D., Yijun C., 2001, Effects of apatite calcium oxyphosphate on speciation and bioavailability of exogenous rare earth elements in the soil-plant system, *Chemical Speciation and Bioavailability*, 13, 49–56.

Zorpas A.A., Vassilis I., Loizidou M. and Grigoropoulou H., 2002, Particle size effects on uptake of heavy metals from sewage sludge compost using natural zeolite clinoptilolite, *Journal of Colloid and Interface Science*, 250, 1–4.

# ASSESSMENT OF THE POPS ISSUES IN THE REPUBLIC OF MOLDOVA

ALA COJOCARU[*] AND ANATOLIE TARITA
*Quality of the Environment Laboratory*
*Institute of Ecology and Geography,*
*1, Academiei Str., Chisinau, MD 2028*
*Republic of Moldova*

**Abstract.** POPs chemicals threat human health and the environment all over the world. The intense use of organochlorinated pesticides (OCPs) in the past, which reached a peak in the 1970s, caused pesticides accumulation in and contamination of soil and crops. Presently it is well-known that some OCPs can persist in the soil for 10–20 years and more. This explains why DDT and its metabolites are still detectable in the Moldovan environment (soils, aquatic sediments), despite the fact that it was banned in 1970 and have not been used in significant amounts since then. Another problem not less important is that in the Republic of Moldova there are more than 6,400 tons (according to 2005 data) of obsolete pesticides. This amount of pesticides (DDT, HCH, Heptachlor, Polychlorcamphen and other obsolete chlororganic pesticides) are stored in inappropriate conditions. These stores and surrounding areas are an "ecological bomb" for environment.

**Keywords:** persistent organic pollutants, organochlorinated pesticides, DDT, HCH, PAH, PCBs, remediation, contaminated sites

## 1. Introduction

During the last decade the problem of persistent organic pollutants (POPs) was placed on the environmental agenda of the Republic of Moldova as part of toxic

---

[*] To whom correspondence should be addressed: environment@ozon.mldnet.com

substances and waste management programmes. It was defined as a separate field of actions, which became a priority one among the main environmental issues of the country.

The national policy regarding POPs, which is an integral part of national environmental policy, is driven by understanding that a comprehensive chemical safety management system needs to be created in the Republic of Moldova.

The use of POPs pesticides (herbicides, fungicides, etc.) for crop protection or other purposes increased environmental hazards (soil pollution and toxic effects on other parts of the environment). They tend to accumulate in the soil and in biota, and residues reach surface water and groundwater through leaching.

Among the pesticides that have been used in Moldova in the past, the organochlorinated pesticides (OCPs), are thought to pose the biggest health and environmental risks due to their toxicity, persistence and bioaccumulation potential. In the 1990s, their use in Moldova almost ceased. However, their intensive use in the past, persistence in the environment and the related health risks still make of them a health and environmental issue.

A preliminary national inventory of POPs was undertaken in order to setting priorities and determining the national objectives in the field of POPs minimization and elimination.

## 2.  Current Pesticides and PCBs-related Information

The Republic of Moldova has never had and does not currently have pesticide producing enterprises or factories; all agrochemicals for plant protection permitted for use in the country have been imported and are imported from abroad. None of the POPs pesticides is presently included in the register of permitted substances for use in agriculture, forestry and households.

In countries of the former Soviet Union, pesticides were manufactured to meet rigid production targets. However, the supply of pesticides often had little relation to the actual demand. Certain pesticides were produced even when it was no longer economically viable to grow the crops they were designed to protect. State farms were obliged to buy the pesticides whether they needed them or not.

In the 1950s to 1990s an estimated total amount of 560,000 tons of pesticides were used in Moldova including 22,000 tons of persistent organochlorinated compounds (OCPs). Pesticides use registered a peak in 1975–1985 but reduced dramatically over the last 10–12 years (from 38,300 tons in 1984 to some 2,800 tons in 2000, as active ingredient). The share of persistent OCPs also decreased, in favor of other pesticide groups (Figure 1).

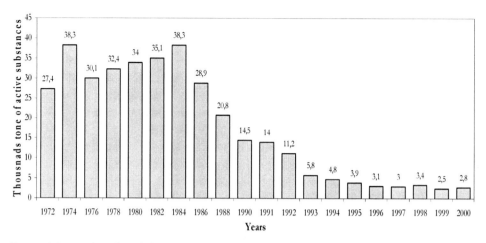

*Figure 1*. Dynamics of pesticides use (Duca et al., 2004, p. 37).

The absence in the past of control on pesticides imports, transportation, storage, and use have resulted in the stockpiling of now banned and useless pesticides which constitute an acute environmental problem/hazard. A pesticide dump was built in 1978 on the territory adjacent to Cismichioi village, in the South of Moldova, in order to find a solution for the ever-increasing amount of obsolete pesticides accumulated in the country. Over a period of ten years (1978–1988) 3,940 tons of pesticides were buried there, including 654.1 tons of DDT.

By the early 1990s, over 1,000 warehouses for pesticide storage have been built in the collective farms. During 1991–2003 about 60% of these were destroyed or dismantled with only 20% of the remaining ones maintained in a satisfactory condition. Significant amounts of obsolete pesticides are stored in the open area. The deteriorated packaging enhances the risk of harmful effect on people's health and environment, some warehouses being situated close to residential areas.[1]

Recent investigations of the State Ecological Inspectorate (SEI) showed a significant pollution of these sites and adjacent areas with OCPs. Maximum allowable concentration (MAC) in soil were exceeded up to 35 times, even at sites as distant as 200 m away from the pesticide stores.[2]

---

[1] National Implementation Plan for the Stockholm Convention on Persistent Organic Pollutants (POPs) Chisinau, Stiinta, 2004, p. 26  http:// www.moldovapops.md

[2] Republic of Moldova: State of the Environment Report 2005:/Inst.of Ecol. and Geography Chisinau, 2006, p. 60.

Currently, the total amount of obsolete pesticides in Moldova is approximately 6,400 tons,[3] including about 3,940 tons buried at the pesticide dump in Cismichioi and 1,712 tons stored in 344 poorly equipped or unfitted facilities, which lack proper monitoring and security. Only 777 tons out of the latter amount are identified compounds and among them POPs pesticides are represented by 80 kg of heptachlor and 1,600 kg of toxaphene. However, no information exists on what amount of POPs pesticides may be among the 935 tons of unidentified obsolete pesticides stored in facilities.[4]

The intense use of OCPs in the past, which reached a peak in the 1970s, caused pesticides accumulation in and contamination of soil and crops.

Existing data on contamination of the environment with OCPs are scaring. During 1976–1990, the soil samples showed pesticide contamination levels exceeding the maximum allowable concentration (MAC) from five times in the Southern zone to 50 times in the Central zone.[5] A research by the Institute for Experimental Meteorology of the State Committee for Meteorology of the former Soviet Union showed that in 1979–1985 about 60% of soil samples were polluted with DDT exceeding the MAC, in spite of the fact that DDT was prohibited in 1970.

Beginning in 1989, due to reduction in pesticide application, investigations showed an anticipated decrease of pesticides-related pressure both in annual and perennial crops. During 1990–1995, the regional Centers of Preventive Medicine of the Ministry of Health analyzed the contents of 28 pesticides' residuals in 10 agricultural crops and foodstuffs. Pesticide residuals were found in 56.4% of the tomato samples and in 40% of the grape samples, but they never exceeded the MAC. This downward trend continued in the last years. The percentage of samples investigated during 1995–2002 showing traces of DDT and HCH has decreased every year and only episodically exceeded the MAC. Traces of heptachlor showed up only once. From the foodstuffs, animal products were found to accumulate most: in 2002, DDT residuals were found in 4.5% of samples and HCH in 1.8% of samples. The contamination level of crops was insignificant.

---

[3] I. Malanciuc, S. Stasiev, A. Veleva, C. Mogoreanu, Management of obsolete pesticides in the Republic of Moldova, the 8th International HCH and Pesticides Forum, 26–28 May, Sofia, Bulgaria, p. 33.

[4] Republic of Moldova: State of the Environment Report 2004:/Inst.of Ecol. and Geography 2005, Chisinau, 2005, p. 94.

[5] A. Tarita, 1995. Organochloro Compounds in soils of the Republic of Moldova. 1995 European year for conservation of nature in the Republic of Moldova: problems, realizations and perspectives, p. 110.

The review of the findings of POPs pesticides assessment revealed the following problems:

- Large amounts of obsolete (including POPs) pesticides are stored in poorly equipped or unfitted storage facilities lacking proper monitoring and security.
- There is no clear ownership and accountability of obsolete pesticides issue related to stockpile and contaminated sites management.
- POPs pesticides are still identifiable in the environment (including soil, surface and ground water and foodstuffs) despite the ban on their use imposed long time ago.
- Large areas around former and existing 340 storage facilities are contaminated with POPs pesticides.

DDT has never been produced in Moldova. Currently DDT is not listed in the official register of permitted substances for use in agriculture, forestry and household. Illegal import of DDT is unlikely, since: (1) it could not be legally used and (2) DDT is no longer seen as an essential pesticide since a number of effective alternative pesticides are in use. Over the last decade, DDT and PAHs concentrations in the environment have constantly decreased, whereas the concentrations of PAHs and PCBs are increasing.[6] The results of the long-term monitoring of DDT residues in soil show a clear downward trend after the peak was reached in the 1980s. In 2005 the average content of DDT residuals in monitored soils was 0.027 ppm, downward from 0.038 ppm the previous year. The MAC (= 0.1 ppm) was exceeded in 8.3% of samples and the highest value registered was 0.372 ppm.[7]

The same tendency has been recorded in the surface waters. According to investigations, the maximum concentrations of DDT residuals detected in surface waters decreased from 8–10 ppb in the 1980s to 0.0–0.01 ppb in 2002.

Concerning DDT stockpiles, a reported amount of 654.1 tons is buried at the Cismichioi pesticide dump. The investigations carried out in 1999 in the framework of the Tacis project "Selected Actions for the Protection of the Danube River Basin" showed that surface soils within and adjacent to the site are contaminated with residues of DDT and its metabolites. The study reported strong evidence that pesticide residues from the dump are being mobilized through exposure of contaminated surface soils and by leaching from the dump into groundwater.

---

[6] K. Terytze, A. Tarita, J. Muller, 2001. Polycyclic Aromatic Hydrocarbons in Soils of the Republic of Moldova. Assessment of the quality of contaminated soils and sites in Central and Eastern European Countries (CEEC) and New Independent Sates (NIS), (Proceedings), International Workshop, Sofia, pp. 199–201.

[7] Republic of Moldova: State of the Environment Report 2005:/Inst.of Ecol. and Geography, Chisinau, 2006, p. 45.

PCBs have never been produced in the Republic of Moldova, all of them being imported. Apparently, no control of the quality of dielectric oils exists at national borders related to the concentrations of PCBs. Their utilization in some sectors has been discontinued or prohibited in the 1980s. However, PCBs continue to be used in power installations and other types of equipment. The major sources of environmental pollution with polychlorinated biphenyls in Moldova are the emissions from the energy sector and industry. The main pathways of environmental pollution are the PCB oil spills and leaks from electric equipment, heat exchangers and hydraulic systems, evaporation from different technical installations, and discharges of industrial liquid waste.

The preliminary inventory identified that about 30,000 tons of dielectric oils are used in electrical power installations, including approximately 23,300 tons in high voltage transformers, 5,400 tons in circuit breakers and 400 tons in capacitors. From the total amount, 95–97% is in the equipment that belongs to power supply entities (producers, transporters, and distributors) and 3–5% in the consumers' electrical installations. The losses of dielectric oils in the energy sector are estimated at 9–10 tons per year.

In contrast to transformer oil, the capacitors used in Moldova contain polychlorinated biphenyls (PCB). The total number of capacitor batteries located at 20 electrical substations throughout the country is almost 20,000, containing a total amount of 365 tons of trichlorobyphenil. Most of the capacitors are concentrated at the Vulcanesti substation in the south of Moldova. Over 12,000 batteries are kept at the substation including many out-of-use capacitors, 56 kg each, containing 19 kg of PCB oil. Thus, a total amount of 230 tons of trichlorobyphenil and 670 tons of PCB-contaminated equipment is stored there.[8]

Analysis of PCB congeners at several electric substations demonstrated the pollution of soils on the territory, sometimes a severe one. The concentration of PCBs in the topsoil collected beneath the captors battery at the Vulcanesti substation reached a level of 7,100 ppm which is exceeding the MAC by five orders of magnitude, and in Donduseni substation, 95.4 ppm that is 1,590 times the MAC value.[9]

Beginning of November 2003, the Moldovan authorities have started repackaging and transportation of obsolete pesticide stockpiles scattered across the country to a limited number of warehouses. Currently about 2,450 tons are

---

[8] National Implementation Plan for the Stockholm Convention on Persistent Organic Pollutants (POPs) Chisinau, Stiinta, 2004, p. 32. http:// www.moldovapops.md

[9] Republic of Moldova: State of the Environment Report 2005:/Inst. of Ecol. and Geography, Chisinau, 2006, p. 46.

stored in 26 storehouses and it is expected by the end of 2007 the remaining about 550 tons will be stored in additional 6 storehouses. Within the WB/GEF POPs Project it is proposed to dispose of 1,150 tons of obsolete pesticide stored in 12 warehouses selected based on their risk assessment. These pesticides are to be incinerated in a licensed facility abroad, in accordance with best environmental practices. The remaining part of obsolete pesticide stockpiles (about 1,850 tons) will be disposed of with the assistance from NATO and Moldovan Government.

Upon the finalization of repackaging and centralization of obsolete pesticides in a limited number of storages, about 340 warehouses will be fully emptied from obsolete pesticides, ensuring the elimination of the most direct threats to health and environment. However, emptied warehouses will remain a significant pollution source because their walls, floors, and adjacent territories are contaminated. In total, the country has about one thousand of pesticide contaminated sites (demolished, abandoned or empty pesticides storehouses, solutions preparation units, and adjacent zones) with an average area of less or about of 1 ha, which require a detailed inventory, risk assessment, development and implementation of remediation measures. The total area of pesticides contaminated lands in the country can presently be roughly estimated rather than 1,000 hectares. This estimation is based on the investigation recently undertaken by the State Ecological Inspectorate on the contamination of the areas adjacent to pesticides storehouses.

If no or limited action is taken in contaminated lands, the severity of POPs impacts to the environment and human health would increase, even considering self-degradation and life-time of obsolete pesticides formulations. It is obviously not the case for POPs pesticides, as they are persistent, bio-accumulative and remaining in the environment and in organisms for a long time. Mitigation of residual effects by developing and promoting affordable and acceptable options for the evacuated warehouses and surrounding areas is an important aspect, since the old warehouses are generally not considered by the local authorities and the population as very dangerous places, finally resulting in using as free construction materials for household needs and the adjacent areas for grazing or agriculture. This may generate more severe impacts on human health.

## Conclusion

POPs are able, due to their various physical and chemical properties to enter all environmental compartments and affect environment and humans through various pathways.

There is a number of specific, short term and targeted researches and there is no established monitoring system on POPs in the country. No information exchange among institutions involved in monitoring. Currently investigations over DDT, HCH and its metabolites, and PCBs are carried out in Moldova.

POPs pesticides are still identifiable in the environment (including soil, surface and ground water and foodstuffs) despite the ban on their use imposed long time ago.

Storing POPs pesticides in inappropriate conditions led to the contamination of adjacent lands. There is no systemic information on the level of contamination of these sites.

The analysis of soil samples taken by the State Hydrometeorological Service in 2006 under capacitors and on the adjacent territories of several substations showed a high contamination with PCBs.

There are no developed regulations, standards and/or guidelines covering contaminated sites assessment procedures, remediation criteria, future site use restrictions and site monitoring.

The investigations provided a better understanding of the situation which allowed for setting priorities and determining the national objectives in the field of POPs minimization and elimination, and remediation measures for contaminated sites.

For solution of the environmental problem of POPs having a global character the systemic approach is necessary and has to include the following technical and scientific measures:

- Establishing of chemical and biological monitoring of POPs
- Risk assessment of POPs to humans and ecosystems
- To assess feasibility, local acceptability and affordability of remediation options
- Development and implementation of on-ground remediation measures:

  - Repackaging and centralisation of obsolete pesticides, and PCBs in the energy sector, at the limited number of storage facilities, identification of most appropriate solution for their final elimination/disposal, low-cost measures to minimise impacts from abandoned storage facilities
  - Improvement of temporary storage conditions (repackaging and labelling, consolidation of stocks of waste pesticides)

- Implementation of remediation measures for contaminated sites:

  - Identification of sources of pollution, size of territory, level of pollution, etc.
  - Assessment of risk of the identified polluted lands and creation of the National register on contaminated sites

- Mapping of contaminated sites
- Establishing of the monitoring of contaminated sites
- Development and implementation of ecologically sound technologies and methods on cleaning up/remediation of contaminated sites
- An extensive and POPs specific technical assistance program should be promoted in the Republic of Moldova by involving potential international financial sources and technology transfer options.

**References**

Duca Gh., Cazac V., Gilca G., Poluanti organici persistenti, Chisinau, 2004, p. 52.

Malanciuc I., Stasiev S., Veleva A., Mogoreanu C., Management of obsolete pesticides in the Republic of Moldova, the 8th International HCH and Pesticides Forum, 26–28 May, 2005, Sofia, Bulgaria, p. 33.

National Implementation Plan for the Stockholm Convention on Persistent Organic Pollutants (POPs), Chisinau, Stiinta, 2004, p. 80. http:// www.moldovapops.md

Republic of Moldova: State of the Environment Report 2004, Institute of Ecology and Geography, Chisinau, 2005, p. 123.

Republic of Moldova: State of the Environment Report 2005, Institute of Ecology and Geography, Chisinau, 2006, p. 81.

Tarita A., Organochloro Compounds in soils of the Republic of Moldova. 1995 European year for conservation of nature in the Republic of Moldova: problems, realizations and perspectives, 1995, p. 110.

Terytze K., Tarita A., Muller J., Polycyclic Aromatic Hydrocarbons in Soils of the Republic of Moldova. Assessment of the quality of contaminated soils and sites in Central and Eastern European Countries (CEEC) and New Independent Sates (NIS), (Proceedings), International Workshop, Sofia, 2001, pp. 199–201.

# THE PROBLEM OF SOIL POLLUTION IN RUSSIA AND ASSOCIATED HEALTH PROBLEMS

ALEXANDER CHERNIH AND DINA SOLODOUKHINA[*]
*Kursk State Medical University*
*Department of General Hygiene*
*\*Department of Public Health,*
*Karl Marx Str. 3, Kursk, Russia*

**Abstract.** The article gives the literature review on situation of soil pollution in Russia from the official state reports of the Sanitary-Epidemiologic Monitoring Department and original research studies conducted in different regions of the country. The pollutants analyzed in the article include heavy metals detected in industrial areas and pesticides prevalent in the soil of agricultural regions. Contribution of these soil pollutants to the development of different diseases is reviewed. Situation in one of the agricultural regions of Russia is described in dynamics from the year 1985. The role of anomalous geomagnetic field in Kursk region is discussed with consideration of its possible negative effect on health and its potential interaction with soil pollutants.

Keywords: soil pollution, Russia, Kursk, anomalous geomagnetic fields

## 1. Soil Pollution in the Russian Federation and Associated Health Risks: Overview of the Problem

Health of a human is influenced by a large number of political, economic, social, ecological, psychological factors. Experts from the WHO estimated that 50–52% of health depends on life style, 20–22% depends on heredity and genetics, 18–20% – environmental conditions, and 7–12% are determined by

---

[*] To whom correspondence should be addressed: solodin_kursk@mail.ru

L. Simeonov and V. Sargsyan (eds.),
*Soil Chemical Pollution, Risk Assessment, Remediation and Security.*
© Springer Science+Business Media B.V. 2008

health care system. Some researchers suppose that 80% of the modern diseases are related to the negative affect of the environment (Lisitcin and Akopyan, 1998).

Environmental safety is one of the major factors of population health. In many regions of Russia there is an alarming ecological situation that is also determined by the sanitary condition of the soil. Soil pollution can be the source of secondary air pollution, water, agricultural products and fodder. Different from other mediums, there is no possibility of rapid purification of the soil, so that absorbed chemicals can be preserved there for many years. Soil pollution in industrial areas can achieve critical level that presents a real threat to the health of people (Kryatov et al., 2006). Though in the last years in the majority of Russian cities there is reduction of harmful substances emission, still a lot of chemical toxicants are detected in the air, soil, food products. It shows that problem of chemical safety is one of the most important for preservation of health of all people not only of those who are employed in special industries (Kryatov et al., 2006).

Russian Federal Center of State Sanitary Epidemiologic Control provides sanitary-hygienic monitoring of soil pollution by detection of major chemicals polluting the soil, measuring intensity of pollution, performing factorial analysis to find the reasons for contents' changes. By the official report of this Center the leading pollutants for the soil of Russia in the last years have been heavy metals like lead, cadmium and zinc. Copper, nickel, chrome and mercury are less common and usually are determined by the certain industrial production (Kryatov et al., 2006; State Report, 2004).

Complex investigations conducted in Moscow by the leading national institutes (headed by the National Research Institute of Human Ecology and Environmental Hygiene named after A.N. Sisin of Russian Academy of Medical Sciences) detected 19 soil anomalies of complex multi-element contents that occupy 29% of the city area. Everywhere there were present lead, copper, zinc, tin, silver. In some areas there were elements released by the certain factory. The chemical pollution is aggravated by biological pollution. There were detected high microbial and viral contamination of soil and sand in sand-boxes of children institutions. We have to consider that the higher chemical pollution of the soil the more the role of the soil as a transmission factor of parasitic and infectious diseases. Pathogenic enterobacteria and helmints are more resistant to chemical soil pollution than truly soil microorganisms – E. coli antagonists (Filimonova, 1985).

In the year 2003 in the Russian Federation the number of soil samples not satisfying the hygienic standards by sanitary-hygienic parameters increased from 12.2% (2002) to 12.4%. On the territory of 12 regions of Russia proportion of samples with elevated contents of dangerous pollutants was 2–4.3 times higher than average in the country. Six regions (Samara, Tula, Chita, Sverdlovsk, Khabarovsk regions, Primorskiy kray) had 29.6–53.4% of such samples that was 11–36% higher than in the year 2002. Particularly dangerous is increasing pollution of build-up areas with heavy metals. In industrial areas the proportion of soil samples with increased amount of heavy metals achieved 67.3% that is 2.3–5.7 times higher than average in Russia. The worst situation was in the Far East, Ural, and Siberia.

In the year 2003 there was increase in number of samples with high contents of mercury from 0.4% to 0.6% (Sverdlovsk, Irkutsk regions, Primorskiy kray), cadmium – from 2.1% to 2.3% (Moscow, Vologda, Chita, Sverdlovsk regions, Primorskiy kray) (Kryatov et al., 2006).

Table 1 is giving the list of regions where the proportion of soil samples with increased amount of chemical pollutants in the inhabited areas, is more than average in the country (Zavistyaeva, 2004).

TABLE 1. Regions of Russia with high level of soil chemical pollution, 2000–2002

| # | Region of Russian Federation | The proportion of soil samples containing chemicals above TLV level, % | |
|---|---|---|---|
| | | year 2000 | year 2002 |
| | Russian Federation | 13.6 | 12.3 |
| 1. | Primorskiy krai (Far East region) | 65.1 | 48.6 |
| 2. | Moscow | 30.4 | 41.1 |
| 3. | Chitinskaya oblast | 14.0 | 39.8 |
| 4. | Vologodskaya oblast | 21.8 | 34.5 |
| 5. | Sverdlovskaya oblast | 36.3 | 34.1 |
| 6. | Krasnoyarskiy kray | 7.6 | 32.1 |
| 7. | Saint-Petersburg | 27.7 | 30.1 |
| 8. | Republic of Mordoviya | 15.4 | 27.3 |
| 9. | Novgorodskaya oblast | 20.2 | 24.7 |
| 10. | Orenburgskaya oblast | 22.0 | 23.1 |
| 11. | Irkutskaya oblast | 25.0 | 23.1 |
| 12. | Republic of Udmurtiya | 20.4 | 17.1 |

Soil of rural areas is mainly polluted by chemicals used in agricultural sector. Eco-toxicological expertise assessment of the soil contents in 1998 by the Ministry of Agriculture of the Russian Federation showed that 1.4 million hectares of the country are polluted by heavy metals, from which 1.7% of the soil is containing high-toxic substances and 3.8% – substances of the 2nd class of toxicity including some pesticides. In spring 5% and in autumn 4.7% of the soil in agricultural areas of 32 regions of Russia were estimated to have increased level of DDT and metaphos. Threshold limit value (TLV) for the mentioned chemicals varied from 2.5 TLV in Primorskiy krai to 28.6 TLV in Novosibirsk region and 50.2 TLV in Kurgansk region (Kryatov et al., 2006).

Table 2 represents the remained amount of pesticides in the soil samples above TLV in the year 2000 in Russia (Chiburaev et al., 2003).

TABLE 2. The leading groups of pesticides detected in the soil, which had remaining amount above TLV, Russia, 2000

| Group of pesticides | Number of samples | Proportion in the total, % |
|---|---|---|
| Mercurous-organic compounds | 149 | 32.7 |
| DDT | 82 | 18.0 |
| Organophosphates | 52 | 11.4 |
| Herbicides of chlorophenyxi-acetic acid | 49 | 10.7 |
| Pyretroids – insecticides | 41 | 9.0 |

In the last years the use of organochlorides, particularly DDT, banned by Stockholm's convention of 2001, dramatically decreased in the country but the problem of their storage and utilization has remained unsolved. DDT was excluded from the list of chemicals eligible for use in agriculture in 1971, but for prevention of some endemic diseases in 1988 there were used 126 tons of DDT. In the south of Russia (Krasnodarskiy kray) the research showed that breast milk of women living in that area contains DDT and its metabolites. There are also more reproductive system disorders in population inhabiting this region (Kryatov et al., 2006).

Despite the fact that in the last years there was significant reduction in use of pesticides, we have to consider their high toxicity and ability to preserve for a long time in the environment, so there can be a large threat from the remaining stores of pesticides. Still the use of hexachlorane, dichloranilin containing dioxins and dibenzofurans, is continued. The total amount of dangerous or unrecognized pesticides stored in 11 regions of Russia exceeds 4,500 tons and can achieve up to 14,000 tons. Improper storage system can lead to long-term pollution by chemical toxicants of the soil and environment in general. Potential effects associated with exposure to organochlorine pesticides include neuro-behavioural disorders, cancer, and spontaneous abortion (Levy and Wegman, 2000). Pollution of agricultural soil in combination with other anthropogenic factors like industrial pollution can lead to summation of risks like lead, cadmium, mercury, organochlorides and others. This can have negative effects on health. The WHO experts emphasize on the significant relationship between the level of toxicants in the soil and food and population health. Epidemiological studies showed that lead can enter the human body with food, soil and air (Kryatov et al., 2006; WHO Report, 2002).

There is direct correlation between increased concentration of toxic elements in the soil and occurrence of pathologic conditions. For instance, in Tula region there was detected association between chromium exposure and malignant neoplasm ($r = 0.7$), diseases of endocrine system ($r = 0.6$) and genitourinary tract diseases ($r = 0.6$). In Novgorod region there was detected association between lead exposure and incidence of neurological and musculoskeletal diseases in children; neurological diseases, diseases of blood and neoplasm in adults. Zink exposure was detected as a factor associated with diseases of the gastrointestinal system and skin (State Report, 2004; Zavistyaeva, 2006).

Generally, in areas with dangerous level of soil pollution there is increased number of children who have frequent cases of diseases of different classes; in areas of very high pollution there are reproductive disorders in women like infertility, premature delivery, miscarriages, newborns with low-birth weight (State Report A, 1999; State Report B, 1999).

Heavy metals detected in biological mediums are particularly common in children population. In the town of Karabash (Chelyabinsk region) 60.3% of children had increased level of lead in the hair (Zavistyaeva, 2006). Systemic exposure to heavy metals determines the certain physiological changes, which have clinical manifestation in the form of neuralgia, cephalalgia, blood changes, liver, skin diseases (Kryatov et al., 2006).

Another aspect of soil pollution in Russia is contamination of land by oil and oil products. At all stages of oil fields development from boring to industrial manufacturing there can be accidental pollution of the environment because of pipeline breaks, oil storage damages, or human mistake. Annually out of 300 mln tons of the oil extracted in the country, 1.5% and more is lost at extraction, transportation or storage that is around 4.5 million tons per year. Today about 800,000 hectares of the soil need to be cleaned from the oil. There is a special concern on this problem because in the present system of epi-demiological monitoring oil is not included in the number of detected major pollutants. There is also no research on the role of oil-polluted soil on health of people (Kryatov et al., 2006).

## 2. Kursk Region as an Area of High Electromagnetic Field and Intensive Agricultural Production

Electromagnetic pollution of the environment is another issue which becomes a large-scale problem in Russia. This problem is particularly important in large cities, which have many various sources of electromagnetic fields (EMF) and large density of population. In the last ten years with the rapid development of telecommunication and electronics intensity of EMF also increased. Electro-magnetic radiation is also used for therapeutical purposes in medical institutions (like micro-wave therapy, diathermia). For these reasons the problem of EMF and their possible negative effects on health makes more and more concern for physicians, as well as for ecologists, hygienists, physicists, biologists (Chernih, 2005). According to the literature the most sensitive to EMF are immune (Belskiy, 2000), nervous (Fathutdinova, 2001), endocrine (Lai and Singh, 1997), and hemopoietic (Jernova et al., 2001) systems. Irreversible changes in the human organism are observed even at short-term repeated exposures to EMF of low intensity. For multiple repetitions of EMF radio-waves biologic effect is cumulated. However, the variety of clinical manifestations and absence of specific characteristics in changes of health condition make it very difficult to distinguish EMF as a health hazard or risk factor for the certain pathologic conditions. Experimental studies showed the potential negative effects of EMF on animals, namely it can lead to death of embryos or malformations, promote development of neoplasm like cancer of mammary glands in rats. Weak association between exposure to EMF and biological effects, absence of dosage effect give the reason to call EMF promoters but not inducing or causal factors of cancers. The latter can be viruses, pesticides, solvents and other physical or biological agents (Adey, 1990).

Ecological and hygienic aspects of electromagnetic fields are not well investigated. It is reasonable to study variation of naturally existing geomagnetic field particularly in the regions of magnetic anomalies. The research has shown that there are several magnetic anomalies on the Earth where geomagnetic field is different from other areas on average by 10%. In Russia there are the two largest ones, which are Angaro-Ilimskaya and Kursk magnetic anomalies (KMA). Horizontal component of GMF induction on the territory of KMA is achieving 80,000 nT, vertical component is up to 170,000 nT, that is 5–6 times more than in other areas. KMA is spreading from the north-west to the south-east from the Baltic Sea to the Azov Sea, the width is 250 kms, the area occupied is around 160 thousand square kms, and it covers 12 regions of Russia and Ukraine (Smolensk, Kaluga, Bryansk, Oryol, Kursk, Berlgord, Voronezh, Sumi, Kharkov, Donetck, Lugansk, Rostov regions). Maximal intensity of anomalous GMF was reordered on the territory of Kursk region in the town Jeleznogorsk, that allows to consider it as a center of KMA and its anomalous geomagnetic field is a natural risk factor for health. Geophysical reason for KMA is the enormous mass of iron quartzite in the earth deposits, magnetization of which under the influence of normal GMF creates the additional field called anomalous. In conditions of the increased sun activity variation of GMF is dramatically increasing, that affects the human organism, particularly it increases the frequency of cardio-vascular diseases. Yu. A. Grigoriev proved that GMF is a risk factor for people working in shielded buildings (Grigoriev, 1995). Several authors detected increased resistance of *Shigella Zonne* to antibiotics in anomalous regions (Belskiy, 2000).

Incidence of acute gastrointestinal infections in the center of KMA is 2–4 times higher than in Kursk, where intensity of GMF is 4–5 times lower than in Jeleznogorsk. Infectious disease process is characterized by the longer duration and higher severity in the area of anomalous GMF.

We raise the problem of GMF and their affect on health because this factor can intervene with other risks like environmental pollutants. Kursk region is predominantly agricultural with fertile black soil. Traditional cultures planted in the region are wheat, barley, maize, sugar beet, potatoes, vegetables, fruits (apples, berries).

In experimental study of the combined affect of EMF and pesticide chlorpyriphos there was detected interaction and synergistic effect on mortality of rats. For this reason there should be more concern about use of agricultural chemicals in the area of KMA. There can be increasing toxicity of chemicals even if their dosage is in the range of maximal admissible concentration. Legal regulations must target the use of pesticides because anomalous GMF is a natural non-modifiable factor. Increase of restrictions in use of agricultural chemicals will provide reliability of existing or newly designed hygienic standards for many agricultural chemicals (Chernih, 2002).

We analyzed the use of pesticides in Kursk region in the last 20 years. There was detected a tendency of decreased use of chemicals for plant protection. There was increase in use of highly active pesticides with low norms of spending, that give minimum harm to the environment. In 1985 there was used 6,621 tons of pesticides with 3,174 tons of active substance. In the year 2002 this figure decreased up to 548 tons with 137 tons of active substances. Rate of reduction is 12.1 for the total weight and 23.2 times for the weight of active substances. The reasons for this trend is mainly economic problem in agricultural sector, as well as the reduction of donations to agricultural companies from public sector. Variety of the used pesticides in Kursk region increased 2.6 times from 64 items in 1985 to 165 in 2002. There were excluded insecticides of disk synthesis, which can cause harmful biologic consequences, high toxic organochlorides and organophosphates. Today organochlorides take 2.7% in the total (versus 59.7% in 1985) and organophosphates take 6.6% (versus 17.2% in 1985). There was increase in use of carbamates, heterocyclic compounds, synthetic pyretroids, avermectines, azoles (Chernih, 2003, 2004).

From the year 1995 the average amount of pesticides per 1 hectare of cultivated land gradually has decreased and in the year 2002 it was 0.24 kg or 0.11 kg of the active substance. This is obviously a positive trend that shows reduction of ecologic danger for the population health and environment.

By the data of Kursk regional center responsible for environmental monitoring in 2001 about 11,300 hectares of agricultural lands were contaminated with pesticides, including 10,000 hectares of ploughed fields (from which 0.900 hectares are moderately dangerous) and 1,300 hectares of perennial plantings (from which 0,200 hectares are moderately dangerous). In the last six years the amount of polluted lands has increased. The major chemicals – pollutants are simazin, bayleton, and DDT. One of the areas famous for fruit production

(Oboyan) has 89% of land polluted with the mentioned chemicals. The remained amount of DDT in the soil was 2.5–2.7 times higher than TLV in spring and 1.9–2.2 times higher than TLV in autumn (Chernih, 2004).

To sum up, though the use of pesticides in Kursk region in the last years has decreased, we observe increase in the list of used chemicals. There must be continued strict monitoring of the use of agrochemicals considering the larger participation of private commercial organizations supplying agrochemicals. In the conditions of anomalous geomagnetic fields in the region there should be paid more attention to environmental safety important for population health.

## 3. Conclusions and Recommendations for Improvement of the Situation

The existing legal basis in Russia is not enough for effective control and organization of monitoring system on soil pollution and its risk assessment. Today there is justification of threshold limit value for more than 500 substances, the majority of which are pesticides and only 39 items are industrial chemical pollutants. Absence of legal norms or their under-development leads to the situation that some of the ministries and administrative departments set their own levels of TLV for different substances. So, one of the major directions in research for preservation and protection of soil should be defining of chemical substances in the soil in concentrations that are safe for health (Kryatov et al., 2006). There should be raised the following questions:

- Improvement of ranking system of chemical pollutants of soil by their toxicity and threat with consideration of similar studies in the developed countries
- Analysis and reconsideration of legal normative basis and harmonization of TLV with international standards
- Development of norms for the major pollutants like oil, heavy metals, etc.
- Development of methodology of prognosis of possible changes in health under the influence of soil pollutants considering the level of soil pollution
- Investigation of quantitative relationships between soil pollution and response of the dependent factor in the chain of effect (plant, air, water, microorganisms, human)
- Improvement of calculation techniques of justification of TLV of different chemicals and their combinations in the soil
- Development of methodology of investigation of health status of people as the integral measurement of soil affect, polluted by chemical and biological factors
- Standardization of methods used for detection of chemical substances in the soil
- Improvement of the national and regional law on soil preservation as the basis of ecological safety and public health (Kryatov et al., 2006; State Report A, 1999; State Report B, 1999)

Preservation of soil safety is one of the major ecologo-hygienic tasks in the Russian Federation. In order to solve it there should be combined the efforts of many professionals – epidemiologists, hygienists, occupational therapists, agricultural chemists, public health specialists. There must be formed the national policy aimed at development of long – term sustainable programs to solve ecological and hygienic problems of soil pollution in the country before it led to catastrophic consequences.

## References

Adey, W.R., Joint actions of environmental nonionizing electro-magnetic fields and chemical pollution in cancer promotion, *Environmental Health Perspectives* Vol. 86, 1990, pp. 297–305.

Belskiy, V.V., L.M. Zakaryan, V.V. Kiselyova, P.V. Kalutckiy, Ecological situation in the region of Kursk magnetic anomaly, Materials of the Regional Conference "Problems of population health in the Region" (Belgorod, 2000), pp. 119–127.

Chernih, A.M., P.V. Kalutckiy, V.M. Emeliyanov, Experimental Study of Combined Affect of the Constant Magnetic Field and Agrochemicals on Organism: Methodical Recommendations (Moscow, 2002), pp. 23–24.

Chernih, A.M. Threats to human health in use of pesticides, *Hygiene and Sanitation* Vol. 5, 2003, pp. 25–29.

Chernih, A.M. Perumal Jaya Chandran, Chernih Yu. A., Monitoring of use of pesticides in Kursk region, *Kursk Vestnik – A Human and His Health* Vol. 2–3, 2004, pp. 87–92.

Chernih, A.M., A.I. Elkin, V.N. Pozdeev, Ecological threats to human health under the influence of electro- and anomalous geomagnetic fields (literature review), *Military Medicine Journal* Vol. 326, 2005, pp. 46–50.

Chiburaev, V.I., Ya. G. Dvoskin, I.V. Bragina, Pollution with pesticides in the RF as a potential threat for population health, *Hygiene and Sanitation* Vol. 3, 2003, pp. 68–71.

Fathutdinova, L.M. Influence of electromagnetic fields on the nervous system", *Occupational Medicine and Industrial Ecology* Vol. 9, 2001, pp. 20–22.

Filimonova, E.V., Scientific Justification of *Enterococcus* identification for evaluation of sanitary and epidemic soil safety: Dissertation Abstract (Moscow, 1985).

Jernova, A.I., L.M. Sharshina, V.A. Chiruhina, Observation of hemoglobin allotropy in magnetic field in vitro, *Bulletin of Experimental Biology* Vol. 9, 2001, pp. 272–274.

Grigoriev, Yu. A. Weakened geomagnetic field as a risk factor for workers employed in shielded buildings, *Occupational Medicine and Industrial Ecology* Vol. 4, 1995, pp. 7–11.

Kryatov, I.A., N.V. Rusakov, N.I. Tonkopy, The ecologo-hygienic problem of soil pollution, *News of RAMN* Vol. 5, 2006, pp. 18–21.

Lai, H., N.P. Singh, Melatonin and a spin-trap compound blocked radiofrequency radiation-induced DNA stand breaks in rat brain cells, *Bioelectromagnetics* Vol. 18, 1997, pp. 446–454.

Levy, B.S., D.H. Wegman, Occupational Health: Recognizing and Preventing Work-Related Disease and Injury. 4th ed. (Philadelphia: Lippincott, Williams & Wilkins, 2000).

Lisitcin, Yu. P., A.S. Akopyan, Panorama of Health Preservation. Reforms in Medical Services and Unsolved Problems of Health Care Privatization (Moscow, 1998).

State Report "About Sanitary-Epidemiologic Situation in the Russian Federation in the year 2003", Moscow, 2004, pp. 30–36, 103–108.

State Report A: About environmental situation in the Russian Federation in the year 1998, Moscow, 1999, pp. 53–56.

State Report B: About Sanitary-Epidemiologic Situation in the Russian Federation in the year 1998, Moscow, 1999, pp. 128–129.

WHO Report: Our Planet is our Health, Geneva, 1992, pp. 60–105.

Zavistyaeva, T. Yu., About the problem of conducting sanitary-hygienic monitoring of soil in Russia, *News of S.-Petersburg state medical university* Vol. 3, 2004, pp. 41–45.

Zavistyaeva, T. Yu., The role of soil as one of the indicators of population health in the system of sanitary-hygienic monitoring, *Population Health and Environment* Vol. 1, 2006, pp. 18–21.

# PRESENT STATE, RISK ASSESSMENT AND TREATMENT OPTIONS FOR THE CONVERSION OF LIQUID BALLISTIC MISSILE PROPELLANTS IN AZERBAIJAN REPUBLIC

AIAZ EFENDIEV AND ELNUR MAMMADOV
*Department of Chemical Sciences*
*Azerbaijan National Academy of Sciences*
*10 Istiglaliyyet Str., 1001 Baku AZ, Azerbaijan*

It is known, that after collapse of the Soviet Union large quantities of the components of liquid ballistic missile propellants were left on the territory of several former republics of the USSR including Azerbaijan. They consist of high energetic fuels and the oxidizers designated as "Melange". Both components are separately stored in special containers. Fuels and oxidizers that are not longer used by the Azerbaijan Military Forces are stored at two sites – military bases at Elet and Mingechevir.

The oxidizers ammount to 1,400 metric tons of 73–80% nitric acid and 20–27% nitric oxides with different inhibitors/additives, while the fuels are 450 metric tons of Samine (50% triethylamine + 50% xylidine) and 25 metric tons of Izonit (isopropyl nitrate).

After the withdrawal of Soviet troops from Azerbaijan these chemicals have been abandoned. They are still stored on military bases and guarded by military staff. The Army does not need it for military use and has no means for a safe storage and handling of these chemicals.

On all storage tanks leakages of chemicals are evident (Figures 1 and 2). In particular oxidizers are continuously leaking into the atmosphere and to the soil. Due to their aggressiveness in combination with water they corrode stainless steel certainness as well as aluminium storage tanks. The leakage of Samine and Izonit is mainly detected by a strong characteristic smell at all storage sites.

## 1. Oxidizers

### 1.1. NITRIC ACID ($HNO_3$)

Fire: Nitric acid is not combustible, but reacts with water or steam to produce heat. Contact of concentrated nitric acid with combustible materials may increase the hazard of fire and may lead to an explosion.

*Figure* 1. Leaking tank.

*Figure 2.* Tanks in contact with groundwater.

Human toxicity: Contact with nitric acid or inhalation of nitric acid fumes will result in severe cauterization of skin, the respiratory system and eyes.

## 1.2. NITRIC TETROXIDE ($N_2O_4$)

Fire: Does not burn itself but supports combustion as a strong oxidising agent. May ignite combustible materials (wood, paper, oil, clothing, etc.), containers may explode when heated.

Toxicity: Inhalation of nitric tetroxide fumes/gas will result in slowly evolving but progressive inflammation of lungs. Dangerous for the skin, respiratory system and eyes.

## 1.3. SAMINE

Samine is a mixture of 50% triethylamine and 50% xylidine. It is used as a fuel for ballistic and air defence missiles.

Triethylamine ($C_6H_{15}N$)

$$C_2H_5-N(-C_2H_5)-C_2H_5$$

Fire: Triethylamine is a flammable/combustible material that may be ignited by heat or sparks of flames. Vapours are heavier than air and form explosive mixtures with air. Vigorous reaction with strong acids.
Toxicity: Mainly local effects. Eye irritant – eye contact causes severe burns. Clothing wet with triethylamine will cause skin burns. Vapours irritate nose, throat, lung, causing coughing, and difficult breathing.

Xylidine ($C_8H_{11}N$ – six isomers)

Fire: Xylidines are flammable/combustible substances that may burn but do not ignite readily. When heated, vapours may form explosive mixtures with air. Contact with metals may evolve hydrogen gas.
Toxicity: Possibly carcirogenic, intoxication may result in headache and dizziness.

## 1.4. IZONIT

Isopropyl Nitrate (2-Propyl Nitrate, $C_3H_7NO_3$)

$$O_2N-O-CH(-CH_3)-CH_3$$

Fire: Highly flammable, self-ignition possible when in contact with organic material, explosive.

## 2. Treatment Options

There have been 2 Workshops in Baku, organized by NATO and 1 Workshop in Kiev organized by OSCE and NATO. As a result of these Workshops the following strategy of utilization of missile propellants has been developed.

### 2.1. TREATMENT OPTIONS FOR OXIDIZERS

For the oxidizers neutralization appears to be the most simple and useful method for destruction, yielding fertilizer for application in agriculture.

Simple method of neutralization of the oxidizer that might reflect local conditions in Azerbaijan is to use carbonate rocks which are available in Azerbaijan. They are used for cement and calcium oxide production. Thus, calcium oxide is fairly available and could be used for the preparation of lime milk, resulting in calcium hydroxide solution. The calcium hydroxide solution could be used for neutralization of the oxidizer, yielding calcium nitrate. Calcium nitrate can be used as a fertilizer.

### 2.2. TREATMENT OPTIONS FOR SAMINE

Samine as a mixture of 50% xylidines and 50% triethylamine can be re-used after separation.

There are many industrial uses for both xylidines and triethylamine. As the boiling points of xylidines (213–226°C) and triethylamine (89.3°C) are considerably different, the separation of Samine into its basic compounds by means of distillation and a subsequent commercial re-use should be considered.

Catalytic incineration of samine should be also considered.

### 2.3. TREATMENT OPTIONS FOR ISOPROPYL NITRATE

Isopropyl nitrate can be hydrolysed with the use of acid solutions. Another option is the use of isopropylnitrate as an additive to dizel fuel in order to increase its cetane number. The third option is catalytic incineration.

## 3.  Results and Conclusion

The proposals for the conversion of propellants in Azerbaijan has been accepted by NATO and American Company UXB has started destroying the propellants, first of all the oxidizers. Up to now 650 tons of oxidizers have been converted into fertilizer and the process is going on. The next stage will be utilization of fuels. After we finish utilization of propellants we will face the problem of remediation of soil. The following picture show how the soil in the place where the missile propellants are kept is polluted (Figure 3).

*Figure 3*. Polluted soil.

**Acknowledgment**

The authors would like to express their sincere gratitude to Prof. Dr. Wolfgang Spyra of Branderburg University of Technology, Germany who made a great contribution to the problem of utilization of Melanj in Azerbaijan Republic for active participation and valuable advices.

# USING *LEPIDIUM* AS A TEST OF PHYTOTOXICITY FROM LEAD/ZINC SPOILS AND SOIL CONDITIONERS

BEATA JANECKA[*] AND KRZYSZTOF FIJALKOWSKI
*Czestochowa University of Technology*
*Brzeznicka 60A, 42 200 Czestochowa, Poland*

**Abstract.** A high sensitivity of *Lepidium sativum* to phytotoxic substances makes it suitable for the biological tests conducted to assess the state of land and water environment. *Lepidium* is commonly used in tentative, quick and inexpensive phytotoxic properties of soil contaminated with heavy metals, pertrochemical compounds and polycyclic aromatic hydrocarbons. The test can also be used to assess the influence of sewage sediments, applied for remediation and agriculture purpose.

The article describes the trials of applying the *Lepidium* test as an guidance of phytoremediation possibilities of spoil dump.

**Keywords:** *Lepidium* test, phytoremediation, lead and zinc dumps

## 1. Introduction

The long-term activities of the non-ferrous metal industry in the southern regions of Poland has resulted in considerable metal contamination of surface soils, particularly in close proximity to metal emission sources such as industrial facilities and waste spoil dumps.

A typical spoils dump created in southern Poland between 1915 and 1930 was chosen for this study. The dump used hydraulic enrichment processing of the zinc-lead ore, containing about 6% zinc, 1.5% lead and 0.1% cadmium. The dump is located less than 20 m from a farm and agriculture fields which is a potential source of considerable human exposure to metals and a threat to crop quality.

Previous revegetation trials using routine methods for heap application proved to be unsuccessful. Most of plants perished within a short time of

---

[*] To whom correspondence should be addressed: bjanecka@is.pcz.czest.pl

L. Simeonov and V. Sargsyan (eds.),
*Soil Chemical Pollution, Risk Assessment, Remediation and Security.*
© Springer Science+Business Media B.V. 2008

introduction to the site and the growth and development of the remaining plants was limited by soil proprieties. Possible explanations for growth limitation may stem from the lack or excess of certain chemical compounds in the soil responsible for vegetative processes, unfavourable physical parameters of the soil or a combination of these two factors operating simultaneously.

The project required a rapid and inexpensive test for screening the impact of various soil additives in different ratios on vegetative growth. Therefore, *Lepidium sativum*, a plant species frequently used as a biological indicator, was applied to the site as a screening indicator of soil toxicity and physical properties (Garbisu and Alkorta, 2001; Sas-Nowosielska et al., 2005). The sensitivity of *Lepidium sativum* to phytotoxic factors enables the use of this species in tentative, quick and cost-effective analyses of the phytotoxicity of wastes and soils contaminated with heavy metals, petrochemical compounds or polycyclic aromatic hydrocarbons. The test can also be used to assess the feasibility of sewage sludge during heap remediation projects (Maila and Cloete, 2005).

The root sensitivity of *Lepidium sativum* to the presence of mutagenic and carcinogenic compounds in the ground allows the use of this species in laboratory studies as a biological indicator for soil toxicity (Ernst, 1996). In research conducted by Walter et al. (2006), *Lepidium* seeds were used to evaluate the toxicity of the sewage sludge applied for agricultural purposes. Similarly, Maila and Cloete (2002) applied *Lepidium sativum* seeds as an indicator of PAH pollution in the soil.

*Lepidium* use as an indicator species requires analysis of its roots. Under the influence of cytotoxic and mitotoxic compounds, inhibition of the dividing processes of the root meristematic cells appears which leads to a reduction of roots length (Broda and Grabias, 1979; Gong et al., 2001).

## 2.   Aim and Scope of the Project

The experiment is part of a comprehensive project focused on non-ferrous dump management. The project is aimed at phytochemostabilization of the spoil dump surface soils, which should abate the resuspension of metal-bearing dust and contribute to the reduction of contaminated leachates from the spoil dump. The aim of the present task is the examination of how spoil material and local biosolids planned for use as soil conditioners may adversely influence the growth of plants used for remediation. The project proposes to use *Lepidium* as a test species for a quick and cost-effective indicator of soil or amendment toxicity.

## 3. Experiment

### 3.1. MATERIALS

Heap material:
The experiments were conducted on material derived from the spoils waste dump from the zinc-lead ore mechanical enrichment process. The dump was established between 1915 and 1930, and consists of fine-grained dolomite (particle size 0–10 mm) containing about: 6% Zn, 1.5% Pb and 0.1% Cd. The plant nutrient content of the soil is rather sparse.

Organic Additives:
It was concluded that one of the possible causes of failure in the previous trials of spoils dump revegetation was a shortage of nutrients and organic matter in the soil. Taking this into consideration, further experiments included the following organic additives were used:

- Garden peat
- Municipal compost
- Sewage sludge
- Subsoil used for mushroom production

All additives met environmental standards for use in metals contaminated soils.

### 3.2. METHODOLOGY OF THE *LEPIDIUM* TEST (Broda and Grabias, 1979)

The experiment was performed using waste originating from the spoils dump, and the additives as listed previously. The *Lepidium* seeds were pre-treated before the experiment by incubating them under controlled temperatures for a period of 17–22 hours. Seeds were selected for use in the study if their roots reached about 1 mm length.

Petri dishes (120 mm diameter) were filled with 20 g of investigated material (waste, garden peat, sewage sludge, sub-soil for mushrooms production). Thirty seeds were placed on the Petri dish surface and then covered with filter paper. The control group of seeds were prepared in a similar way and kept moist using distilled water. The experiment was conducted in triplicate. The dishes were placed into an incubator (25°C) for 24 hours. After incubation, the roots were stained with methylene blue. The length of the roots were measured and evaluated.

In the next phase of the experiment, the phytotoxicity of the waste and the additives (peat, compost, sewage sludge and soil used for growing mushrooms) were evaluated.

In the final phase of the experiment, six variants of the ground/material mixtures were studied (Table 1) for their phytotoxic effects.

The methodology of the experiment is described in previous sections.

TABLE 1. Experiment design

| Variant I | Variant II | Variant III | Variant IV | Variant V | Variant VI |
|-----------|-----------|-------------|------------|-----------|------------|
| Control | waste | waste + 1.5% peat | waste + 1.5% compost | waste + 1.5% sewage sludge | waste + 1.5% mushroom subsoil |
| | | waste + 2.5% peat | waste + 2.5% compost | waste + 2.5% sewage sludge | waste + 2.5% mushroom subsoil |
| | | waste + 5% peat | waste + 5% compost | waste + 5% sewage sludge | waste + 5% mushroom subsoil |
| | | waste + 10% peat | waste + 10% compost | waste + 10% sewage sludge | waste + 10% mushroom subsoil |

## 4. Results

The first part of experiment (root elongation test) examined the spoils and additive materials used for the presence of nutrients and toxic compounds. The results are shown in Table 2.

TABLE 2. Average length of *Lepidium sativum* roots in different spoil/additive variants

| Experiment | Average length of Lepidium roots [mm] |
|------------|---------------------------------------|
| Control | 14.5 |
| Waste | 18.7 |
| Peat | 20.6 |
| Compost | 7.7 |
| Sewage sludge | 4.0 |
| Mushroom subsoil | 3.6 |

The subsequent experimental phases assessed the various amendment mixture compositions within the heap material. The ratios of particular mixtures were based upon the results obtained from preliminary experiments (Table 3).

## Discussion

The experiments conducted during the course of this project have confirmed the applicability of the *Lepidium* test as a screening method for determining the toxicity of the materials being remediated as well as the soil amendments added as a source of plant nutrients.

The methodology described above allows for a quantification of the phyto-toxic or phytopromotional impact of various amendments used for revegetation purposes. The best results were observed with the use of compost, where root elongation was highest when compared with the control group. The addition of peat, sewage sludge and mushroom subsoil, resulted in shortening of Lepidium roots by 30–50%.

TABLE 3. Average length of *Lepidium* roots in different variants of the experiment

| Variant of the experiment | Length of *Lepidium* roots in relation to the soil conditioners added [mm] |
|---|---|
| Control | 21.1 |
| Waste | 21.4 |
| Waste + 1.5% peat | 15.7 |
| Waste + 2.5% peat | 15.5 |
| Waste + 5% peat | 18.9 |
| Waste + 10% peat | 19.0 |
| Waste + 1.5% compost | 21.7 |
| Waste + 2.5% compost | 22.3 |
| Waste + 5% compost | 22.7 |
| Waste + 10% compost | 11.9 |
| Waste + 1.5% sewage sludge | 12.6 |
| Waste + 2.5% sewage sludge | 11.9 |
| Waste + 5% sewage sludge | 7.4 |
| Waste + 10% sewage sludge | 5.1 |
| Waste + 1.5% mushroom subsoil | 15.1 |
| Waste + 2.5% mushroom subsoil | 13.1 |
| Waste + 5% mushroom subsoil | 14.7 |
| Waste + 10% mushroom subsoil | 10.9 |

Actually, the results of this experiment points to the conclusion that the plant toxicity from heap material is negligible. This observation is limited, however, to the conditions of this particular test. Special care should be taken to control the quality and ratio of soil amendment material introduced to improve the heap material structural, physical and chemical properties.

An analysis of the data showed that some of the additives should be added in carefully measured quantities, as an excess of some additives may result in adverse effect, as observed in the case of mushroom subsoil.

## Summary

The pilot study of heap material properties, based on physico-chemical analysis and biological test (*Lepidium*) have shown the potential for using plants for the remediation of spoils waste dumps. The *Lepidium* test did not indicate the presence of inorganic phytotoxic substances in dump material. The occurrence of excess phytotoxic organic compounds was similarly excluded, due to the high sensitivity of *Lepidium* to organic toxicants. Soil additives introduced to improve heap material agricultural quality should be examined carefully in order to avoid undesirable effects due to unknown or unforeseen impacts on plant growth.

Plans for continued research of the *Lepidium* test on a lab scale using pot experiments and various ratios of biosolids (sewage sludge, peat) will be based on the results of previous *Lepidium* testing.

The pot experiments will be conducted in parallel with studies investigating the effects of soil moisture and soil texture variations on the process of plant growth and development. To positively confirm the results, the duration of the experiment should be expanded.

Biological tests, in certain controlled instances, are competitive with classic and instrumental analytical methods. Furthermore, these experiments take less time, are cheaper and in many cases provide sufficient amounts of data needed to guide remediation decisions concerning polluted grounds (Kuczyńska et al., 2004).

## Acknowledgments

The authors wish to express their thanks to Mr. Tyler Lane for his contribution to this paper.

## References

Broda B., Grabias B., Ocena biologiczna aktywności cytotoksycznej niektórych związków chemicznych występujących w środowisku człowieka za pomocą testów roślinnych I. Test Lepidium (Biological estimation of cytotoxic activity of some chemical compounds present in human environment by plants. I. Lepidium test – in Polish) Farmacja Polska XXXV, no.12, s. 707–709, 1979

Ernst W.H.O., Bioavailability of heavy metals and decontamination of soil by plants, Geochemistry, no. 11, pp. 163–167, 1996

Garbisu C., Alkorta I., Phytoextraction: a cost-effective plant-based technology for the removal of metals from the environment, Bioresource Technology, Elsevier Science, pp. 229–236, 2001

Gong P., Wilke B.-M., Strozzi E., Fleischmann S., Evaluation and refinement of a continuous seeds germination and early seedling growth test the use in the ecotoxicological assessment of soil, Chemosphere, no. 44, pp. 491–500, 2001

Kuczyńska A., Wolska L., Namieśnik Ł., Szybkie testy biologiczne (ekotesty) – nowe podejście do oceny stopnia zanieczyszczenia środowiska. (Rapid biological tests – a new approach to assess the contamination of the environment Mikrozanieczyszczenia w środowisku człowieka, Konferencje 55, Wyd. Politechniki Częstochowskiej, Częstochowa 2004

Maila M.P., Cloete T.E., The use of biological activities to monitor the removal of fuel contaminants – perspective for monitoring hydrocarbon contamination: a review, International Biodeterioration and biodegradation, no. 55, pp. 1–8, 2005

Maila M.P., Cloete T.E., Germination of Lepidium sativum as a method to evaluate polycyclic aromatic hydrocarbons (PAHs) removal from contaminated soil, International Biodeterioration and biodegradation, no. 50, pp. 107–113, 2002

Sas – Nowosielska A., Kucharski R., Małkowski E., Feasibility studies for phytoremadiation of meta – contaminated soil, in Manual for soil analysis – analysis – monitoring and assessing soil bioremediation, Springer, Berlin – Heidelberg, 2005

Walter I., Martínez R., Cala V., Heavy metal speciation and phytotoxic effect of three sewage sludges for agricultular uses, Enviromental Pollution, no. 139, pp. 507–514, 2006

# HYDROGEOPHYSICAL TECHNIQUES FOR SITE CHARACTERIZATION AND MONITORING: RECENT ADVANCES IN GROUND-PENETRATING RADAR

SÉBASTIEN LAMBOT[1,2*], EVERT SLOB[3],
MARNIK VANCLOOSTER[1], JOHAN A. HUISMAN[2]
AND HARRY VEREECKEN[2]
[1]Université catholique de Louvain, Belgium
[2]Forschungszentrum Jülich, Germany
[3]Delft University of Technology, The Netherlands

**Abstract.** We introduce ground penetrating radar (GPR) basic principles and applications in environmental engineering, with emphasis on quantitative methods for soil water content estimation. The main limitations of these techniques are discussed. Then, we summarize our recent advances on the development and use of advanced off-ground GPR for shallow subsurface characterization. The proposed method is based on full-waveform forward and inverse modelling of the radar signal, thereby maximising inherently information retrieval capabilities from the radar measurements.

**Keywords:** agricultural research, contamination of surface waters, ecosystems, environmental research, ground-penetrating radar, hydro-geophysical, hydrological

## 1. Introduction

A significant increase in global food and biofuel demands, predicted for the next 50 years, poses huge challenges for the sustainability and productivity of both terrestrial and aquatic ecosystems and the services they provide to society (Tilman et al., 2002). Sustainable and optimal management of water and land resources strongly relies on knowledge of the soil hydrological properties, which are essential in agricultural and environmental research and engineering as they control hydrological processes, contamination of surface and subsurface water,

---

[*] To whom correspondence should be addressed: sebastien.lambot@uclouvain.be

plant growth, sustainability of ecosystems and biodiversity, and climate. The soil constitutes an interface between the earth and the atmosphere and governs all important key processes of the hydrological cycle such as infiltration, runoff, evaporation, as well as partitioning of energy at the earth's surface into sensible and latent heat exchange with the atmosphere. For instance, soil moisture has been shown to play a crucial role in land-atmosphere interactions and feedbacks to climate change (Seneviratne et al., 2006).

The characterization and monitoring of the distribution of the soil properties in the environment is therefore needed to develop site-specific management practices that match human activities with local environmental requirements (Stafford, 2000; Zhang et al., 2002). Obtaining soil information with the required spatiotemporal resolution is however complicated by the inaccessibility of the subsurface and its inherent variability. In addition, the interconnectivity of different subsystems at different scales requires a holistic approach. Common techniques to characterize soil hydrological properties are either suited to small areal scales (<0.1 m), such as reference sampling methods, capacitive sensors, and time domain reflectometry (TDR) (Robinson et al., 2003), or to large areal scales (>10–100 m), such as airborne and spaceborne passive microwave radiometry and active radar systems (de Rosnay et al., 2006; Dobson and Ulaby, 1986; Jackson et al., 1996). Small scale techniques are usually invasive, sometimes requiring boreholes, and may not be representative of the soil properties at the management scales. For instance, common methods of characterization for contaminated soils still often rely on the use of drilling and soil sampling, although there is evident concern about the limitations of such methods. The limited understanding of biogeochemical-hydrological processes during remediation operations and the inadequacy of conventional observation approaches for monitoring these processes limits our ability to guide contamination remediation. For the large scale techniques, the characterization is limited to the top first centimeters of soil and temporal resolution is relatively poor. More appropriate techniques are needed to measure the variability of soil hydrological properties at intermediate scales (0.1–100 m), which is crucial in applications that include agricultural water management and soil and water conservation, and to bridge the present scale gap between ground truth measurements and remote sensing (Famiglietti et al., 1999). As an example, the lack of suitable characterization techniques has been cited as the most critical factor preventing the wider implementation of precision agriculture in the 21st century (Bouma et al., 1999; Kirchmann and Thorvaldsson, 2000; Stafford, 2000).

Non-invasive and proximal sensing techniques can be used for mapping and monitoring key hydrological or pedological variables at intermediate or field scales. Soil electric sounding is commonly used as a technique to generate proxies of soil moisture and salinity or contaminants. Soil electric sounding can be performed by classical geo-electric (Kemna et al., 2000; Kemna et al., 2002;

Vanderborght et al., 2005) or electromagnetic induction (EMI) (Lesch et al., 1998; Sudduth et al., 2001) techniques. Soil electric conductivity is multivariate and depends strongly on soil water content, water salinity, and soil texture and structure (Mualem and Friedman, 1991; Rhoades et al., 1976). As these parameters partly vary independently within a field, their respective correlation with electric conductivity can not be interpreted uniquely and is subject to large uncertainties. Yet, electric soundings have been particularly successful to qualitatively delineating management zones or detecting contaminant plumes.

Over the last decade, ground-penetrating radar (GPR) has gained an increasing interest in environmental research and engineering applications (Annan, 2005). Several GPR methods are available to identify soil dielectric permittivity, usually from the GPR wave propagation velocity determination (Huisman et al., 2003). As the dielectric permittivity of liquid water overwhelms the permittivity of other soil components, soil dielectric permittivity constitutes an accurate surrogate measure of soil water content (Tabbagh et al., 2000; Topp et al., 1980). For instance, it is well known that capillary pressure helps to regulate groundwater interactions with the surface, and other processes that deal with buried gas or water. However, measuring or calculating the capillary pressure is difficult because the continuum between drainage (decreasing water saturation) and imbibation (increasing water saturation) exists in a system that reacts slowly to change where parameters may not return to their initial state. This makes it difficult to predict the distributions of multiple fluids in a porous medium. Plug et al. (2007a, b) hypothesized that the ability of surrounding rock to transmit electrical current provides new insight into the physical behavior of the groundwater capillary pressure. Through laboratory experiments where the authors varied the degree of water saturation present in sand samples, it was found that perturbations during drainage conditions were slow to take affect, while perturbations during imbibation were processed more quickly. These results show that the capillary pressure is a unique function of water saturation and dielectric permittivity; the latter can be measured for buried sediments at the Earth's surface, allowing scientists insight into processes at the boundary between groundwater and overlying material. This is important for assessing the process of $CO_2$ uptake or delivery by the soil from and to the atmosphere.

In this chapter, we introduce GPR basic principles and summarize the state-of-the-art of GPR for environmental applications with emphasis on quantitative characterization methods. Then, we present our recent advances on the development and use of mono-static off-ground GPR for shallow subsurface characterization. This chapter intends to provide a general overview of GPR in hydrogeophysics, with its potentialities and limitations, as a basis to guide the scientist or engineer through the various existing GPR methods to solve particular environmental problems.

## 2.  Ground-Penetrating Radar

GPR is a geophysical technique which is particularly appropriate to image the soil in two or three dimensions with a high spatial resolution, up to a depth of several meters or even up to a few tens of meters. GPR operates by transmitting high frequency (typically in the range 10–2000 MHz) electromagnetic waves into the soil (see Figure 1). Wave propagation is governed by the soil dielectric permittivity $\varepsilon$ (determining wave velocity), electric conductivity $\sigma$ (determining wave attenuation), magnetic permeability $\mu$, and their spatial distribution. Electromagnetic contrasts create partial wave reflections and transmissions that are measured by a receiving antenna, depending on the mode of operation (reflection or transmission). For non-magnetic materials as prevalent in the environment, $\mu$ is equal to the free space magnetic permeability $\mu_0$. Detailed descriptions of the fundamental principles of GPR can be found in (Daniels, 2004).

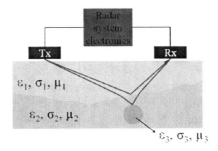

*Figure 1.* Ground penetrating radar (GPR) basic principles. Tx is the transmitting antenna. Rx is the receiving antenna. Slid lines represent wave ray paths.

GPR has been primarily used to image the subsurface and detect buried objects (Annan, 2002). In the areas of unsaturated zone hydrology and water resources, GPR has been used to identify soil stratigraphy (Boll et al., 1996; Bristow et al., 2000; Davis and Annan, 1989; Kung and Lu, 1993; Vaughan et al., 1999), to locate water tables (Nakashima et al., 2001), to follow wetting front movement (Vellidis et al., 1990), to assist in subsurface hydraulic parameter identification (Gloaguen et al., 2001; Hubbard et al., 1997), to measure soil water content (Chanzy et al., 1996; Greaves et al., 1996; Huisman et al., 2001; Schmalholz et al., 2004; Serbin and Or, 2003; van Overmeeren et al., 1997; Weiler et al., 1998), to assess soil salinity (al Hagrey and Müller, 2000), and to support the monitoring of contaminants. This last application requires that the electromagnetic properties of the contaminants are distinctly different from other solid and liquid soil components (Brewster and Annan, 1994; Darayan et al., 1998; Knight, 2001; Yoder et al., 2001).

Figure 2 illustrates a GPR image of the subsurface, where measurements have been made along a transect. The profile is recorded with a PulsEkko 1000 system using fixed offset 250 MHz center frequency shielded antennas with a transmitter-receiver separation of 25 cm. The horizontal step size was 5 cm and a time step of 0.2 ns was used. The recording is made in the central part of the Netherlands in a partial consolidated sand environment with an unconsolidated top layer. For large parts the surface response is constant, but near 3 m and 4.5 m, uplift in the direct arrival can be seen, which is due to near surface objects. A strong surface feature can be seen between 40 m and 50 m where it seems that amplitude inversion has occurred. Several other discontinuities occur in the first arrivals. Many subsurface reflections can be seen due to layering, which do not appear to be very continuous. This can be due to a discontinuous interface, due to strong variation in water content, or due to strong scattering at a shallower depth level, as can be observed near 45 m where the surface produces a strong signal and there appears to be scattering from an object with an apex at 10 ns below which almost all signals are masked. Strong moisture variations are also possible because of the presence of small bushes and trees. Several strong hyperbolic events can be seen, coming from relatively small objects in the direction of the measured line, e.g., at positions of 11.5 m and 4 ns and 37 m and 6 ns.

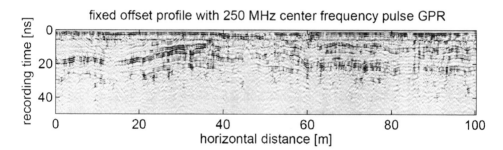

*Figure 2.* GPR image of the subsurface, illustrating soil stratigraphy.

In addition to providing images of the subsurface reflections, quantitative information can be derived from the GPR data. Generally, GPR signal analysis is performed using ray-tracing approximations (as shown in Figure 1) and tomographic inversion (Cai and Mc-Mechan, 1995; Goodman, 1994). Several methodologies are generally adopted for determining wave propagation velocity and retrieve soil water content from GPR data (Huisman et al., 2003):

- Determination of the wave propagation time to a known interface using single-offset surface GPR (Grote et al., 2003; Lunt et al., 2005; van Overmeeren et al., 1997; Weiler et al., 1998)

- Detection of the velocity-dependent reflecting hyperbola of a buried object using single-offset surface GPR along a transect (Vellidis et al., 1990; Windsor et al., 2005)
- Extraction of stacking velocity fields from multi-offset radar soundings at a fixed central location (common midpoint method) (Garambois et al., 2002; Greaves et al., 1996)
- Determination of the ground-wave velocity for surface water content retrieval using multi- and single-offset surface GPR (Chanzy et al., 1996; Du and Rummel, 1994; Galagedara et al., 2003, 2005a, b; Grote et al., 2003; Huisman et al., 2001, 2002)
- Determination of the surface reflection coefficient using single-offset off-ground GPR (Chanzy et al., 1996; Redman et al., 2002; Serbin and Or, 2003, 2004)
- Determination of the two-dimensional spatial distribution of water between boreholes using transmission tomography (Alumbaugh et al., 2002; Binley et al., 2001; Rucker and Ferre, 2005; Zhou et al., 2001)

In particular, time-lapse GPR measurements have recently permitted to monitor soil water dynamics between boreholes and infer the soil hydraulic properties governing water flow (Binley et al., 2001; Cassiani and Binley, 2005; Kowalsky et al., 2005; Linde et al., 2006; Rucker and Ferré, 2004; Tsoflias et al., 2001). GPR can also be used to monitor remediation amendments and processes (Hubbard et al., 2005). Remediation approaches, such as in situ chemical treatments and biostimulation, usually result in transformations in the soil, such as the dissolution and precipitation of minerals, gas evolution, redox variations, biofilm generation, and changes in permeability and porosity. Provided sufficient sensitivity to these changes, time-lapse GPR can be used for remotely monitoring hydrological-biogeochemical soil properties during remediation operations, at both the laboratory and the field scale.

Although these techniques are well established in the scientific and engineering communities and have been relatively successful in an impressive number of applications, they still suffer from major limitations originating from the strongly simplifying assumptions on which they rely with respect to electromagnetic wave propagation phenomena. As a result, a bias is introduced in the estimates due to limited GPR model adequacy and, moreover, only a part of the information contained in the radar data is used, generally the propagation time (and for ground coupled antennas, this is still ambiguous (Yelf, 2004)). Common methods are moreover not adapted for automated real-time monitoring over long periods (multiple days).

Resorting to the physical and mathematical basis of GPR wave propagation is necessary to estimate simultaneously both the depth dependent soil dielectric permittivity and electric conductivity. The relation between the subsurface constitutive parameters and the measured electromagnetic field is governed by Maxwell's equations. Reconstruction of the unknown constitutive parameters from the known field implies inverse modeling. Inverting electromagnetic data has been a major challenge in applied geophysics for many years (Gentili and Spagnolini, 2000; Lazaro-Mancilla and Gomez-Treviño, 2000; Spagnolini, 1997). Successful inversion is challenging since it involves rigorous forward modeling of the three-dimensional (3-D) GPR-subsurface system, which is furthermore computationally very time-consuming. Moreover, the inverse problem should satisfy elemental well-posedness conditions. Nevertheless, inversion is becoming a rational choice due to the ever increasing power of computers (Sasaki, 2001).

## 3.  Full-Wave Analysis of Off-Ground GPR Data

Recently, we have developed a full-wave electromagnetic model for the particular case of mono-static off-ground GPR, i.e., a single antenna plays simultaneously the role of emitter and receiver and is situated at some distance (e.g., 20 cm) above the soil (Lambot et al., 2004c, d). The model includes internal antenna and antenna-soil interaction propagation effects (this is usually not accounted for using common GPR methods) and considers an exact solution of the 3-D Maxwell's equations for wave propagation in multilayered media (1-D solutions are usually considered). Both phase and amplitude information is inherently used for model inversion, thereby maximizing information retrieval from the available radar data, both in terms of quantity and quality. The technique has been successfully validated in a series of hydrogeophysical applications (Lambot et al., 2004a, b, 2005, 2006a, c). Recently, we integrated the method with hydrodynamic modeling to retrieve the soil hydraulic properties from time-lapse proximal radar data and to monitor the dynamics of water content profiles (Lambot et al., 2006b). In addition, the radar model permits to enhance subsurface imaging, by suppressing antenna effects, which are ambiguous otherwise (Lopera et al., 2007a, b). We summarize below the proposed method in order to draw the attention of the reader on essential GPR aspects and scientific methodology for the advanced interpretation of GPR data.

### 3.1.  GPR FORWARD MODELING

#### 3.1.1.  *Antenna Equation in the Frequency Domain*

We use a monostatic ultra-wideband stepped-frequency continuous wave (SFCW) radar system that is set up using a vector network analyzer (VNA), combined with a directive horn antenna to be operated off the ground. The advantage of VNA technology over traditional GPR systems is that the measured quantities constitute an international standard and are well defined physically. The GPR signal to be modeled consists of the frequency-dependent complex ratio $S_{11}(\omega)$ between the returned signal and the emitted signal, $\omega$ being the angular frequency. The VNA reference plane where $S_{11}(\omega)$ is actually measured is established at the connection between the antenna feed point and the VNA cable.

The antenna is modeled using the block diagram depicted in Figure 3 (Lambot et al., 2004d). It relies on Maxwell's equation linearity and assumes that the spatial distribution of the electromagnetic field measured by the antenna does not depend on the subsurface, i.e., is constant. This is expected to be a valid assumption if the antenna is not too close to the ground and assuming the soil to be described by a horizontally multilayered medium. The model consists of a linear system composed of elementary model components in series and parallel, all characterized by their own frequency response function accounting for specific electromagnetic phenomena.

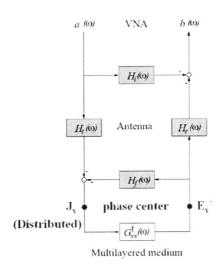

Figure 3. Block diagram representing the VNA-antenna-multilayered medium system modeled as linear systems in series and parallel (Lambot et al., 2004d).

The resulting transfer function relating $S_{11}(\omega)$ measured by the VNA to the frequency response $G_{xx}^{\uparrow}(\omega)$ of the multilayered medium (air-subsurface system) is expressed in the frequency domain by:

$$S_{11}(\omega) = \frac{b(\omega)}{a(\omega)} = H_i(\omega) + \frac{H_t(\omega)G_{xx}^{\uparrow}(\omega)H_r(\omega)}{1 - H_f(\omega)G_{xx}^{\uparrow}(\omega)}, \tag{1}$$

where $b(\omega)$ and $a(\omega)$ are, respectively, the received and emitted signals at the VNA reference plane; $H_i(\omega)$, $H_t(\omega)$, $H_r(\omega)$, and $H_f(\omega)$ are, respectively, the complex return loss, transmitting, receiving, and feedback loss transfer functions of the antenna; and $G_{xx}^\uparrow(\omega)$ is the transfer function of the air-subsurface system modeled as a multilayered medium (see below).

Due to inherent variations in the impedance between the antenna's feed point, antenna aperture, and air, multiple wave reflections (ringing) occur within the antenna. The return loss transfer function $H_i(\omega)$ represents the part of this ringing, measured at the reference plane that is independent of the backscattered electromagnetic field $G_{xx}^\uparrow(\omega)$. The transmitting and receiving transfer functions, or global transmittances, describe the antenna gain and phase delay between the measurement point and the source and receiver virtual point. The positive feedback loop with transfer function $H_f(\omega)$, similarly to $H_i(\omega)$, accounts for impedance variations between the antenna feed point, antenna aperture, and air, which cause a part of the backscattered field to be reflected again towards the subsurface. This leads to multiple wave reflections between the antenna and the soil. Transfer functions $H_i(\omega)$ and $H_f(\omega)$ play the role of global reflection coefficients. We refer to (Lambot et al., 2004d) and (Lambot et al., 2006a) for additional details on the determination of these characteristic antenna transfer functions.

### 3.1.2. Zero-Offset Green's function for Multilayered Media

The air-subsurface system is modeled as a 3-D stratified dielectric and conductive planar system consisting of $N$ horizontal layers separated by $N-1$ interfaces parallel to the horizontal $\hat{x}\hat{y}$ plane. The $n$ th layer is homogeneous and characterized by a dielectric permittivity $\varepsilon_n$, electric conductivity $\sigma_n$, magnetic permeability $\mu_n$, and thickness $h_n$. The emitting part of the horn antenna is approximated by an infinitesimal horizontal $x$-directed electric dipole (second subscript $x$ in $G_{xx}^\uparrow(\omega)$), whereas the receiving part of the antenna is emulated by recording the horizontal $x$-directed component of the backscattered electric field (first subscript $x$ in $G_{xx}^\uparrow(\omega)$). The point source is located at the origin S of the coordinate system, in the upper half-space. The use of a 3-D model is essential to account for the spherical divergence in wave propagation, and is especially necessary for correct signal amplitude modeling. Thanks to the zero-offset, source-receiver mode of operation, the dipole approximation is expected to be valid, as the measured waves have mainly propagated vertically. The radiation pattern of the antenna, different from that of an infinitesimal dipole, plays therefore a negligible role and may be disregarded. It could however be readily accounted for by considering a distributed source over the antenna aperture, using the superposition principle, thereby emulating the radiation pattern of any specific antenna. This would however increase the total number of unknowns in the antenna model.

The theoretical basis for GPR wave propagation is given by Maxwell's equations. Green's functions, i.e., the point-source solutions of Maxwell's equations for electromagnetic waves propagating in multilayered media, are well known. Following the approach of (Slob and Fokkema, 2002) and (Lambot et al., 2004d), the analytic expression for zero-offset Green's function in the spectral domain (2-D spatial Fourier domain) is found to be:

$$\tilde{G}_{xx}^{\uparrow} = \frac{1}{2}\left( \frac{\Gamma_n R_n^{TM}}{\eta_n} - \frac{\xi_n R_n^{TE}}{\Gamma_n} \right)\exp(-2\Gamma_n h_n),\tag{2}$$

where the subscript $n$ equals 1 and denotes here the first interface and first layer, $R_n^{TM}$ and $R_n^{TE}$ are, respectively, the transverse magnetic (TM) and transverse electric (TE) global reflection coefficients accounting for all reflections and multiples from subsurface interfaces, $\Gamma_n$ is the vertical wave number defined as $\Gamma_n = \sqrt{k_\rho^2 + \xi_n \eta_n}$, $k_\rho$ is a spectral domain transformation parameter, $\xi_n = j\omega\mu_n$, $\eta_n = \sigma_n + j\omega\varepsilon_n$, and $j = \sqrt{-1}$.

The global TM-mode and TE-mode reflection coefficients at interface $n$ ($n = 1, \cdots, N-1$) are given by:

$$R_n^{TM} = \frac{r_n^{TM} + R_{n+1}^{TM}\exp(-2\Gamma_{n+1} h_{n+1})}{1 + r_n^{TM} R_{n+1}^{TM}\exp(-2\Gamma_{n+1} h_{n+1})}\tag{3}$$

$$r_n^{TM} = \frac{\eta_{n+1}\Gamma_n - \eta_n\Gamma_{n+1}}{\eta_{n+1}\Gamma_n + \eta_n\Gamma_{n+1}}\tag{4}$$

$$R_n^{TE} = \frac{r_n^{TE} + R_{n+1}^{TE}\exp(-2\Gamma_{n+1} h_{n+1})}{1 + r_n^{TE} R_{n+1}^{TE}\exp(-2\Gamma_{n+1} h_{n+1})}\tag{5}$$

$$r_n^{TE} = \frac{\mu_{n+1}\Gamma_n - \mu_n\Gamma_{n+1}}{\mu_{n+1}\Gamma_n + \mu_n\Gamma_{n+1}}\tag{6}$$

where $r_n^{TM}$ and $r_n^{TE}$ denote the local-plane wave TM and TE mode reflection coefficients, respectively, at interface $n$. These expressions are in a recursive form, with $n = N-1, \cdots, 1$. The recursion is initiated by the observation that there are no upgoing waves from the lower half-space, such that $R_{N-1}^{TM} = r_{N-1}^{TM}$ and $R_{N-1}^{TE} = r_{N-1}^{TE}$. The local reflection coefficients determine that part of the electromagnetic wave which is reflected at a dielectric interface, while the other part is transmitted.

The transformation of (2) from the spectral domain to the spatial domain is carried out by employing the 2-D Fourier inverse transformation:

$$G_{xx}^{\uparrow} = \frac{1}{4\pi} \int_0^{+\infty} \tilde{G}_{xx}^{\uparrow} k_\rho dk_\rho , \tag{7}$$

which reduces to a single integral in view of the invariance of the electromagnetic properties along the $x$ and $y$ coordinates and the zero-offset configuration. The integral can be computed with standard integration routines using Gaussian-like quadrature. Both equations (1) and (7) relate the raw radar data to the soil electromagnetic properties and their vertical distribution.

### 3.1.3. *Model Inversion*

Subsurface parameter identification by inverse modeling is a nonlinear optimization problem which consists in finding the parameter vector $\mathbf{b} = [\varepsilon_n, \sigma_n, h_n]^T$, $n = 1 \cdots N$, such that an objective function $\phi(\mathbf{b})$ is minimized (the magnetic permeability is disregarded here). For the particular case where no prior information about the parameters is used and assuming that observation errors are normally distributed with mean zero and covariance matrix $\mathbf{C}$, and are independent and homoskedastic, the maximum likelihood theory reduces to a weighted least squares problem. The objective function to be minimized is accordingly defined as follows:

$$\phi(\mathbf{b}) = \left| \mathbf{G}_{xx}^{\uparrow*} - \mathbf{G}_{xx}^{\uparrow} \right|^T \mathbf{C}^{-1} \left| \mathbf{G}_{xx}^{\uparrow*} - \mathbf{G}_{xx}^{\uparrow} \right|, \tag{8}$$

where $\mathbf{G}_{xx}^{\uparrow*} = G_{xx}^{\uparrow}(\omega)$ and $\mathbf{G}_{xx}^{\uparrow} = G_{xx}^{\uparrow}(\omega, \mathbf{b})$ are vectors containing, respectively, the observed and simulated radar measurements, from which major antenna effects have been filtered using (1). Since the Green's function is a complex function, the difference between observed and modeled data is expressed by the amplitude of the errors in the complex plane.

Objective function (8) indirectly relates the response of the multilayered medium to its constitutive parameters. However, as for most electromagnetic inverse problems, this function is highly nonlinear and characterized by an oscillatory behavior and a multitude of local minima, depending on the dimensionality of the inverse problem. This complex topography necessitates the use of a robust global optimization algorithm. Following the approach of (Lambot et al., 2002), we use the global multilevel coordinate search (GMCS) algorithm (Huyer and Neumaier, 1999) combined sequentially with the classical Nelder-Mead simplex algorithm (NMS) (Lagarias et al., 1998).

### 3.1.4. *Model Validation and Applications*

We performed radar measurements above a sand box containing two sand layers subject to various water contents, under which a metal sheet was installed as perfect electric conductor to control the bottom boundary condition in the electromagnetic model (Lambot et al., 2004d). Radar measurements were performed in the range 1–2 GHz, with a frequency step of 4 MHz. Figure 4 represents the modeled and measured Green's function in the frequency domain for a specific value of the water content. Figure 5 shows the estimated relative dielectric permittivity as a function of the top layer water content, obtained by inverse modeling. Apart from the remarkable agreement between the measured and modeled data, we observe that the inversely estimated dielectric permittivity is very consistent with the different water contents and compare favorably to time domain reflectometry (TDR) measurements. These results demonstrate

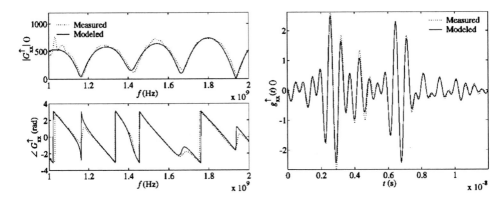

*Figure 4.* Measured and modeled Green's function for the antenna being above a two-layered medium, with a perfect electric conductor as bottom boundary condition (Lambot et al., 2004d). Radar data are presented in both the frequency (left) and time (right) domains.

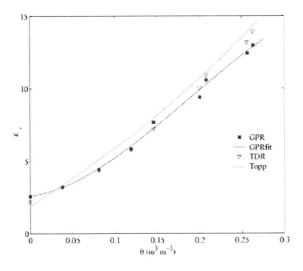

*Figure 5.* Inversely estimated relative dielectric permittivity as a function of water content. GPR results are compared to reference TDR measurements. Topp's model is shown (Lambot et al., 2004d).

the adequacy of the GPR forward model, the uniqueness of the inverse solution for this specific inverse problem, and the stability of the inverse solution with respect to actual modeling and measurement errors. Differences were partly attributed to the different operating frequency ranges of both systems (TDR and GPR) and to the different measurement scales. The estimation of the electric conductivity and its frequency dependence was performed simultaneously, which is particularly analyzed in (Lambot et al., 2005). We observed that for the specific sand under investigation and in the limited frequency range 1–3 GHz, the frequency dependence of the dielectric permittivity was negligible and the frequency dependence of the electric conductivity could be described by a linear model, parameterized using only two parameters to invert for.

In (Lambot et al., 2004a), we applied that GPR approach to monitor water content as a function of time and depth during a free drainage event in a 2-m high laboratory sand column (see Figure 6). The sand column was also equipped with TDR probes at two different depths. We subsequently used the obtained water content time series to identify the soil hydraulic properties, described by the Mualem-van Genuchten parameterization, using hydrodynamic inverse modeling (Lambot et al., 2002). We observed in particular that GPR was less sensitive than TDR to the small scale soil heterogeneities (sedimentation layering) and the water dynamics was better described at the GPR characterization scale.

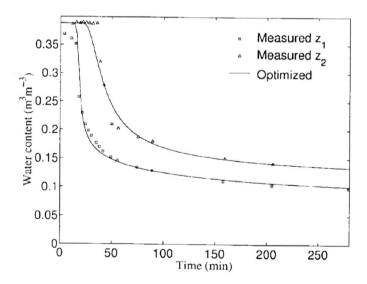

*Figure 6.* GPR monitoring of water content as a function of time at two different depths of a laboratory lysimeter subject to free drainage (Lambot et al., 2004a). Markers represent GPR-derived soil moisture at two different depths and solid lines represent modeled moisture time series using hydrodynamic modeling.

In an other study, we identified from a single radar measurement performed at a controlled outdoor test site (TNO facilities, The Hague, The Netherlands) a continuous vertical water content profile in hydrostatic equilibrium with a water table (Lambot et al., 2004b). In that case the profile could be constrained using the Mualem-van Genuchten parameterization as it corresponds to the characteristic water retention curve. As a result, only four parameters had to be inverted for in order to reconstruct the whole continuous profile. Continuous profiles may be emulated using sufficiently thin layers compared to the minimal wavelength in the electromagnetic model. Synthetic experiments were performed as well to demonstrate the well-posedness of the inverse problem. We observed that continuous profiles are not appropriate in terms of GPR reflections, thereby reducing information content in the radar data.

In (Lambot et al., 2006a), we investigate the effect of soil surface roughness on the measured GPR Green's function and on the inverse estimates. Indeed, this may be critical as the electromagnetic model assumes only smooth surfaces. We observed that the effect of roughness is negligible when Rayleigh's criterion is satisfied, namely, when the roughness amplitude is inferior to at least one height of the minimal wavelength. The inverse solution appeared to be stable with respect to roughness.

In (Lambot et al., 2006c), we analyze the proposed approach for mapping surface water content in field conditions. For that specific purpose, inversion of the Green's function is performed in the time domain, on a time window focused on the surface reflection only. We showed that compared to the traditional surface reflection method, filtering antenna effects and performing full-wave inversion provides valuable advantages compared to other existing techniques. In particular, the antenna distortions effects are filtered out (resulting in increased accuracy), the antenna elevation should not be known a priori as it is inversely estimated as well, and finally the measurements above a calibrating perfect electric conductor situated at exactly the same height as the field measurements are not required.

Finally, we tackled the issue of retrieving continuous electromagnetic profiles and soil hydraulic properties from time-lapse GPR data (Lambot et al., 2006b). The proposed inverse modeling flowchart is presented in Figure 7. Electromagnetic inverse modeling is constrained by hydrodynamics and petrophysical laws, reducing significantly the complexity of the inverse problem. We demonstrated using numerical experiments the uniqueness of the inverse problem. The inversion is however complicated by the large computing resources required. In that respect, we used the JUMP supercomputer of the Forschungszentrum Jülich and we operated with 16 processors in parallel. The computation time for that inversion was about 7 hours.

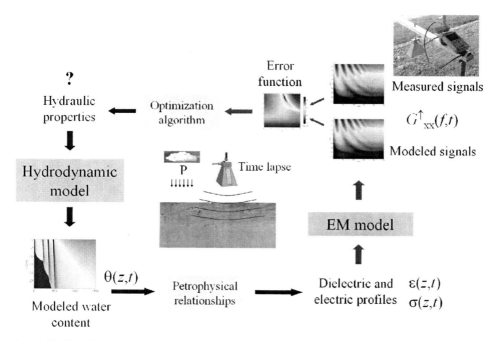

*Figure 7.* Flowchart representing the integrated electromagnetic and hydrodynamic inversion of time-lapse radar measurements for estimating soil hydraulic properties and electric profiles, where $t$ is time. $z$ is depth, and $f$ is frequency (Lambot et al., 2006b).

## Conclusions

This chapter has emphasized the need for non-invasive techniques for soil hydrological characterization and monitoring. In that respect, GPR is increasingly used for such applications as it permits both high resolution subsurface imaging and quantitative characterization. Although traditional GPR techniques have been useful in many applications, more advanced technology and processing are still needed to benefit from the full potential of GPR for subsurface investigations. We have illustrated such an advanced approach for characterizing the shallow subsurface from proximal measurements. Future research will focus on similar full wave inverse modeling approaches for other GPR configurations and applications ranges. In addition, the simultaneous integration of several sensors, such as GPR, EMI, and microseismic, will permit to go a step further in soil characterization by providing complementary information in a data fusion framework. Finally, further understanding of the petrophysical relationships linking the soil geophysical (electromagnetic) properties to the soil physico-chemical properties is still needed.

## Acknowledgments

This work was supported by a Marie Curie Intra-European Fellowship within the 6th European Community Framework Programme, Delft University of Technology (The Netherlands), the Université catholique de Louvain (Belgium), the Fonds National de la Recherche Scientifique (FNRS, Belgium), and the Forschungszentrum Jülich GmbH (Germany).

## References

al Hagrey, S.A. and Müller, C., 2000. GPR study of pore water content and salinity in sand. Geophysical Prospecting, 48: 63–85.

Alumbaugh, D., Chang, P., Paprocki, L., Brainard, J., Glass, R.J. and Rautman, C.A., 2002. Estimating moisture contents in the vadose zone using cross-borehole ground penetrating radar: A study of accuracy and repeatability. Water Resources Research, 38: 1309.

Annan, A.P., 2002. GPR - History, Trends, and Future Developments. Subsurface Sensing Technologies and Applications, 3(4): 253–270.

Annan, A.P., 2005. GPR methods for hydrogeological studies. In: Y.R.a.S.S. Hubbard (Editor), Hydrogeophysics, Springer, New York, pp. 532.

Binley, A., Winship, P., Middleton, R., Pokar, M. and West, J., 2001. High-resolution characterization of vadose zone dynamics using cross-borehole radar. Water Resources Research, 37(11): 2639–2652.

Boll, J., van Rijn, R.P.G., Weiler, K.W., Steenhuis, T.S., Daliparthy, J. and Herbert, S.J., 1996. Using ground penetrating radar to detect layers in a sandy field soil. Geoderma, 70: 117–132.

Bouma, J., Stoorvogel, J., van Alphen, B.J. and Booltink, H.W.G., 1999. Pedology, Precision Agriculture, and the Changing Paradigm of Agricultural Research. Soil Science Society of America Journal, 63: 1763–1768.

Brewster, M.L. and Annan, A.P., 1994. Ground penetrating radar monitoring of a controlled DNAPL release: 200 MHz radar. Geophysics, 59: 1211–1221.

Bristow, C.S., Bailey, S.D. and Lancaster, N., 2000. The sedimentary structure of linear sand dunes. Nature, 406(6791): 56–59.

Cai, J. and Mc-Mechan, G.A., 1995. Ray-based synthesis of bistatic ground penetrating radar profiles. Geophysics, 60: 87–96.

Cassiani, G. and Binley, A., 2005. Modeling unsaturated flow in a layered formation under quasi-steady state conditions using geophysical data constraints. Advances in Water Resources, 28(5): 467–477.

Chanzy, A., Tarussov, A., Judge, A. and Bonn, F., 1996. Soil water content determination using digital ground penetrating radar. Soil Science Society of America Journal, 60: 1318–1326.

Daniels, D.J., 2004. Ground Penetrating Radar, 2nd Edition. The Inst. Electrical Eng., London.

Darayan, S., Liu, C., Shen, L.C. and Shattuck, D., 1998. Measurement of electrical properties of contaminated soils. Geophysical Prospecting, 46: 477–488.

Davis, J.L. and Annan, A.P., 1989. Ground penetrating radar for high resolution mapping of soil and rock stratigraphy. Geophysical Prospecting, 37: 531–551.

de Rosnay, P. et al., 2006. SMOSREX: A long term field campaign experiment for soil moisture and land surface processes remote sensing. Remote Sensing of Environment, 102(3–4): 377–389.

Dobson, M.C. and Ulaby, F.T., 1986. Active microwave soil moisture research. IEEE Transactions on Geoscience and Remote Sensing, 24: 23–36.

Du, S. and Rummel, P., 1994. Reconnaissance studies of moisture in the subsurface with GPR. In: M.T.v.G.a.F.J.L.a.L. Wu (Editor), Proceedings of the Fifth International Conference on Ground Penetrating Radar, Waterloo cent. for Groundwater Res., Univ. of Waterloo, Waterloo, Ont., Canada, pp. 1241–1248.

Famiglietti, J.S., Devereaux, J.A., Laymon, C.A., Tsegaye, T., Houser, P.R., Jackson, T.J., Graham, S.T., Rodell, M. and van Oevelen, P.J., 1999. Ground-based investigation of soil moisture variability within remote sensing footprints during the Southern Great Plains 1997 (SGP97) Hydrology Experiment. Water Resources Research, 35(6): 1839–1851.

Galagedara, L.W., Parkin, G.W. and Redman, J.D., 2003. An analysis of the GPR direct ground wave method for soil water content measurement. Hydrological Processes, 17: 3615–3628.

Galagedara, L.W., Parkin, G.W., Redman, J.D. and Endres, A.L., 2005a. Field studies of the GPR ground wave method for estimating soil water content during irrigation and drainage. Journal of Hydrology, 301: 182–197.

Galagedara, L.W., Redman, J.D., Parkin, G.W., Annan, A.P. and Endres, A.L., 2005b. Numerical modeling of GPR to determine the direct ground wave sampling depth. Vadose Zone Journal, 4: 1096–1106.

Garambois, S., Sénéchal, P. and Perroud, H., 2002. On the use of combined geophysical methods to assess water content and water conductivity of near-surface formations. Journal of Hydrology, 259: 32–48.

Gentili, G.G. and Spagnolini, U., 2000. Electromagnetic inversion in monostatic ground penetrating radar: TEM horn calibration and application. IEEE Transactions on Geoscience and Remote Sensing, 38(4): 1936–1946.

Gloaguen, E., Couteau, M., Marcotte, D. and Chapuis, R., 2001. Estimation of hydraulic conductivity of an unconfined aquifer using cokriging of GPR and hydrostratigraphic data. Journal of Applied Geophysics, 47: 135–152.

Goodman, D., 1994. Ground penetrating radar simulation in engineering and archeology. Geophysics, 59: 224–232.

Greaves, R.J., Lesmes, D.P., Lee, J.M. and Toksov, M.N., 1996. Velocity variations and water content estimated from multi-offset, ground-penetrating radar. Geophysics, 61: 683–695.

Grote, K., Hubbard, S.S. and Rubin, Y., 2003. Field-scale estimation of volumetric water content using GPR ground wave techniques. Water Resources Research, 39(11): 1321, doi:10.1029/2003WR002045.

Hubbard, S., Chen, J., Williams, K., Peterson, J. and Rubin, Y., 2005. Environmental and agricultural applications of GPR. In: S.L.a.A.G. Gorriti (Editor), Proceedings of the 3rd International Workshop on Advanced Ground Penetrating Radar, Delft University of Technology, Delft, The Netherlands, pp. 45–49.

Hubbard, S.S., Rubin, Y. and Majer, E., 1997. Ground-penetrating-radar-assisted saturation and permeability estimation in bimodal systems. Water Resources Research, 33(5): 971–990.

Huisman, J.A., Hubbard, S.S., Redman, J.D. and Annan, A.P., 2003. Measuring soil water content with ground penetrating radar: A review. Vadose Zone Journal, 2: 476–491.

Huisman, J.A., Snepvangers, J.J.J.C., Bouten, W. and Heuvelink, G.B.M., 2002. Mapping spatial variation in surface soil water content: comparison of ground-penetrating radar and time domain reflectometry. Journal of Hydrology, 269: 194–207.

Huisman, J.A., Sperl, C., Bouten, W. and Verstraten, J.M., 2001. Soil water content measurements at different scales: accuracy of time domain reflectometry and ground penetrating radar. Journal of Hydrology, 245: 48–58.

Huyer, W. and Neumaier, A., 1999. Global optimization by multilevel coordinate search. Journal of Global Optimization, 14(4): 331–355.

Jackson, T.J., Schmugge, J. and Engman, E.T., 1996. Remote sensing applications to hydrology: soil moisture. Hydrological Sciences, 41(4): 517–530.

Kemna, A., Binley, A., Ramirez, A. and Daily, W., 2000. Complex resistivity tomography for environmental applications. Chemical Engineering Journal, 77(1–2): 11–18.

Kemna, A., Vanderborght, J., Kulessa, B. and Vereecken, H., 2002. Imaging and characterisation of subsurface solute transport using electrical resistivity tomography (ERT) and equivalent transport models. Journal of Hydrology, 267(3–4): 125–146.

Kirchmann, H. and Thorvaldsson, G., 2000. Challenging targets for future agriculture. European Journal of Agronomy, 12: 145–161.

Knight, R., 2001. Ground penetrating radar for environmental applications. Annual Review of Earth and Planetary Sciences, 29: 229–255.

Kowalsky, M.B., Finsterle, S., Peterson, J., Hubbard, S., Rubin, Y., Majer, E., Ward, A. and Gee, G., 2005. Estimation of field-scale soil hydraulic and dielectric parameters through joint inversion of GPR and hydrological data. Water Resources Research, 41: W11425, doi:10.1029/2005WR004237.

Kung, K.J.S. and Lu, Z.B., 1993. Using ground penetrating radar to detect layers of discontinuous dielectric constant. Soil Science Society of America Journal, 57: 335–340.

Lagarias, J.C., Reeds, J.A., Wright, M.H. and Wright, P.E., 1998. Convergence properties of the Nelder-Mead Simplex method in low dimensions. Siam Journal on Optimization, 9(1): 112–147.

Lambot, S., Antoine, M., van den Bosch, I., Slob, E.C. and Vanclooster, M., 2004a. Electromagnetic inversion of GPR signals and subsequent hydrodynamic inversion to estimate effective vadose zone hydraulic properties. Vadose Zone Journal, 3(4): 1072–1081.

Lambot, S., Antoine, M., Vanclooster, M. and Slob, E.C., 2006a. Effect of soil roughness on the inversion of off-ground monostatic GPR signal for noninvasive quantification of soil properties. Water Resources Research, 42: W03403, doi:10.1029/2005WR004416.

Lambot, S., Javaux, M., Hupet, F. and Vanclooster, M., 2002. A global multilevel coordinate search procedure for estimating the unsaturated soil hydraulic properties. Water Resources Research, 38(11): 1224, doi:10.1029/2001WR001224.

Lambot, S., Rhebergen, J., van den Bosch, I., Slob, E.C. and Vanclooster, M., 2004b. Measuring the soil water content profile of a sandy soil with an off-ground monostatic ground penetrating radar. Vadose Zone Journal, 3(4): 1063–1071.

Lambot, S., Slob, E.C., van den Bosch, I., Stockbroeckx, B., Scheers, B. and Vanclooster, M., 2004c. Estimating soil electric properties from monostatic ground-penetrating radar signal inversion in the frequency domain. Water Resources Research, 40: W04205, doi:10.1029/2003WR002095.

Lambot, S., Slob, E.C., van den Bosch, I., Stockbroeckx, B. and Vanclooster, M., 2004d. Modeling of ground-penetrating radar for accurate characterization of subsurface electric properties. IEEE Transactions on Geoscience and Remote Sensing, 42: 2555–2568.

Lambot, S., Slob, E.C., Vanclooster, M. and Vereecken, H., 2006b. Closed loop GPR data inversion for soil hydraulic and electric property determination. Geophysical Research Letters, 33: L21405, doi:10.1029/2006GL027906.

Lambot, S., van den Bosch, I., Stockbroeckx, B., Druyts, P., Vanclooster, M. and Slob, E.C., 2005. Frequency dependence of the soil electromagnetic properties derived from ground-penetrating radar signal inversion. Subsurface Sensing Technologies and Applications, 6: 73–87.

Lambot, S., Weihermüller, L., Huisman, J.A., Vereecken, H., Vanclooster, M. and Slob, E.C., 2006c. Analysis of air-launched ground-penetrating radar techniques to measure the soil surface water content. Water Resources Research, 42: W11403, doi:10.1029/2006WR005097.

Lazaro-Mancilla, O. and Gomez-Treviño, E., 2000. Ground penetrating radar in 1-D: an approach for the estimation of electrical conductivity, dielectric permittivity and magnetic permeability. Journal of Applied Geophysics, 43: 199–213.

Lesch, S.M., Herrero, J. and Rhoades, J.D., 1998. Monitoring for temporal changes in soil salinity using electromagnetic induction techniques. Soil Science Society of America Journal, 62: 232–242.

Linde, N., Binley, A., Tryggvason, A., Pedersen, L.B. and Révil, A., 2006. Improved hydro-geophysical characterization using joint inversion of cross-hole electrical resistance and ground-penetrating radar traveltime data. Water Resources Research, 42: W12404, doi:10.1029/2006WR005131.

Lopera, O., Milisavljevic, N. and Lambot, S., 2007a. Clutter reduction in GPR measurements for detecting shallow buried landmines: a Colombian case study. Near Surface Geophysics, 5(1): 57–64.

Lopera, O., Slob, E.C., Milisavljevic, N. and Lambot, S., 2007b. Filtering soil surface and antenna effects from GPR data to enhance landmine detection. IEEE Transactions on Geoscience and Remote Sensing, 45(3): 707–717.

Lunt, I.A., Hubbard, S.S. and Rubin, Y., 2005. Soil moisture content estimation using ground-penetrating radar reflection data. Journal of Hydrology, 307(1–4): 254–269.

Mualem, Y. and Friedman, S.P., 1991. Theoretical predictions of electrical conductivity in saturated and unsaturated soil. Water Resources Research, 27: 2771–2777.

Nakashima, Y., Zhou, H. and Sato, M., 2001. Estimation of groundwater level by GPR in an area with multiple ambiguous reflections. Journal of Applied Geophysics, 47: 241–249.

Plug, W.J., Slob, E., Bruining, J. and Tirado, L.M.M., 2007. Simultaneous measurement of hysteresis in capillary pressure and electric permittivity for multiphase flow through porous media. Geophysics, 72(3): A41-A45.

Plug, W.J., Slob, E., van Turnhout, J. and Bruining, J., 2007. Capillary pressure as a unique function of electric permittivity and water saturation. Geophysical Research Letters, 34(13): 5.

Redman, J.D., Davis, J.L., Galagedara, L.W. and Parkin, G.W., 2002. Field studies of GPR air launched surface reflectivity measurements of soil water content. In: L. Steven Koppenjan and Hua (Editor), Proceedings of the Ninth International Conference on Ground Penetrating Radar, Santa Barbara, CA., USA, pp. SPIE 4758: 156–161.

Rhoades, J.D., Raats, P.A.C. and Prather, R.J., 1976. Effects of liquid-phase electrical conductivity, water content, and surface conductivity on bulk soil electrical conductivity. Soil Science Society of America Journal, 40: 651–655.

Robinson, D.A., Jones, S.B., Wraith, J.M., Or, D. and Friedman, S.P., 2003. A review of advances in dielectric and electrical conductivity measurement in soils using time domain reflectometry. Vadose Zone Journal, 2: 444–475.

Rucker, D.F. and Ferre, T.P.A., 2005. Automated water content reconstruction of zero-offset borehole ground penetrating radar data using simulated annealing. Journal of Hydrology, 309(1-4): 1–16.

Rucker, D.F. and Ferré, T.P.A., 2004. Parameter estimation for soil hydraulic properties using zero-offset borehole radar: Analytical method. Soil Science Society of America Journal, 68(5): 1560–1567.

Sasaki, Y., 2001. Full 3-D inversion of electromagnetic data on PC. Journal of Applied Geophysics, 46: 45–54.

Schmalholz, J., Stoffregen, H., Kemna, A. and Yaramanci, U., 2004. Imaging of water content distributions inside a lysimeter using GPR tomography. Vadose Zone Journal, 3: 1106–1115.

Seneviratne, S.I., Luthi, D., Litschi, M. and Schar, C., 2006. Land-atmosphere coupling and climate change in Europe. Nature, 443(7108): 205–209.

Serbin, G. and Or, D., 2003. Near-surface water content measurements using horn antenna radar: methodology and overview. Vadose Zone Journal, 2: 500–510.

Serbin, G. and Or, D., 2004. Ground-penetrating radar measurement of soil water content dynamics using a suspended horn antenna. IEEE Transactions on Geoscience and Remote Sensing, 42: 1695–1705.

Slob, E.C. and Fokkema, J., 2002. Coupling effects of two electric dipoles on an interface. Radio Science, 37(5): 1073, doi:10.1029/2001RS2529.

Spagnolini, U., 1997. Permittivity measurements of multilayered media with monostatic pulse radar. IEEE Transactions on Geoscience and Remote Sensing, 35: 454–463.

Stafford, J.V., 2000. Implementing precision agriculture in the 21st century. Journal of Agricultural Engineering Research, 76 (3): 267–275.

Sudduth, K.A., Drummond, S.T. and Kitchen, N.R., 2001. Accuracy issues in electromagnetic induction sensing of soil electrical conductivity for precision agriculture. Computers and Electronics in Agriculture, 31: 239–264.

Tabbagh, A., Camerlynck, C. and Cosenza, P., 2000. Numerical modeling for investigating the physical meaning of the relationship between relative dielectric permittivity and water content of soils. Water Resources Research, 36: 2771–2776.

Tilman, D., Cassman, K.G., Matson, P.A., Naylor, R. and Polasky, S., 2002. Agricultural sustainability and intensive production practices. Nature, 418(6898): 671–677.

Topp, G., Davis, J.L. and Annan, A.P., 1980. Electromagnetic Determination of Soil Water Content: Measurements in Coaxial Transmission Lines. Water Resources Research, 16: 574–582.

Tsoflias, G.P., Halihan, T. and Sharp, J.M., 2001. Monitoring pumping test response in a fractured aquifer using ground-penetrating radar. Water Resources Research, 37(5): 1221–1229.

van Overmeeren, R.A., Sariowan, S.V. and Gehrels, J.C., 1997. Ground penetrating radar for determining volumetric soil water content: results of comparative measurements at two test sites. Journal of Hydrology, 197: 316–338.

Vanderborght, J., Kemna, A., Hardelauf, H. and Vereecken, H., 2005. Potential of electrical resistivity tomography to infer aquifer transport characteristics from tracer studies: A synthetic case study. Water Resources Research, 41(6): 23.

Vaughan, D.G., Corr, H.F.J., Doake, C.S.M. and Waddington, E.D., 1999. Distortion of isochronous layers in ice revealed by ground-penetrating radar. Nature, 398(6725): 323–326.

Vellidis, G., Smith, M.C., Thomas, D.L. and Asmussen, L.E., 1990. Detecting wetting front movement in a sandy soil with ground penetrating radar. Trans. ASAE, 33: 1867–1874.

Weiler, K.W., Steenhuis, T.S., Boll, J. and Kung, K.J.S., 1998. Comparison of ground penetrating radar and time domain reflectometry as soil water sensors. Soil Science Society of America Journal, 62: 1237–1239.

Windsor, C., Capineri, L., Falorni, P., Matucci, S. and Borgioli, G., 2005. The estimation of buried pipe diameters using ground penetrating radar. Insight, 47(7): 394–399.

Yelf, R., 2004. Where is true time-zero? In: E.C.S.a.A.Y.a.J. Rhebergen (Editor), Proceedings of the Tenth International Conference on Ground Penetrating Radar, Delft University of Technology, Delft, The Netherlands, pp. 279–282.

Yoder, R.E., Freeland, R.S., Ammons, J.T. and Leonard, L.L., 2001. Mapping agricultural fields with GPR and EMI to identify offsite movement of agrochemicals. Journal of Applied Geophysics, 47: 251–259.

Zhang, N., Wang, M. and Wang, N., 2002. Precision agriculture: a worldwide overview. Computers and Electronics in Agriculture, 36: 113–132.

Zhou, C., Liu, L. and Lane, J.W., 2001. Nonlinear inversion of borehole-radar tomography data to reconstruct velocity and attenuation distribution in earth materials. Journal of Applied Geophysics, 47: 271–284.

# GEOCHEMICAL ASSESSMENT OF RECLAIMED LANDS

# IN THE MINING REGIONS OF UKRAINE

MYKOLA KHARYTONOV[*]
*Soil Science & Ecology Department*
*State Agrarian University, 25 Voroshilov Str., 49027*
*Dnipropetrovsk, Ukraine*

**Abstract.** The land territories of Ukraine are remarkable for its fertility on the one hand and the existence of the mining sites useful minerals considerable by the amount and assortment on the other hand. In most cases they are mined in the open cut mining method. By the type of bedding in the earth's crust the deposits are differed in flat ones (as Nikopol manganese ore deposit, Kerch iron ore deposit, Pridneprovsky brown coal field, Verkhnedneprovsk deposit of the non-ferrous metal ore, etc.) and steeply dipping ones (as Krivoy Rog and Kerch iron ore deposits, etc.). According to the mechanical-and physical properties of the useful minerals and the overburden of the rock qualitative composition and its usefulness for the utilization the land territories are breached by the quarries, outside dump, tailings, industrial areas of a working and transporting lines. Considering in the dynamics the mining process of the useful minerals it is possible to motivate the rational order of the mine take for the quarries, the use of their worked-out face for the impoundment of the tailings for the creation of the technical deposits from the simultaneously mined minerals and also the different types of the land reclamation. The planned out and prepared properly surface has to be usable for the utilization in the agriculture, in the industrial and civil construction (the tops and platforms of the high waste heaps), for the summer – cottage community building (the small flat areas on the filled – in quarries dumps and tailings, for the creation of the rest zones with the use of tree plantations (the steep slopes of the rock dumps). Landscapes and land present the experimental objects reclamation stations situated in Nikopol manganese ore basin, iron ore, coal deposits in Kerch, Krivoy Rog, Western Donbass mining regions (Kharytonov, 1996; Masyuk and Kharytonov, 2000; Masyuk et al., 1997). The valuation of the lands disturbed by mining and premises for the realization of the land reclamation was given, the technology of the creation

---

[*] To whom correspondence should be addressed: mykola_kh@yhoo.com

L. Simeonov and V. Sargsyan (eds.),
*Soil Chemical Pollution, Risk Assessment, Remediation and Security.*
© Springer Science+Business Media B.V. 2008

of the breach lands soil forming cover was motivated, the chemical composition was estimated.

**Keywords:** geochemical assessment, land fertility, ore deposits, reclaimed lands

## 1. Land Reclamation in the Nikopol Manganese Ore Basin

The rocks of the Nikopol manganese basin in the southeast part of Ukraine are presented the holocene, postpliocene, neogene and paleogene deposits. Taken on day surface they are problem for environment management for the future. There are 9,000 ha of mined land with soil replaced in the region, but most has had ineffective reclamation. Another 10,000 ha or more exists without soil replacement (mostly from mining operations more than 30 years ago), and an estimated additional 150,000 ha or more in Nikopol region suffers significant adverse affects from the non-restored mined land The reclamation of disturbed land is conducted in one technological cycle with the process of manganese ore mining. The soil mass is taken off, piled up and heaped onto the land after the rock has been replaced.

Accepted conventional meanings for the substrata as following:

BS – black soil
LS – loamy soil
RBLS – red-brown loamy soil
RBC – red-brown clay
GGC – green-grey clay

TABLE 1. Trace elements content in the plant meliorated rocks of the Nikopol manganese Ore Basin, ppm (numerator-AAB pH 4,8, denominator-1NHCl)

| Substrate | Co | Ni | Pb | Mn | Zn | Cu | Fe | Cr |
|---|---|---|---|---|---|---|---|---|
| Black soil | 0.97 | 1.82 | 4.76 | 140 | 16.0 | 1.86 | 5.0 | 1.0 |
| | 4.88 | 8.4 | 16.6 | 535 | 62 | 8.3 | 1,747 | 3.0 |
| Loamy soil | 1.85 | 3.6 | 10.1 | 69.7 | 10.0 | 2.5 | 5.0 | 2.0 |
| | 5.56 | 7.9 | 13.8 | 343 | 81 | 6.6 | 1,283 | 3.1 |
| Red-brown loamy soil | 1.7 | 3.75 | 9.36 | 62.0 | 11.3 | 2.76 | 4.3 | 2.19 |
| | 5.68 | 8.43 | 12.2 | 348 | 28.3 | 6.8 | 1,243 | 3.0 |
| Red-brown loamy Soil | 1.41 | 3.19 | 8.06 | 68.0 | 9.5 | 2.18 | 4.0 | 1.9 |
| | 6.85 | 11.3 | 12.6 | 467 | 35 | 8.57 | 1,848 | 3.75 |
| Green Montmorillonite clay | 0.87 | 1.49 | 5.78 | 49.7 | 11.7 | 2.14 | 4.7 | 1.27 |
| | 2.88 | 3.5 | 11.87 | 291 | 37.2 | 7.9 | 1,330 | 3.3 |
| Grey-green clay | 1.09 | 2.7 | 4.3 | 51.3 | 5.9 | 2.2 | 4.4 | 1.4 |
| | 2.43 | 4.0 | 11.0 | 155.0 | 16.3 | 7.3 | 1,082 | 3.5 |
| Ancient-alluvial sand | 0.87 | 3.26 | 3.75 | 29.0 | 9.5 | 1.85 | 9.8 | 1.21 |
| | 3.0 | 6.5 | 6.2 | 52.0 | 19.8 | 4.0 | 577 | 1.58 |
| Dark-grey schist clay | 0.8 | 5.75 | 6.24 | 206.3 | 8.7 | 2.22 | 11.2 | 1.33 |
| | 3.33 | 9.2 | 11.7 | 2,053 | 35.8 | 8.22 | 2,005 | 3.0 |

GMC – green motmorillonite clay
AAS – ancient-alluvial sand
DGSC – dark-grey schist clay

The case when weathered rocks accumulate the high level of manganese was fixed for dark-grey schist clay (Table 1, Figure 1).

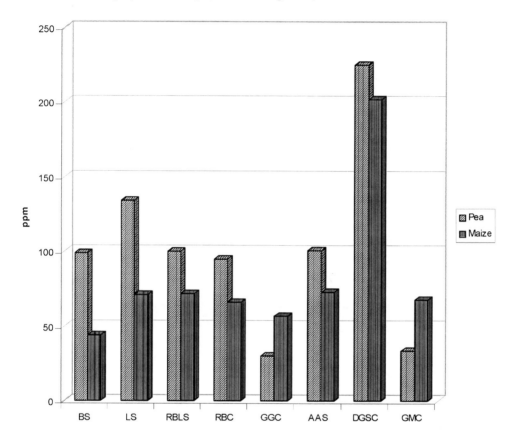

*Figure 1.* Manganese content in maize and pea growing in the rock substrata of the Nikopol manganese Ore Basin, ppm.

## 2.   Land Reclamation in the Kerch Iron Ore Basin

The experimental materials were following rocks and substrata: loam soil (LS), grey-green clay (GGC), sand-clay deposits (SCD) and such technogenic substrates as iron ore including rock (IOR), iron ore waste (IOW).

Iron Content in the Soil and Rock Substrata Water Solution Extraction is present in the Figure 2.

The field experiments used disturbed rocks separately and mixtures with technogenic substrata. The experiments were carried out with farm crops (alfalfa, pea, wheat), having different requirements for substrate fertility.

Farm crops yield on the phytomeliorated rock mass mixture was compared to variants with additional 30–50cm of soil stratum added.

The preliminary long phytomelioration of rock mass may be recommended as technological model of land reclamation in similar mining regions.

The heavy metals migration through rock-soil-plant system was researched for two mentioned land reclamation stations both the Nikopol manganese and the Kerch Iron Ore basins (Figure 1, Table 2).

TABLE 2. Heavy metals concentrations in wheat grain (Kerch Iron Ore Deposit), ppm

| Trial | Cr | Ni | Cd | Pb | Mn | Zn | Cu | Fe |
|-------|-----|-----|-----|-----|------|------|-----|-----|
| Rocks mixture (RM) | tr. | tr. | tr. | tr. | 42.0 | 29.3 | 3.5 | 180 |
| RM + 30 cm soil mass | tr. | tr. | tr. | tr. | 53.0 | 32.4 | 3.6 | 138 |
| RM + 50 cm soil mass | tr. | tr. | tr. | tr. | 44.0 | 43.0 | 5.5 | 124 |
| RM + 80 cm soil mass | tr. | tr. | tr. | tr. | 40.0 | 29.8 | 6.3 | 156 |
| Loam soil (LM) | tr. | tr. | tr. | tr. | 44.0 | 40.3 | 4.1 | 228 |
| LM + 50 cm | tr. | tr. | tr. | tr. | 40.0 | 31.8 | 3.1 | 116 |
| Sand-clay deposits (SCD) | tr. | tr. | tr. | tr. | 41.0 | 24.9 | 5.6 | 184 |
| SCD + 50 cm soil mass | tr. | tr. | tr. | tr. | 46.0 | 33.0 | 5.4 | 254 |
| Iron Ore including rock | tr. | tr. | tr. | tr. | 45.0 | 43.0 | 4.8 | 226 |

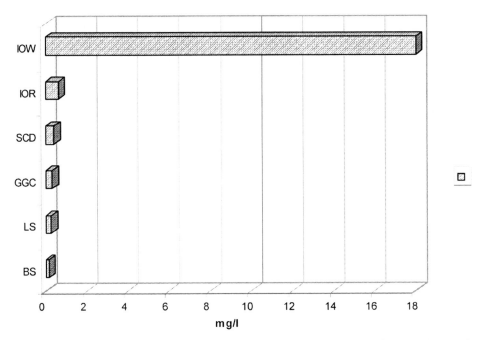

*Figure 2.* Iron content in the soil and rock substrata water solution extraction (Kerch Iron Ore Deposit), ppm.

## 3. Land Reclamation in the Western Donbass Coal Region

It was established that geochemical situation in the Western Donbass coal region is more risked. Every year during the coal mining up to the 7 millions attendant rocks (aleurolites and sand rocks, argillites and clays, pyrite, etc.) are emerged in the earth surface. The lowering of the day surface reaches to 4 m. This is a reason to oak-grove destruction and plant successions changes within this territory. Additional environmental problem is exit of high-mineralized attendant ground waters. To protect the Samara River floodplain from flooding with high mineralized attendant ground waters was proposed to realize several preventive measures including the creation of depositories for the ground water (Yevgrashkina, 2003).

Meantime $FeS_2$ content is 4–5% in the coal wastes. Following weathering of rocks taking out on the earth surface is connected with the process of heavy metals and some another toxic ions migration. That is why the loam soil was offered as geochemical barrier to toxic salts and heavy metals migration (Masyuk and Kharytonov, 2000).

The different models of land reclamation have researched in the field experiments of long-term standing on the following scheme:

1. Mining rocks (MR) +30 cm black soil (30BS)
2. MR+50BS
3. MR+70BS
4. MR+100BS
5. MR+ 70 cm meadow soil (MS70)
6. MR+50cm loam soil (50LS) +30cm BS
7. MR+50LS +50 BS
8. MR+50LS+70BS

The geochemical estimation of the heavy metals migration has been done along new created profiles of the different land reclamation versions in the Western Donbass (Tables 3 and 4).

TABLE 3. Heavy metals content in the surface and neighboring with toxic rocks stratums (Western Donbass Coal Mining), ppm

| Trial | Stratum, cm | Cu | Mn | Zn | Pb | Cd | Fe | Ni | Co |
|---|---|---|---|---|---|---|---|---|---|
| MR | 0–10 | 0.34 | 75.3 | 7.8 | 0.91 | 0.25 | 1.6 | 9.0 | 2.75 |
| | 10–30 | 0.3 | 78.3 | 8.65 | 0.85 | 0.28 | 1.6 | 8.95 | 2.75 |
| MR + 70 cm | 55–70 | 0.18 | 16.0 | 0.6 | 1.46 | 0.1 | 0.78 | 2.22 | 0.5 |
| MS | 70–85 | 0.29 | 41.0 | 1.65 | 0.73 | 0.16 | 1.56 | 4.1 | 1.13 |
| MR + 100 cm | 80–100 | 0.13 | 0.42 | 0.13 | 0.43 | 0.05 | 0.63 | 0.65 | 0.5 |
| BS | 100–120 | 0.34 | 28.8 | 2.6 | 1.1 | 0.1 | 6.0 | 3.95 | 0.63 |

TABLE 4. Heavy metals content in wheat grain (Western Donbass Coal Mining), ppm

| Version | Co | Ni | Pb | Zn | Cu | Fe | Cd | Cr | Mn |
|---|---|---|---|---|---|---|---|---|---|
| MR + 30BS | tr. | tr. | tr. | 30.5 | 4.8 | 70.5 | 0.60 | 1.9 | 34.2 |
| MR + 50BS | tr. | tr. | tr. | 32.7 | 5.1 | 68.0 | 0.75 | 1.2 | 32.0 |
| MR + 70 BS | tr. | tr. | tr. | 31.9 | 4.8 | 28.0 | 0.44 | 1.2 | 32.0 |
| MR + 50LS + 30BS | tr. | tr. | tr. | 26.7 | 4.6 | 52.0 | 0.50 | 1.0 | 30.4 |
| MR + 50LS + 50BS | tr. | tr. | tr. | 23.6 | 4.2 | 36.5 | 0.63 | 0.8 | 28.4 |
| MR + 50LS + 70BS | tr. | tr. | tr. | 23.8 | 4.0 | 29.5 | 0.50 | 0.5 | 31.2 |

## Conclusion

Thus, land reclamation applies as preservation measure to restore disturbed lands, to avoid environment pollution.

Triple-stratum land reclamation model (MR+50LS+70BS) with protective screen of loamy soil is the more hopeful for mould board coal mining rocks negative environmental impact mitigation.

## References

Kharytonov, N. Ecotoxicological problems under mining at the Ukrainian steppe, 30th International Geological Congress, Beijing, China, 1996.

Masyuk, N., Kharytonov, N. Environmental Impact Assessment of Different Reclamation Models in The Brown Coal Region of Ukraine, 5th International conference on Precision Agriculture, July 16–19, Bloomington, MN, USA, 2000.

Masyuk, N., Kharytonov N., Mitsik A. Prospects of use of Kerch iron ore layer rock for land recultivation, 4th International Conference on Biogeochemistry of trace elements, Abstracts Proceedings, CA, USA, 1997.

Yevgrashkina, G.P. Mining Branch Activity Impact on the hydro-geological and soil-meliorative territory conditions. Monograph. Dnepropetrovsk, 2003, 200 p. (in Russian).

# URBAN SOILS POLLUTION IN BELARUS: PRIORITY POLLUTANTS AND LEVELS

VALERY KHOMICH[*], TAMARA KUKHARCHYK
AND SERGEI KAKAREKA
*Institute for Problems of Natural Resources Use & Ecology of
National Academy of Sciences of Belarus, Skoriny Str. 10, Minsk,
220114, Belarus*

**Abstract.** In the paper the results of soil pollution investigation in cities of Belarus are discussed. Detected levels of contamination by heavy metals, mineral oil, water soluble compounds and POPs (PAH and PCBs) are given. Main sources of soil pollution, peculiarities of pollutants distribution, and spatial structure of anomalies are shown.

**Keywords:** urban soil, soil pollution, sources of pollution, heavy metals, water soluble compounds, PCBs, PAH

## 1. Introduction

One of the most important ecological problems of nowadays is chemical pollution of soils of cities and industrial centers, where, on the one hand, a great amount of pollutants are concentrated, and on the other hand, a significant part of population lives, being constantly subjects to the risk of their influence.

As it is well-known, the consequences of soil pollution are multiple: change of soil behavior; possibility of water, air and vegetation contamination; entrance of pollutants into human body. For this reason considerable attention is given in many countries to the investigation of soil pollution, assessment of contamination level, and revelation of the most polluted sites (brownfields, hot spots) (Ferguson, 1998; Clarinet, 2002). According to the European Environment Agency, there are around 2 million polluted sites in Europe, 10 thousand of which require to be remediated (The European Environment..., 2005).

---

[*] To whom correspondence should be addressed: valery_khomich@mail.ru

By now quite extensive factual material on soil pollution in Belarusian cities is collected. However they are studied highly unevenly. Most cities have been investigated using a sparse set of sampling points (within the system of soil pollution monitoring). And some of the cities underwent a detailed testing of soils pollution – Minsk, Gomel, Svetlogorsk, Mogilev, Pinsk, Brest, and Mozyr belong to the most investigated ones (Environment of Belarus…, 2002; Khomich et al., 2004; The Forecast of Environmental…, 2004).

Among the pollutants heavy metals (Pb, Cd, Ni, Cu, Zn etc.), soluble compounds and mineral oils are tested. Recently, however, persistent organic pollutants such as PAH and PCBs are measured in soils, as well.

## 2.  Methods and Objects

In the paper the results of eco-geochemical investigation of some cities of Belarus (Minsk, Gomel, Svetlogorsk and Pinsk) and impact zones of local sources are summarized. For assessment of soil pollution by heavy metals and water soluble compounds regular grid of soil sampling with interval 0.5–1.0 km was applied. Upper soil layer (0–10 cm) was tested, and mixed samples of soil were collected. During the process of soil sampling functional zones and urban landscapes were taken into account. Up to now more than 3,000 soil samples were tested on heavy metals and water soluble compounds.

Investigations of soil pollution by PCBs and PAH were carried out at the territory of industrial enterprises and substations of electricity distribution, where there is a danger of these pollutant discharges into environment with leakages and wastes. Soil sampling in impact zones was performed on the basis of the several factors: sources location, pathways of pollutants discharges into environment, visually identification of polluted sites, relief of the territory, etc. In total, 60 substations and 15 enterprises were investigated, around 250 soil samples were tested by PCBs content and around 100 – by PAH content.

## 3.  Results and Discussions

Heavy metals. It was detected that in Belarusian cities such heavy metals, as lead and zinc are priority soil pollutants. Practically everywhere their concentrations exceed background concentrations in 1.5 times and more (Table 1). Besides, copper and nickel belong to accumulating elements, although levels of their accumulation are determined by local sources. In a number of cities high concentrations of cadmium and mercury are revealed. As a rule, their main sources are industrial and communal wastes.

TABLE 1. Heavy metals content in soils of Minsk and Gomel cities

| Indicator | Minsk (1,340)* | | | | Gomel (695) | | | |
|---|---|---|---|---|---|---|---|---|
| | Pb | Zn | Cu | Ni | Pb | Zn | Cu | Ni |
| Average, mg/kg | 20.5 | 39 | 13.3 | 8.8 | 19 | 49 | 22.6 | 12.7 |
| Maximum, mg/kg | 785 | 1,077 | 716 | 217 | 320 | 7,367 | 3,618 | 102 |
| Standard deviation | 31 | 46 | 29 | 10 | 26 | 302 | 147 | 8 |
| Background level, mg/kg | 10 | 20 | 5 | 5 | 11.5 | 13.4 | 12 | 10.5 |

\* – in the brackets the number of soil samples is given

Spatial structure of polluted areas depends on the distribution of pollution sources. Generally, soils are more polluted in central (historical) parts of cities: this is characteristic for Gomel, Minsk, Pinsk, and Brest. The analysis of genesis of Pb and Zn anomalies has demonstrated that in large cities anomalies are caused by vehicle emissions, emissions from stationary sources, dispersion of wastes etc. In several cities the largest polluted areas are mainly located at industrial areas, waste water treatment facilities, industrial landfills. Generally, types of pollutants depend on the specialization of enterprises. In Gomel, for example, large anomaly of chromium was revealed in the impact zone of *Centrolit enterprise*, and that of strontium – near *Gomel Chemical Plant*. In Svetlogorsk great zinc anomaly is connected with industrial zone where landfill of industrial wastes, waste water treatment facilities and zinc-containing sludge depot are located.

There is quite a distinct link between heavy metals accumulation in urban soils, on the one hand, and functional prescription of the territory, on the other. The highest levels of copper, manganese, nickel, lead, and zinc accumulation are found in soils of industrial zone (Figure 1). Thus, the average content of copper in soils of industrial zone of Minsk city is about 2.8 times higher than in those of green and residential zones; lead – in 1.7–1.8 times; zinc – in 1.5–1.8 times.

According to (The Forecast of Environmental…, 2002), the total area of soils in Belarus with the content of heavy metals exceeding Maximum Permissible Levels makes up to 78,600 ha. The area of soils polluted by heavy metals with dangerous concentrations is different. Thus, in Gomel the area with average lead content in soils above Maximum Permissible Levels is about 12% of the city territory, zinc – about 9%.

Water soluble compounds belong to typical pollutants of urban soils (Environment of Belarus…, 2002; Khomich et al., 2004). It was revealed that soils of gardens are polluted by nitrates, potassium and phosphates; soils of industrial areas – by sulphates and fluorine. Average water soluble compounds content in urban soil exceeds the background level in 2.5 times.

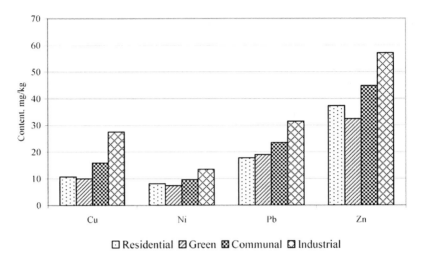

□ Residential ▨ Green ▨ Communal ▢ Industrial

*Figure 1.* Comparison of average content of heavy metals in soils of different functional zones of Minsk city.

The formation of the largest area of salt-affected soils in Belarus (more than 1,000 ha) in the impact zone of potassium enterprises in Soligorsk is related to the production of large volumes of solid and liquid wastes. Up to date more than 550 million tons of solid wastes and about 65 million tons of clay salty slimes have been accumulated here on the ground surface. The distance of soil salinization is within 3–4 km. The content of salt in soil reaches up to 3–4% (Toorenkov et al., 1986).

Vast area of salt-affected soils is located in the impact zone of *Gomel Chemical Plant*, which produces phosphate fertilizers. Soil salinization is caused by large amount of solid waste – phosphorus gypsum formation and its further wind dispersion. Phosphorus gypsum dumps occupy more than 60 ha of soil. Soils in the adjacent areas are polluted by sulphates, phosphates and fluorine.

Alkali soils near some chemical enterprises have been also revealed. One of the anomalies was formed near the waste water treatment facility of *Svetlogorsk Khimvolokno Amalgamation*. The anomaly originated as a result of highly mineralized sewage failure flood at the beginning of 1970 and its subsequent replenishment by infiltrated waters from the sludge and waste water storage. High content of salts in subsurface horizons is stipulated by intensive soil water evaporation and salts deposition. Here, in a salt crust formed on the soil surface, sulphate content makes up to 3–10 g/kg. Ground waters are highly mineralized (sum of ions – 1,000–2,000 mg/l), and are sulphate natrium in composition.

Persistent organic pollutants (POPs). There are two main groups of POPs by origin: industrial chemicals and by-products. It should be noticed that investigation of environment pollution by PCBs in Belarus has started recently (Kakareka et al., 2003; Kukharchyk, 2006).

PCBs are one of the most serious problems for Belarus because they have been used for over 40 years. According to the results of the inventory (Kukharchyk et al., 2005; National plan, 2006), the total of roughly 1,500 tons of PCBs were revealed in Belarus, 99% of which is concentrated in power transformers and capacitors. It was found that PCB-containing equipment is used at over 760 industrial enterprises. Largest volumes of PCB-containing equipment are found at substations of transmission network where power capacitors are installed.

Practically all soils at investigated substations are polluted with PCBs. In most cases "hot spots" with high concentration of PCBs were revealed. PCBs content in the soils of the territories, where electrical equipment is used and stored, makes up to milligrams or sometime grams per kilogram. Maximum concentrations of PCBs (2–21 g/kg) were tracked in the soil near destroyed capacitors and transformers as a consequence of PCB leakage.

The highest concentrations of PCBs are found in top soil layer. In some cases high PCB concentrations are fixed at the depth of 50 cm. In several cases "hot spots" of PCB leakage can be easily noticed due to dark color and "burnt out" vegetation. Low-chlorinated PCBs prevail in the soils of the places where capacitors are installed or stored, and high-chlorinated PCBs – where transformers are stored.

As a rule, contaminated areas are rather local (typical spot occupies less than a square meter). Nevertheless, they are potential sources of secondary ground and surface water, as well as bottom sediments, pollution. Moreover, spreading of PCBs beyond the places of PCB-containing equipment usage or storage occurs: the sum of 6 PCBs at the distance of 100–150 m from capacitor battery makes up to 0.35–4.6 mg/kg in the soil.

Among POPs, which represent a by-product of various industrial and thermal processes, investigation of PAH is the most urgent in Belarus because of a large number of emission sources (Kakareka et al., 2003). It should be noticed that PAH discharges into environment are possible with raw materials or additional materials (shale and coal oils, technical carbon, etc.).

The obtained results show, that PAH concentrations in urban soils are highly dispersed: starting from values below detection limit method up to 26.3 mg/kg (the sum of 16 PAH compounds). High PAH concentrations are found in different functional zones including industrial, residential and even green zones. Three main sources of PAH discharges into urban soils can be singled out: with emissions; with ashes and soot remains from solid fuel combustion; and finally with bonfire ashes in green zones. In 20% of cases the sum of PAH exceeds Maximum Permissible Level.

The study of PAH distribution in soils of industrial enterprises of various types has allowed to detect great variability of meanings and high level of soil pollution by PAH practically in all cases where thermal processes are present, or (raw) materials contaminated by PAH are used (Table 2).

TABLE 2. PAH content in soils of enterprises, mg/kg

| Compounds | Machine-building (44)* | | Tire production (10) | |
|---|---|---|---|---|
| | Maximum | Average | Maximum | Average |
| Naphthalene | 0.37 | 0.04 | 0.34 | 0.12 |
| Acenaphthylene | 0.82 | 0.02 | 0.31 | 0.07 |
| Acenaphthene | 0.16 | 0.03 | 0.13 | 0.02 |
| Fluorene | 0.16 | 0.01 | 0.17 | 0.04 |
| Phenanthrene | 3.32 | 0.28 | 8.89 | 2.08 |
| Anthracene | 0.08 | 0.01 | 1.01 | 0.19 |
| Fluoranthene | 4.12 | 0.27 | 3.03 | 0.73 |
| Pyrene | 3.42 | 0.2 | 4.02 | 0.86 |
| Benz[a]anthracene | 3.61 | 0.3 | 5.51 | 1.18 |
| Chrysene | 2.18 | 0.15 | 3.04 | 0.61 |
| Benzo[b]fluoranthene | 2.55 | 0.16 | 1.2 | 0.27 |
| Benzo[k]fluoranthene | 1.78 | 0.11 | 0.94 | 0.18 |
| Benzo[a]pyrene | 2.24 | 0.12 | 0.48 | 0.1 |
| Indeno[1,2,3-cd]pyrene | 1.32 | 0.07 | 1.56 | 0.32 |
| Dibenz[a,h]anthracene | 0.56 | 0.02 | 0.27 | 0.04 |
| Benzo[ghi]perylene | 1.38 | 0.09 | 0.51 | 0.11 |
| Sum of 16 compounds | 27.16 | 1.9 | 31.41 | 6.91 |

* – in the brackets the number of soil samples is given

Maximum PAH concentrations, which reach up to 31 mg/kg (the sum of compounds), have been revealed in the soils of enterprise for tire production. In the technological process technical carbon (soot) is used, which represents the main source of soil pollution. Maximum Permissible Levels are exceeded in dozens and hundreds times, while samples that exceed Maximum Permissible Level for the majority of compounds are found in 80–100% of cases.

High PAH concentrations have been revealed in soils of industrial enterprises of machine-building and metallurgic profile, at which cast-iron and steel castings are realized, and other thermal processes take place. The average contents of 16 PAH compounds makes up to 1.9 mg/kg, maximum contents – 27.2. It has been detected that the highest levels of soil pollution are characteristic for the industrial areas of old enterprises, including those where casting shops are closed.

## Conclusion

The obtained results confirmed the contamination of soils in Belarus by heavy metals, water soluble compounds, PCBs and PAH. The levels and area of pollution depend on many factors among which the types of sources and pathways of pollutants discharges are the most important ones. Taking into account the obtained data, such international obligation like Stockholm convention on POPs, and current situation with chemical and analytical equipment in the country, the following future tasks are determined: to investigate industrial enterprises; to reveal high polluted areas (hot spots); to create a data base of polluted sites; to develop standards for soil investigation and soil quality assessment; to assess risk connected with soil pollution; to choose and apply the technology for soil remediation.

## References

CLARINET. Brownfields and Redevelopment of Urban Areas. A report from the Contaminated Land Rehabilitation Network for Environmental Technologies/U. Ferber, D. Grimski. – Version: August 2002. – Wien: Umweltbundesamt GmbH (Federal Environmental Agency – Austria).

Environment in Belarus/ed. by V. F. Loginov. Minsk, Belarus, 2002 (in Russian).

Ferguson C. Assessing Risks from Contaminated Sites: Policy and Practice in 16 European Countries/C. Ferguson/Land Contamination & Reclamation. – 1999.– No 7 (2). pp. 33–54.

Kakareka S. V., Kukharchyk T. I. and V. S. Khomich. Persistent Organic Pollutants: Sources and Emission Estimation. Minsk: Minsktipproekt, 2003 (in Russian).

Khomich V. S., Kakareka S. V. and Kukharchyk T. I. Ecogeochemistry of Urban Landscapes of Belarus. Minsk: RUP Minsktipproekt, 2004 (in Russian).

Kukharchyk T. Polychlorinated Biphenyls in Belarus. Minsk: Minsktipproekt, 2006 (in Russian).

State of Environment in Belarus. Ecological Bulletin, 2000–2005/ed. by V. F. Loginov. Minsk, Belarus, 2006 (in Russian).

The European Environment State and Outlook, 2005/EEA. – Denmark, 2005.

The Forecast of Environmental Changes in Belarus for 2010–2020/ed. by V. F. Loginov. Minsk, RUP Minskproekt, 2004 (in Russian).

The National plan of the Republic of Belarus for the implementation of its obligations under the Stockholm Convention on persistent organic pollutants for the period of 2007–2010 and until 2028. Ministry of natural resources and environmental protection of the Republic of Belarus, Global Environment Facility, World Bank. Minsk, Belarus, 2006.

Toorenkov N., P. Zhigarev, and L. Yakoobovich. Protection of soils from technogenic pollution. BelNIINTI, Minsk, 1986 (in Russian).

# IDENTIFICATION OF POLLUTANTS IN SOILS AROUND THE OBSOLETE PESTICIDES STOCKS IN UKRAINE

VICTORIA LOKHANSKA[*], SERGIY MELNYCHUK
AND YURIY BARANOV
*Ukrainian Laboratory of Quality and Safety of Agricultural Products by National Agricultural University, 15 Geroiv Oborony Str., 03041 Kiev, Ukraine*

**Abstract.** One of the problems, connected with obsolete pesticides stocks, is the pollution of adjacent soils. The contamination of soil and groundwater with different pesticides, mainly forbidden nowadays, has caused serious environmental problems. Since 2004, the Ukrainian Laboratory of Quality and Safety of Agricultural Products has carried out a series of studies of pesticide residues in soils, adjacent to obsolete pesticides stocks, in the frames of a joint project with EPA USA. The methods used were GLC/FID, GLC/ECD, GLC/MS.

**Keywords:** obsolete pesticides, soils, multiresidues GLC determination, methods GLC/FID, GLC/MS

## 1. Introduction

According to data of FAO at the beginning of 2005 in the world about 500,000 tons of obsolete pesticides, including 20,000 tons in Ukraine, are accumulated. There is more than 2,000 tons belong to persistent organic pollutants (POPs) (Ligostaeva and Antonov, 2005). POPs belong to different chemical classes, but all of them are highly-toxic, stable to decomposition, they can be carried on long distances by air and water, and they are accumulated in fatty tissue (Stockholm Convention, 2004).

On the descriptions of 12 POPs, registered in Stockholm convention, they are divided by 3 groups:

- High-toxic pesticides (DDT, dieldrin, aldrin, heptachlor, mireks, toksafen, endrin, chlordan, hexachlorbenzol)

---

[*] To whom correspondence should be addressed: V.Lokhanska@nauu.kiev.ua

L. Simeonov and V. Sargsyan (eds.),
*Soil Chemical Pollution, Risk Assessment, Remediation and Security.*
© Springer Science+Business Media B.V. 2008

- Industrial products (PCB)
- Dioxins as by-products of some manufactures

In Ukraine the problem of management of obsolete pesticides requires an urgent decision, because about 20,000 tons of these dangerous substances are distributed across the country. Through proper storage and packing such pesticides can be rendered less harmful to human health and the environment. Research showed the presence of hazardous substances in the environment, including possible ingress into the food chain. Ownership of many of such stocks has passed to farmers, and new proprietors frequently do not know what to do with such stocks, the level of danger associated with these substances, and some proprietors are poorly informed about the legislation related to these stocks. Although in Ukraine recommendations on work with obsolete pesticides exist, the pace of funding and decision-making remains unsatisfactory.

Ukraine actively joined the forming of transnational nature protection policy and practical elaboration principles of steady development which requires introduction of weighed ecologically friendly approaches to nature management.

The partner project with US Environmental Protection Agency (EPA USA) has been started in the Ukrainian Laboratory of Quality and Safety of Agricultural Products of the National Agricultural University since August, 2004. Researches are complex with bringing in of a number of institutes of NAN of Ukraine and specialists of Ministry of Environment Protection of Ukraine. Placed conducting of pilot researches were chosen 3 storages in Zhidachev district of the L'vov area (v. Grusyatichi, v. Kniselo and v. Bakivtsi) and 3 storages in the Korsun'-Shevchenkiv district of the Cherkassy area (v. Kvitki, v. Petrushki and v. Komarivka). The purpose of this article is an attempt to present the analytical decision of task of determining the residue amounts of multi-component mixture of persistent pesticides in soils, adjoining to the places of storage of obsolete pesticides, and describe the method which was elaborated at our laboratory.

## 2.  Results and Discussion

Multiresidue methods of pesticides determining are most acceptable to the aims of monitoring and established in many countries (Tchmil, 2002). In particular, the methods of determination 29 chlorine-organic pesticides (COP) in drinking water are well known (Official Methods of analysis of A.O.A.C., 1999a). Another well known method is method of determination of 44 pesticides from the groups of FOS and COP in water, and complex of toxicants, containing a residue of 91 pesticides, in fruit and vegetables (Official Methods of analysis of A.O.A.C., 1999b) In connection with wide distribution of methods of GLC/MS and LC/MS in combination with solid state extraction the examined methods of analysis got new development (Becker, 1985; Official Methods of analysis of A.O.A.C., 2005).

In Ukraine at different times the prototypes of modern methods of determination of multiresidual methods were developed in one sample (Klissenko et al., 1999), although principles of development some differ from presented higher. In the real work the method of determination of 20 persistent pesticides is offered in one test of soil. It is based on extraction of complex of pesticides from the sample of soil by an acetone by insisting, shaking and subsequent filtration, to cleaning by a redistribution in the binary system: water acetone-n-hexane, concentration of extracts and determination of capillary gas chromatography a method with selective detectors: TID and ECD. In the case of necessity, TFE with a cartridge for non-ionic connections of octadecyl C 18, Oasis, etc. can be used for the additional cleaning of extracts.

For the increase of reliability of identification of microquantity of the analyzed pesticides in accordance with principles of quality of the analytical measuring management, capillary GLC/MS was used with an ionic trap. At possibility not only to avoid fallaciousness of results of determination (especially at the level of track amounts) but also identify unknown peaks is arrived at, using the presence of function of RTL GLC/MS.

Terms of chromatography:
Gas-chromatograf type Crystal-4000 with an ECD and TID

- The capillary column of Zebron ZB-1 (30 m x 0.32 mm x 0.5 mkm)
- Liquid phase – 100% methyl polysiloxane
- Temperature condition:
     column – 190°C (TID), 200°C (ECD)
     vaporizer – 220°C (TID), 250°C (ECD)
     detector – 250°C (TID), 310°C (ECD)
- Charges of gases: nitrogen – 40 cm$^3$/min
     hydrogen – 15 cm$^3$/min
     air – 250 cm$^3$/min

GC of Finigan TRACE GC Ultra with the mass-selective detector of Polaris Q (Thermo Electron Corporation, USA).

- The capillary column of DB-5ms (30 m x 0.32 mm x 0.5 mkm).
- Immobil phase – 95% methyl polysiloxane + 5% phenyl polisiloxane
- Temperature condition:
     evaporator – 250°C, splitless transfer line – 300°C
     detector – 200°C
     column – 40°C/1 min, 25°C/min up to 300°C (3–20 min)
- Gas (helium) charges – 20 cm$^3$/min

V. LOKHANSKA ET AL.

The aforementioned terms of chromatography allow attaining the high degree of division of components of pesticides mixture (Table 1).

The advantage of the capillary column (GLC/TID) is ability to divide chlorine-triazine (simazine, atrazine and propazin) and also isomers of HCH ($\alpha, \beta, \gamma$), what is hard arrived on the packing columns.

During the chromatography process of the standard mixture of the pesticides the satisfactory division of the components is also arrived in the mode of the programming of the temperature of the column (GLC/MS). The metrological performance of the method of the pesticides residue determination in soil, were got by the method of the introduction to the analyzed test of the added amounts of the pesticides (three concentrations on each pesticide, recovering the linear range of the determination, were entered) (Table 1), to meet with sanitary-hygienic control, in particular MAC (maximum allowance concentration) of the

TABLE 1. Pesticide residues complex determination conditions

| Pesticide | Rt, GCL, min | LOQ, ppm | Rec., % | Rt, GLC/MS, min | GC/MS confirmation | | |
|---|---|---|---|---|---|---|---|
| | | | | | m/z1 | m/z2 | m/z3 |
| α-HCH | 3.95 | 0.001 | 78.9 | 8.94 | 183 | 181 | 217 |
| β-HCH | 4.19 | 0.001 | 75.6 | 9.18 | 183 | 181 | 219 |
| γ-HCH | 4.53 | 0.001 | 68.4 | – | 183 | 181 | 219 |
| Trifluraline | 4.56 | 0.05 | 79.6 | 8.65 | 306 | 264 | 43 |
| Simazine | 5.12 | 0.05 | 78.8 | – | 201 | 186 | 44 |
| Atrazine | 5.3 | 0.05 | 79.3 | 9.06 | 200 | 215 | 173 |
| Propazine | 5.42 | 0.05 | 87.9 | 9.08 | 214 | 229 | 172 |
| Semeron | 7.51 | 0.05 | 84.2 | 9.59 | 213 | 198 | 58 |
| Prometryn | 8.8 | 0.05 | 84.8 | 9.76 | 241 | 184 | 226 |
| Dimethoate | 4.98 | 0.01 | 84.5 | 9.03 | 87 | 93 | 125 |
| Diazinon | 6.28 | 0.01 | 83.9 | 9.24 | 179 | 137 | 152 |
| Parathion-methyl | 8.12 | 0.01 | 86.8 | 9.73 | 109 | 263 | 125 |
| Malathion | 10.25 | 0.01 | 92.3 | 9.97 | 125 | 173 | 93 |
| Aldrin | 8.45 | 0.001 | 81.4 | 10.15 | 66 | 263 | 79 |
| Heptachlor | 6.95 | 0.001 | 69.5 | 9.86 | 100 | 272 | 274 |
| Pendimethalin | 13.67 | 0.05 | 81.3 | 10.36 | 252 | 162 | 281 |
| DDD | 17.86 | 0.001 | 72.3 | 11.20 | 235 | 237 | 165 |
| DDE | 14.23 | 0.001 | 77.4 | 10.87 | 246 | 248 | 318 |
| DDT | 23.1 | 0.001 | 81.6 | 11.49 | 235 | 237 | 165 |
| Phosalone | 55.44 | 0.05 | 82.8 | 12.23 | 182 | 121 | 97 |
| Chlorpyriphos | 11.34 | 0.01 | 77.8 | 10.20 | 97 | 197 | 199 |
| Propiconasol | 33.85 | 0.01 | 76.9 | 11.55 | 69 | 173 | 259 |
| Propiconasol | 35.36 | 0.01 | 75.9 | 11.61 | 69 | 173 | 259 |

pesticides in soils, more, than on an order exceed the lower limit of the determination in soil. The developed method is simple on execution, allows to save the far of solvents and is reproduced quickly (duration of analysis does not exceed 4th hours, but at the use of the speed-up methods of the extraction yet less than).

As an example in Table 2 the results of the application of the method are resulted for the multiresidual of the pesticides in the soils, selected on the territories adjoining to obsolete pesticides stocks. Soils were selected in the different areas of Ukraine. In view of the storage terms of pesticides in these facilities, the high level of adjacent soils contamination does not cause a surprise, but requires giving the fact of the extraordinary value.

Data analysis of the Table 2 shows that residue of COP ($\beta$, $\gamma$ - HCH, DDT and DDE), treflane and atrazine considerably exceed MAC.

TABLE 2. Pesticides residue in soils adjacent to obsolete pesticide stocks

| Pesticides residue, mg/kg | L'viv region, v. Grusiatychy | | Cherkassy region, v. Kvitky | |
|---|---|---|---|---|
| | Distance from the stock | | Distance from the stock | |
| | 1–3 m | 20 m | 1–3 m | 20 m |
| A-HCH | nd | 0.01 | 0.04 | nd |
| B-HCH | 0.25 | 0.44 | 4.24 | 0.28 |
| Г-HCH | 0.05 | 0.26 | 0.03 | 0.03 |
| Treflan | nd | nd | 113.0 | nd |
| Simazine | nd | nd | nd | nd |
| Atrazine | nd | nd | 31.82 | 0.10 |
| Propazine | nd | nd | nd | nd |
| Semeron | nd | nd | nd | nd |
| Promethrine | nd | nd | nd | nd |
| Dimethoate | 0.01 | nd | 0.28 | 0.01 |
| Diazinone | 0.07 | nd | 0.05 | 0.01 |
| parathion-methyl | nd | nd | nd | nd |
| Malation | 0.01 | nd | nd | nd |
| Aldrine | nd | nd | nd | 0,01 |
| Heptachore | nd | nd | nd | nd |
| Pendimethaline | nd | nd | nd | nd |
| DDD | 0.01 | nd | 0.06 | 0.10 |
| DDE | 0.18 | 0.01 | 0.11 | 2.67 |
| DDT | 0.15 | 0.01 | 0.28 | 0.32 |
| phosalone | nd | nd | nd | nd |

nd – non detected

## Conclusion

The multiresidues method for determination about 20 pesticides from different chemical groups (organochlorine, organophosphate, triazine, dinitroaniline) in the soil was elaborated. The contaminations of the soil adjacent to obsolete pesticides stocks were investigated.

## Acknowledgments

The authors gratefully acknowledge funding though a grant P-169 from the EPA USA. We also acknowledge Ph.D. V. Tsvilichovsky, Mrs. O. Zemtsova, Ing. V. Pavlinchuk and O. Skripnik for the technical assistance in our work.

## References

Becker G. Organohalohen, Organophosphorus and Triazine Compounds, German ver., 1985, S8, pp. 2–8.

Klissenko et al.: Клісенко М.А., Александрова Л.Г., Демченко В.Ф., Макарчук Т.Л. Аналітична хімія залишкових кількостей пестицидів.-К., 1999, 238 с.

Ligostaeva Ye.V., Antonov A.V. Analysis of the Current Status of Treatment of Prohibited and Obsolete Pesticides in the Context of the Development of the National Plan for Implementation of the Stokholm Convention on Persistant Organic Pollutants. – Environmentally Sound Management (ESM) practices on Cleaning up Obsolete Stockpiles of Pesticides for Central European and EECCA Countries. – Abstracts, 8th International HCH and Pesticides Forum, 26–28 May, 2005, Sophia, Bulgaria, pp. 22–23.

(Stockholm Convention): Стокгольмська конвенція про стійкі органічні забруднювачі. – Офіційний переклад на українську мову. Затверджено Правовим департаментом МЗС України. – Вид. СПД, Вальд", К., 2004, 47 с.

Official Methods of analysis of A.O.A.C. International, ed. P. Gunnif, 16 ed., 1999, MD, USA, v.1, ch.10, pp. 13–17.

Official Methods of analysis of A.O.A.C. International, ed. P. Gunnif, 16 ed., 1999, MD, USA, v.1, ch.10, pp. 26–31.

Official Methods of analysis of A.O.A.C. International, ed. W. Horwitz, 18 ed., 2005, MD, USA, v.1, ch.10, pp. 10–17.

Tchmil V: Чмиль В.Д. Состояние и перспективы использования современных инструментальных методов анализа пестицидов в Украине.- Современные проблемы токсикологии, 2002, № 2, pp. 56–62.

# DESIGN OF NEW "IONIC LIQUIDS" FOR LIQUID/LIQUID EXTRACTION OF PERSISTENT TOXIC SUBSTANCES

FLIUR MACAEV[*], EUGENIA STINGACI
AND VIORICA MUNTEANU
*Institute of Chemistry*
*Academy of Sciences of the Republic of Moldova*
*Academy Str.3, MD-2028, Kishinev, Republic of Moldova*

**Abstract.** New ionic liquids bearing an imidazolium core have been prepared in an attempt to design new "liguids" for liquid/liquid extraction of Persistent Toxic Substances. Trends in the properties of these liquid compounds are discussed. The synthesized salts represent a novel class of solvents and may be considered as a new medium for liquid/liquid extraction.

**Keywords:** ionic liquids, persistent toxic substances, liquid/liquid extraction

## 1. Introduction

The knowledge about Persistent Toxic Substances (PTSs) is quite extensive, as they have been studied for some 30–45 years. To some extent this knowledge can also be extrapolated to similar PTSs in Moldova but this always has to be done with care. Currently in Moldova a lot of PTSs are stored in various former collective agricultural warehouses or disposed in uncontrolled dumps. The main environmental problems are those that relate to water pollution (particularly ground water pollution, hazardous wastes, soil degradation and biodiversity conservation.

"Ionic liquids" are a relatively new class of compounds that have been receiving increased attention in recent years as "green" designer solvents that may potentially replace many conventional volatile organic solvents in reaction and separation processes. These unique compounds are organic salts that are liquid over a wide range of temperatures near and at room temperature. Ionic

---

[*] To whom correspondence should be addressed: flmacaev@cc.acad.md

L. Simeonov and V. Sargsyan (eds.),
*Soil Chemical Pollution, Risk Assesment, Remediation and Security.*
© Springer Science+Business Media B.V. 2008

liquids have no measurable vapor pressure; hence, there has been considerable interest in using them in place of volatile organic solvents that can emit problematic vapors. There has been considerable interest in the potential of room temperature ionic liquids for various separation processes. In Moldova there are more than 5,650 tons (according to 2004 data) of obsolete pesticides. Such pesticides till now are kept improperly impartial destroyed, without guard stores at former collective farms. It can be easily to predict that area around indicated stares including rivers, likes are polluted different kind of persistent toxic substances. Generally, water pollution is any chemical, physical or biological change in the quality of water that has a harmful effect on any living thing that drinks or uses or lives (in) it. When humans drink polluted water it often has serious effects on their health. Water pollution can also make water unsuited for the desired use. More over, water pollution is usually caused by human activities. Water pollution is detected in laboratories, where small samples of water are analyzed for different contaminants. Liquid/liquid extraction is a very effective analytical method for the recovery and separations of both metal species and organic compounds. The removal of toxic heavy metal ions from wastewater is of great concern in the environmental field of waste and pollution reduction.

## 2. Results and Discussion

Room temperature "ionic liquids", based imidazolium cation containing carboxyl, 2,3-epoxypropyl and allyl chains have been synthesized in an attempt to design new functionalized liquids for liquid/liquid extraction of PTSs from water.

### Scheme 1

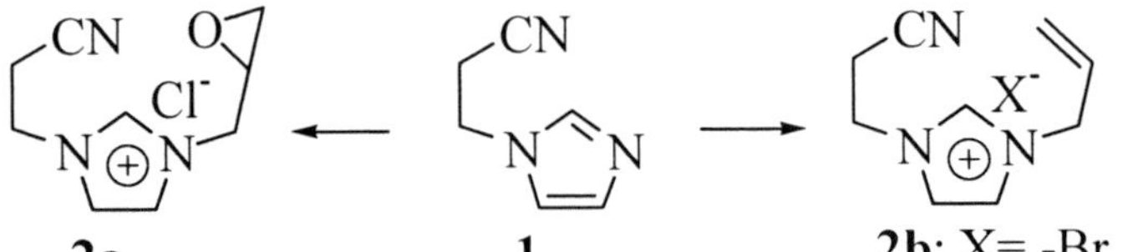

2a        1        2b: X= -Br
                                      2c: X= -BF$_4$
                                      2d: X= -PF$_6$

1$H$-imidazole-1-propanenitrile 1 used in this study was reported previously (Horvath, 1994). The chloride (bromide) salts 2a,b were prepared by quaternization of 1 analogously (Pernak, 2001). The next step was the metathesis of alkylimidazolium salts with the appropriate inogranic salt (KPF$_6$ or KBF$_4$) in water or acetone solution as proposed by Dupont, et al. (2002). Notable when we have prepare this paper, the development of nitrile-functionalized ionic liquids for C-C coupling reaction was reported by Fei et al. (2007).

On the other side we were looked for specific ionic liquids composed of imidazolium cations with a carboxy group like **5**. In the beginning this approach we repeated the method of Wang et al. (2005) but obtained mixture of products.

Herein, we wish to report the regio-specifically synthesis of new compounds **4, 5** with a carboxy group starting from N-methylimidazole **3** and *tert*-butyl 2-bromoacetate according Scheme 2.

### Scheme 2

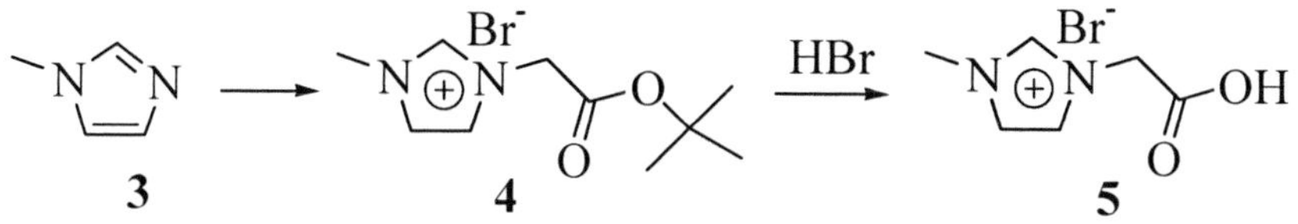

Commercially available imidazole **3** was transformed into 3-(*tert*-butyloxy-carbonylmethyl)-1-methyl-1*H*-imidazolium bromide **4**, which upon reaction with water solution of HBr afforded the 3-carboxymethyl-1-methyl-1*H*-imidazolium bromide **5** in excellent yield. The structures of new salts were established based on their spectral data and elemental analysis.

During the last few years, extraction of kind of metal ions by using ionic liquids have been developed and reviewed by Zhao et al. (2005) and Lee (2006) as well. Mercury compounds extraction from aqueous solutions was carried out us analogously published method (Germani et al., 2007). Liquid/liquid extraction was carried out by contacting equal volumes of ionic liquid and an aqueous solution of $HgCl_2$ under stirring at room temperature. Synthesized ionic liquids showed a remarkable ability to extract Hg(II) ions from the aqueous phase and different trends depending different ionic liquids/Hg(II) molar rations. For example, use nitrile containing ionic liquid produced a corresponding increase of Hg(II) ion percentage partitioned in the organic phase then use 5:1 molar ratio. On the contrary, in the case of carboxyl group containing ionic liquids, a 2:1 molar ratio is high enough to achieve the complete transfer of Hg(II) in the organic phase. Moreover, comparable results were gained with allyl side chain group with combination of 2,3-epoxypropyl chains, pointing out that further modifications in the hydrocarburic moiety of imidazolium salt did not signifi-cantly affect the efficiency of Hg(II) ion extraction. It should be mentioned that our ionic liquids shown practically the same "extractors" properties as reported by Germani, et al. (2007).

## Conclusion

These results open new possibilities in construction of new type specific "extractors" and could be use in the future for PTSs monitoring as well.

## Acknowledgments

The authors gratefully acknowledge funding though a grant 06.21 CRF from the Moldavian-Russian Grants Program. We also acknowledge Professor Gavrilov K. for providing the preliminary results of "extractors" properties and NMR spectroscopic data our ILs.

## Experimental Methods

All the solvents used were reagent quality, and all commercial reagents were used without additional purification. Removal of all solvents was carried out under reduced pressure. Analytical TLC plates were Silufol® UV-254 (Silpearl on aluminium foil, Czecho-Slovakia). IR spectra were recorded on a Specord 75 IR instrument. $^1$H and $^{13}$C NMR spectra were recorded for d$_6$-DMSO 2–3% solution on a Bruker AC-80 (80 and 20 MHz) and on a Varian XL-400 spectrometer (399.95 MHz) apparatus (s-singlet, d-doublet, t-triplet, m-multiplet). N-methylimidazole 3 and *tert*-butyl 2-bromoacetate were reagents from Aldrich Chemical Company.

### Preparation of 1-(2-Cyanoethyl)-3-(2-oxiranylmethyl)-1*H*-imidazolium chloride 2a

Mixture of **1** (4.54 g, 0.036 mol) and 2-chloromethyloxirane (3.26 g, 0.036 mol) in 15 ml acetone was stirred for a 25 hours. Solvent was removed to give 5.68 g of viscously yellow oil **2a**. Yield 75%. IR (v/cm$^{-1}$): 3130, 1450 (C = C), 2250 (CN). $^1$H NMR ($\delta$, ppm): 3.20 t (2H, CH$_2$CN, $J$ = 6.62 Hz), 3.5–3.86 m (1H, CH = C), 3.63–3.69 m (2H, CH = CH$_2$), 3.81–4.19 m (2H, CH$_2$CH), 4.27 t (2H, CH$_2$N, $J$ = 6.4 Hz), 6.94 s, 7.27 s, 7.73 s (3H, imidazole). Mol. For. C$_9$H$_{12}$ClN$_3$O Cal. C 50.59; H 5.66; N 19.67; Find C 50.88; H 5.71; N 5.72.

### Preparation of 1-allyl-3-(2-cyanoethyl)-1*H*-imidazolium bromide 2b

Bromide **2b** (4.19 g) has synthesized by use **1** (2.13 g, 0.018 mol) and 3-bromo-1-propene (2.13 g, 0.018 mol) by the same procedure as for preparation of **2a**. Yellow oil. Yield 75%. IR (v/cm$^{-1}$): 3130, 1450 (C = C), 2250 (CN). $^1$H NMR ($\delta$, ppm): 3.35 t (2H, CH$_2$CN, $J$ = 6.4 Hz), 4.61 t (2H, CH$_2$N, $J$ = 6.4 Hz), 4.94 d (2H, CH = CH$_2$, $J$ = 5.71 Hz), 5.14–5.35 m (2H, CH$_2$CH), 5.8–6.21

m (1H, CH = C), 7.91 s, 8.05 s, 9.56 s (3H, imidazole). Mol. For. $C_9H_{12}BrN_3$
Cal. C 44.65; H 5.00; N 17.36;  Find C 44.76; H 5.81; N 17.33.

### Preparation of 3-allyl-1-(2-cyanoethyl)-1*H*-imidazolium tetrafluoroborate 2c

*Method* A: Mixture of bromide **2b** (2.35 g, 0.0097 mol) and KBF$_4$ (1.23 g, 0.0097 mol) in 2 ml water was stirring at room temperature for 56 hours. Compound **2c** (1.4 g) was obtained as a pale yellow oil. Yield 87%. IR (v/cm$^{-1}$): 3100, 1420 (C = C), 2250 (CN). $^1$H NMR ($\delta$, ppm): 3.20 t (2H, CH$_2$CN, $J$ = 6.27 Hz), 4.50 t (2H, CH$_2$N, $J$ = 6.27 Hz), 4.87 d (2H, CH$_2$CH, $J$ = 5.53 Hz), 5.15–5.42 m (2H, CH = CH$_2$), 5.6–6.3 m (1H, CH = C), 7.77 s, 7.84 s, 9.24 s (3H, imidazole). Mol. For. $C_9H_{12}BrN_3$ Cal. C 43.41; H 4.86; N 16.87; Find C 43.79; H 4.89; N 16.99.

*Method* B: Mixture of bromide **2b** (2.35 g, 0.0097 mol) and 1.26 g (0.0097 mol) of KPF$_6$ in 5 ml acetone was stirred for 40 hours. Solids were filtered, and the solvent removed to give 2.26 g of oil **2c**. Yield 94%.

### Preparation of 3-allyl-1-(2-cyanoethyl)-1*H*-imidazolium hexafluorophosphate 2d

*Method* A: Compound **2b** (2.35 g, 0.0097 mol) has reacted with KPF$_6$ (1.77 g, 0.0097) according to the procedure described for the preparation of tetrafluoroborate **2c** (See method A) to give salt **2d** (1 g) as a yellow oil. Yield 53%. IR (v/cm$^{-1}$): 3185, 1450 (C = C), 2260 (CN). $^1$H NMR ($\delta$, ppm): 3.20 t (2H, CH$_2$CN, $J$ = 6.62 Hz), 4.52 t (2H, CH$_2$N, $J$ = 6.62 Hz), 4.86 d (2H, CH$_2$CH, $J$ = 5.75 Hz), 5.15–5.39 m (2H, CH = CH$_2$), 5.8–6.21 m (1H, CH = C), 7.73 s, 7.84 s, 9.26 s (3H, imidazole). Mol. For. $C_9H_{12}PF_6N_3$ Cal. C 35.19; H 3.94; N 13.68; Find C 35.19; H 3.88; N 13.59.

*Method* B: Salt **2d** (2.93 g) was synthesised starting from **2b** (2.35 g, 0.0097 mol) and KPF$_6$ (1.77 g, 0.0097) in acetone solution by the same procedure as for **2b** (See method B). Yield 97%.

### Preparation of 3-(*tert*-butyloxycarbonylmethyl)-1-methyl-1*H*-imidazolium bromide 4

Bromide **4** was prepared by mixing of equal molar amounts of N-methylimidazole **3** and *tert*-butyl 2-bromoacetate without any solvent. Notable, reaction is exothermic. The resulting viscous (pure according spectral and TLC data) liquid **4** was allowed to cool to room temperature. Yield 94%. $^1$H NMR ($\delta$, ppm): 1.37 s (9H, tert-butyl), 3.92 s (3H, Me-N), 5.29 s (2H, CH$_2$), 7.86 s, 7.90 s, 9.39 s (3H, imidazole). Mol. For. $C_{10}H_{17}BrN_2O_2$ Cal. C 43.34; H 6.18; N 10.11; Find C 43.24; H 6.00; N 9.87.

## Preparation of 3-carboxymethyl-1-methyl-1*H*-imidazolium bromide 5

To prepare the ionic liquid **5**, several drops of 45% solution of HBr in water was added to a solution of 2.77 g (0.001 mol) of **4** in MeCN (5 ml) in a round-bottomed flask fitted with a reflux condenser by refluxing and stirring for 3 hours. The residue was then heated under vacuum at 70°C to remove any excess water and HBr. Yield 82%. M.p. 88–90°C from ethanol. $^1$H NMR ($\delta$, ppm): 3.88 s (3H, Me-N), 4.97 s (2H, CH$_2$), 4.97 s (1H, CO$_2$H), 7.68 s, 7.70 s, 9.10 s (3H, imidazole). Mol. For. C$_6$H$_9$BrN$_2$O$_2$ Cal. C 32.60; H 4.10; N 12.67; Find C 32.54; H 4.00; N 12.88.

## References

Dupont J., Consorti C.S., Suarez A.Z., Souza R.D. *Org. Synthesis*, 2002, *79*, 236–243.

Fei Z., Zhao D., Pieraccini D., Ang W.H., Geldbach T.J., Scopelliti R., Chiappe C., Dyson P.J. *Organometallics*, 2007, *26*, 1588–1598.

Germani R., Mancini M.V., Savelli G., Spreti N. *Tetrahedron Lett.*, 2007, *48*, 1767–1769.

Horvath, A. *Synthesis*, 1994, 102–105.

Lee S. *Chem. Commun.*, 2006, 1049–1063.

National Implementation Plan for the Stockholm Convention on Persistent Organic Pollutants (POPs) Chisinau, Stiinta, 2004.

Pernak J., Czepukowicz. *Ind. Eng. Res.*, 2001, *40*, 2379–2383.

Wang Z., Wang C., Bao W., Ying T. *J. Chem. Res.*, 2005, *6*, 388–390.

Zhao H., Xia S., Ma P. *J. Chem. Technol. Biotechnol.*, 2005, *80*, 1089–1096.

# LAND AND SOIL RESOURCES IN ARMENIA: STATE OF THE ART AND POLICY MEASURES

YURI SUVARYAN AND VARDAN SARGSYAN[*]
*Yerevan State Institute of Economy*
*128 Nalbandyan Str., 375025 Yerevan, Armenia*

**Abstract.** According to the National Action Plan to Combat Desertification in Armenia, about 24,353 km$^2$ of the territory of the Republic, 81.9% (excluding the surfaces of Lake Sevan and water reservoirs), are to different extents exposed to desertification: 26.8% of the total territory of Armenia faces extremely severe desertification; 26.4%, severe desertification; 19.8%, moderate desertification; and 8.8%, slight desertification. Only 13.5% (400 km$^2$) of the territory is not exposed to desertification. In the period between 1950 and 1999, the area of arable land in Armenia decreased by 166,600 ha: meadows by 15,600 ha and pastures by 136,500 ha. There are different problems resulting desertification in Armenia. The paper presents the state of the art of land and soil recourses, the set of problems of soil degradation in Armenia and describes policy measures and shortcomings.

Keywords: desertification, arable land, land management, soil degradation, soil resources

## 1. Background

Development processes in post soviet area, enlargement of volumes of production, as well as intensive involvement of natural resources in the economic turnover are emphasizing importance of the problem of secure environmental management and especially rational use of land and soil resources. The problem is more important for the countries with scarcity of land recourses. Nowadays land degradation and desertification are key problems for the Republic of Armenia. Armenia is situated in the South of the Caucasus and occupies only 29.8 km$^2$. The process of desertification includes 80% of the Republic's territory. In addition to the above-mentioned negative tendency, the new factors came to the

---

[*] To whom correspondence should be addressed: varsarg@mail.ru

L. Simeonov and V. Sargsyan (eds.),
*Soil Chemical Pollution, Risk Assesment, Remediation and Security.*
© Springer Science+Business Media B.V. 2008

229

science over the last few years, such as ineffective land utilization as a result of privatization, cut backs of financial resources directed against droughts and for building of water construction objects, decrease of necessary activities controlling the sphere of urban construction and planning, and measures aimed at prevention of land degradations due to the intensive exploitation of mines. Taking into account a crucial importance of preventing desertification and the UN Convention on "Fighting Desertification....," on May 28th, 2002 the Armenian Government has adopted the "National Action Plan on Desertification" which includes the main strategies for a further prevention of this phenomenon.

## 2.   Land and Soil Resources

### 2.1.   LAND ESTATE AND LAND USES

The total land in Armenia, according to 1997 data, is 2,974,300 ha. The area of specially protected areas of the republic is about 311,000 ha. Despite the scarcity of land, there is a diversity of soils in Armenia. The territory of the Republic of Armenia is divided to the following zones: Semi-desert – total area 236,000 ha, Dry steppe – 242,000 ha, Forest – 712,000 ha, Mountainous – meadows – 629,000 ha.

Soils are mainly distributed in the basins of Dzoraget, Mastara, Arpa, Araks, Eraskh, Agstev, Pambak, Meghri and Vokhchi, in tributary basins of Akhurian, Hrazdan, Kasakh, Azat and Vorotan and in small river basins located in Sevan-Areguni mountain range.

Agriculture is a major land use in Armenia, amounting to about 19% of total land area. The majority of such lands are located in plain areas. These lands produce almost the entire agricultural output in Armenia. Agriculture in Armenia is the main "working" branch with 30% of domestic product and 42% of employment. As in Armenia exists the problem of food security overexploitation of soil recourses is evident. Agriculture is in the intersection of most environmental and development related problems in Armenia.

The shortage of agricultural lands is particularly acute in mountainous regions. Most of arable lands in Armenia are located in the Ararat Valley. Traditionally, cultivation of cereals, fodder, fruit, and vegetable gardening were major agricultural sectors.

In the 1970s and 1980s, highly subsidized large-scale collective farms, either for livestock raising or land cultivation, produced the total agricultural output. Increased productivity was achieved by the use of huge quantities of foodstuff for livestock raising, and the intensive use of fertilizers and other agricultural chemicals for crop production.

Since the break-up of the Soviet Union, land use, agricultural production and trade patterns have dramatically changed in Armenia, as in other FSU regions

(Figure 1). The breakdown of traditional economic ties among the Soviet republics caused the loss of markets for both agricultural inputs (chemicals, food grain for livestock, fuel, machinery and spare parts) and outputs, leading to reduced amounts of arable lands and livestock and hence, a general fall in agricultural output. Large-scale collective farms were no longer sustainable and began to disappear. Individual farmers gradually became the main producers of agriculture output, changing land uses, agriculture practices and adapting to local markets. The natural (subsistence) economy has become stronger in agriculture and brought about increased grazing and hay production. Almost all collective livestock farms have stopped functioning. This had a detrimental effect on pastures near villages, promoting erosion and land degradation of lowlands (IUCN, 2001). Land privatization process has fostered the establishment of private enterprises and small farms.

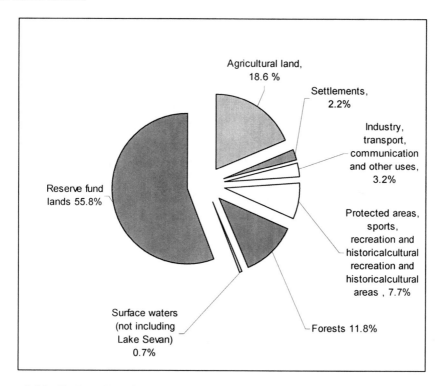

*Figure 1.* Distribution of land in Armenia (National Statistical Service of RA).

Urban land development is not the major land uses. Urban territories occupy small areas of the region. Major concentration is Yerevan, the capital of Armenia. Historically, many environmental problems of the 1970s and 1980s in urban areas were related to poor town planning/town-building and land zoning system. Environmental considerations were largely neglected during the planning and construction processes. An even less controlled situation exists now. Illegal construction of residential blocks and commercial buildings, even in green zones, are not rare in the cities.


## 2.2. LAND DEGRADATION

Degradation and pollution of land resources rank high among the major environmental issues in Armenia. These priorities are underlined in National Environmental Action Plans (NEAPs) of the country. At present, it remains very difficult to take preventive or corrective measures, since severe budgetary constraints do not allow for planning and/or taking large-scale land reclamation and soil protection measures.

Both natural and anthropogenic pressures contribute to land degradation. Among the natural factors, wind and water erosion, landslides, mudflows, flooding, etc. are important driving forces in the region, since the whole region is prone to active geo-dynamic processes. Among anthropogenic factors, bad agricultural practices (intensive land cultivation, over-use of agricultural chemicals, intensive irrigation, over-grazing) as well as unsustainable forestry practices, urbanization and other activities affect land resources.

The following factors contribute to soil quality changes:

- Landslides, which occur on 2% of the territory of the Republic or 0.5 thousand $km^2$, mainly in the Akhuryan river valley and the basins of the Debed, Vedi, Getik and Vorotan rivers
- Mud-flows: about 200 settlements in the Republic are affected by mud-flows; in the Ararat plain they affect 30% of the territory, most of which is agricultural land
- Soil salinization occurs in the Ararat Valley where about 10% of lands are salinated
- Deforestation reaches significant levels in Armenia
- Earthquakes
- Other factors

Soil erosion is one of the most widespread natural phenomena in Armenia and is the most dangerous for the republics' short in arable lands. Erosion here is connected with climate and relief peculiarities as well as anthropogenic factors: irregular woodcutting, unsustainable irrigation and drainage practices, open-pit mining, intensive grazing, land cultivation (especially on steep slopes), etc. Erosion results in reduction of land fertility and degradation of vast land areas. According to data from soil studies (1980–1985), about 44% of the lands of the Republic are to varying extents exposed to erosion and 80% of the area is prone to it. These lands are located mainly in the Marzes of Aragatsotn, Kotayk, Lori, Syunik and Vayots Dzor.

Due to overgrazing, the area of natural meadowlands has decreased from 1.4 million ha in 1940 to 804,500 ha in 2002. In Ararat valley the erosion connected with irrigation reaches 5–6%, whereas erosion taking place in the pastures reaches 60% and over. The Brown soils are exposed to erosion most of all. Technogenic activities such as open-pit mining operations also have adverse effects on land resources, causing land degradation and depletion. Some regions, such as Aragatsotn, Ararat and Siunik have undergone erosion up to 60–90%.

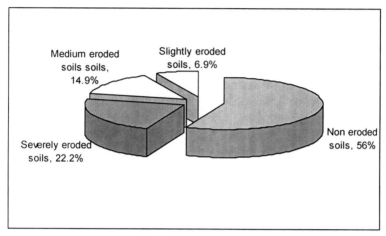

*Figure 2.* Soil characteristics (Action Plan to Combat Desertification).

It may be presumed that erosion processes are one of the reasons for the degradation of environment in highland zones, where a considerable number of pastures and hayfields are concentrated (Figure 2).

The total area of eroded lands in Armenia has been increasing since the 1980s, a 1.9% increase in total eroded area was observed during the last 20 years and the damage from land erosion amounted to 7.5% of the gross agriculture product (UN-ECE/MNP of Armenia, 2000). According to another investigation annual loss of fertile lands makes up 8 million tons (MNP of Armenia, 2001; UNDP, Armenia 1999).

The variety of the relief is one of the reasons of soil erosion in Armenia. Another reason contributing to the development of soil erosion in Armenia is the rigid climatic variation. The fluctuation of air temperatures varies from +42° in Ararat valley to -22 to -44°. Prevalence of continental climate, various elevation of slopes and their extreme fragmentation, combined with scarce vegetative cover create favourable conditions for the development of soil erosion.

Soil salinization is another major issue pertaining to land resources in Armenia. Soils in dry steppe and semi-desert zones in the region are naturally saline. Hence, cultivated soils in such zones need intensive irrigation and drainage.

Unfortunately, since the break-up of the Soviet Union the total area of irrigated lands in Armenia has been declining, irrigated areas have declined from 311,000 ha in 1985 to 280,000 ha in 1995 and 217,000 ha in 2000 (UN-ECE/MNP of Armenia, 2000).

In the region, most irrigation systems are inadequately lined. In addition, they are not properly maintained and need major repairs and/or replacement. Water losses are high, although it is impossible to give exact numbers, and contribute to an increase in the water table and hence, soil salinization. Regretfully, the country lack finances to rehabilitate the systems or plan for new irrigation projects. Irrigation systems need in proper drainage as well, without which water logging and secondary soil salinization can occur. Many irrigation systems do not have drainage systems or have inefficiently operating ones. The systems were destroyed during the last decade and there is a lack of funds to repair or rehabilitate them. Thereby, the secondary salinization of soils is a serious problem at present. In Armenia, salinised soils occupy approx. 42,000 ha in the Ararat Valley.

Lands degraded due to mining activities are found in 281 communities of 11 Marzes of the Republic. According to inventory data, over the period 1978–1998 there were 640 tracts of land degraded by mining, with a total surface area of 7,530 ha, of which 3,780 ha were used as agricultural lands before degradation. Over the total surface area of these degraded tracts, mining activities have been halted on 3,037 ha. These lands should be restored for agricultural use. The remaining 4,493 ha are still being mined.

## 2.3. SOIL POLLUTION

Soil pollution is a serious concern for Armenia. Sources of soil pollution in Armenia include the agriculture, industry, energy, transport, and municipal sectors. During the Soviet era, such pressures as intensive use of mineral fertilizers and agricultural chemicals together with industrial activities, mining, traffic emissions, and the dumping of municipal and toxic solid wastes affected the soil quality in both urban and rural areas. Presently, despite the general decline in use of agro-chemicals, the problem of topsoil pollution still exists. First, agro-chemicals do not easily degrade and heavy metals are still accumulated in soils. Chemical fertilizers and chemical herbicides and pesticides, especially chlorine-organic compounds that remain in the soil for 15-20 years, have long been used in agriculture in Armenia. Second, the uncontrolled import and use of fertilizers and chemicals by individual farmers pose a threat to environmental quality, along with obsolete pesticides stored in inadequate warehouses.

There are limited data on soil pollution by agrochemicals. Historically, the Hydro-meteorological Services (HMSs) in the Republic conducted soil sampling and analysis. Measurements were sporadic and the methods of sampling and analysis employed by HMSs might include unacceptable errors. For example, 3,560 soil samples were taken in 1977–1983 from arable lands and orchards, and only 21 samples showed high pesticide concentrations. High concentrations of DDT and DDE were found in 20% of soil samples taken from arable lands of the Ararat Valley (UN-ECE/MNP of Armenia, 2000, UNEP/MNP of Armenia, 2000).

Soil pollution by heavy metals and oil products is a concern in urban and industrial areas. Before the transition, road traffic accounted for about 60% of soil pollution in urban areas. At present, this figure exceeds 85%, since industries work at a minimum level. Additionally, copper and gold mining operations in the republic were heavily polluting soils. In Armenia, about 30,000 ha of land is polluted by copper, lead and molybdenum due to mining operations in Northeast Armenia. Mining and metallurgy enterprises pollute the soil with heavy metals and chemical compounds. The volume of accumulated industrial wastes reaches several hundred millions cubic meters. Land surrounding the Alaverdi copper-molybdenum plant, in a radius of 3 km, is polluted by heavy metals, with concentrations 20–40 times above norms. Land adjacent to the Ararat gold plant is polluted by heavy metals. Similar enterprises are located in Kadjaran, Kapan, Meghri and Agarak, and their surroundings are also polluted by heavy metals (statistical data are not available).

The city of Yerevan is heavily contaminated. USAID studied soil samples from the area surrounding a thermopower utility and found contamination by polychlorinated biphenyls (UN-ECE/MNP of Armenia, 2000). Almost all settlements are polluted by industrial and household wastes.

## 3. Policy Measures

During the Soviet era, all the lands were public property and belonged to the "United State Land Fund". The Land Fund was divided into several categories based on land use: agriculture, state forestry farms, state land fund, non-agricultural lands (industrial areas, resorts and urban areas, etc.).

Many of land-related problems of the 1970s and 1980s were caused by poor land use planning. Land use planning was a part of central planning system consisting of strictly centralized territorial and sector planning. The planning was conducted at all-union (central) and national levels. State Planning and Building Committees ("Gosplan" and "Gosstroy" respectively) with subordinated branches in the Soviet Republic, were the responsible bodies at the central level. In addition, similar national bodies operated in the sister republics. The

State Planning Committee developed master plans for the entire Soviet Union and provided the major territorial planning guidance for national republics. This agency also worked out short to long-term sector development and industry distribution plans for the entire Soviet Union. Based on these plans, similar national bodies developed national branch development plans.

In essence, the Soviet planning system was ineffective. There was no co-ordination between industrial and land use planning, local conditions were ignored, and many plans were infeasible. Master plans for urban development were based on uniform approaches and characterized by under-valuation of land, lowland development at the expense of agriculture lands and green zones, intensive industrialization, monotonous housing projects etc.

Following their independence, all of the FSU countries, including Armenia, began developing national legal-institutional capacities. In the land resources management field, new land codes, providing land classification according to planned uses, and rules and procedures for land ownership, etc. were adopted. Environmental media-specific statutes on soil protection were also passed in some of these countries.

At present, land resources management and protection responsibilities are widely spread among different agencies. In Armenia the Ministry of Nature Protection is responsible for land resources protection. The Ministry of Agriculture is responsible for planning and management of agricultural land resources. At the same time, the State Committee of the Real Property Cadastre under the Government of Armenia is responsible for the planning and management of all lands other than agricultural. These three agencies are all responsible for some aspect of land resources planning and management. They develop regulations and general policies for land resources planning and management. The Ministry for Environment conducts monitoring of land use and is responsible for inventory of lands affected by geo-dynamic processes.

Overall, all agencies in the land planning and management field experience similar financial and institutional difficulties, as do others involved in environmental and natural resources management. Current legislation is imperfect, especially in the field of land ownership, spatial planning and zoning, etc. Town planning practices are still based on Soviet approaches and do not reflect modern urban concepts or the special nature of transitional economies. Whereas various state plans, programs and projects pertaining to land resources management do exist, financial and implementation mechanisms are lacking or absent.

Armenia NEAP identify priority issues pertaining to all environmental fields, including land resources, and suggest legal-institutional and investment measures for solving these issues. Some of these activities are currently being implemented, but according to Appendix 2 almost no measures for the protection and rehabilitation of soils have not been implemented in recent years, due to the blockade, as well as the economic and energy crises.

## 4. Suggestions

There is a wide variety of measures projected at the "National Action Plan on Desertification in Armenia" aimed to improve land use, to implement a beneficial urban planning, industrial, transport, and energy technologies fighting desertification, as well as directed to an efficient use of the Earth entrails and water resources, however, these measures contain the following shortcomings:

- There are no clearly defined financial mechanisms and responsible authorities for implementation of the above-defined measures.
- Suggested measures for improving mechanisms of environmental economics are of a general character and do not include leverages for economic stimulation or sanctions which would influence the efficient utilization of natural resources and environmental protection.
- There are no proper justifications and recommendations for developing those directions of forestry that are crucial for preventing desertification.
- According to the forecast, 4–6% should be allocated from the Republic's budget for these purposes (in 2000 this index was 2%) and 2% of GDP which would correspond to the developed countries' indexes.

## References

Armenia's Statistical Annual. Yerevan, 2002, p. 310; "Armenia: Economic Tendencies," Yerevan, 2001, p. 163

Constitution of the Republic of Armenia, Yerevan, "Mkhitar Gosh Publishing House," 1997, p. 11

Current Economic Problems of the RA Agriculture Development, Yerevan, 2003, pp. 51–52

Environmental Performance Review of Armenia, 2000 (EPRA)

NEAP, WG6, 1997. National Environmental Action Plan. Land Resources. Use and Protection Management

State of the Environment Report of Armenia, 1998

Strategy of a Sustainable Development of the RA Agriculture, Yerevan, 2002 pp. 39–40

APPENDIX 1. Land requiring improvement or restoration

| Type of land according to purpose | Total surface area (ha) | Land, by type of degradation (ha) | | | | | | | | |
| --- | --- | --- | --- | --- | --- | --- | --- | --- | --- | --- |
| | | Exposed to erosion | | Salinated | Secondary salinization | Degraded | Over humid | Rocky and polluted by wastes | Water-logged | Desertification |
| | | Wind erosion | Water erosion | | | | | | | |
| 1. Agricultural and forest lands (total) | 1,762,438.7 | 4,275 | 9,170 | 864 | 700 | 1941 | 1163 | 3,3742 | 8080 | 3,498 |
| 1.1. Arable lands | 464,261.6 | 1,765 | 2,816 | 790 | 700 | 119 | 528 | 3,477 | 8060 | 1,395 |
| 1.2. Perennial plantings | 42,896.0 | – | – | – | – | – | – | – | – | 450 |
| 1.3. Meadows | 136,892.4 | 2 | 1,572 | 74 | – | 1356 | 620 | 5,540 | 20 | 153 |
| 1.4. Pastures | 633,532.7 | 2,412 | 4,714 | – | – | 466 | – | 2,4660 | – | 1,500 |
| 1.5. Other | 484,856.0 | 96 | 68 | – | – | – | 15 | 65 | – | – |
| 2. Irrigated agricultural lands | 179,209.0 | 440 | 186 | 270 | – | – | 80 | – | 2 | – |
| 3. Specially protected areas | 233,324.0 | 82 | 22 | – | – | – | – | 210 | – | – |
| 3.1. Nature protection lands | 226,518.0 | 82 | 22 | – | – | – | – | 210 | – | – |
| 3.2. Recreational lands | 910.0 | – | – | – | – | – | – | – | – | – |
| 3.3 Historic-cultural lands | 1,912.0 | – | – | – | – | – | – | – | – | – |
| 3.4. Other | 3,984.0 | – | – | – | – | – | – | – | – | – |
| 4. Forest land | 371,326.0 | 116 | 68 | – | – | – | 15 | 65 | – | – |
| 5. Water bodies | 149,114.0 | – | 5 | – | – | – | – | – | – | – |
| 6. Reserve lands | 963,343.0 | 155 | 12 | – | – | 331 | – | 1,350 | – | 12 |

APPENDIX 2.  Soil protection and recovery mneasures (Source: National Statistical Service of RA)

|  | Measure | Total for 2002 |
|---|---|---|
| Agro-technical and hydro-technical measures, ha | Irrigation | 42,514 |
|  | Drainage | – |
|  | Measures for mud-flow and erosion mitigation | 722 |
|  | Cleaning of bushes, stones and wastes | 1,347 |
|  | Restoration of vegetation | 163 |
|  | Restoration of fertile layer | – |
|  | Soil desalination | – |
| Chemical and biological measures, kg/ha | Application of fertilizers | 9,154.3 |
|  | Application of chemical weed-killers and pest-killers | 3.6 |
| Expenditures for the protection and recovery of soils, thousand drams (1,000 dram in 2002 was 2 USD) | Hydro-technical constructions to prevent erosion | 75,000 |
|  | Construction to prevent mud-flows | 8,225 |
|  | Construction of terraces | 25,300 |
|  | Establishment of field-protective forest belts | 200 |
|  | Restoring vegetation | 6,500 |

# SORPTION-DESORPTION TESTS TO CHARACTERIZE SOILS CONTAMINATED BY HEAVY METALS: IMPLICATIONS FOR RISK ASSESSMENT

MIQUEL VIDAL[*] AND ANNA RIGOL
*Departament de Química Analítica*
*Universitat de Barcelona*
*1–11 Martí i Franquès, 3ª Planta, 08028 Barcelona, Spain*

**Abstract.** Sorption-desorption tests at laboratory scale are required to assess the risk derived from the incorporation of heavy metals in soils after a continuous or an accidental release. However, the quantification of derived parameters that can serve as input data for models and related Environmental Decision Support Systems (i.e., the solid-liquid distribution coefficient, $K_d$, and the reversibly sorbed fraction) are operationally dependent and they strongly vary because of a number of factors. This fact may provoke under or overestimations of the assessed risk in various orders of magnitude. In this work examples are shown to illustrate several factors that must be considered and controlled for a better and more reliable risk assessment.

**Keywords:** heavy metal, soil, sorption isotherm, sorption reversibility, extraction tests, risk assessment

## 1. Introduction

Various are the steps to be considered to assess the risk derived from a metal incorporation in soils. It is required the qualitative and quantitative description of the source term (e.g., to ascertain whether the source is particulate or soluble; whether it is formed by a single metal or it is multimetallic), the knowledge of soil properties, as well as the establishment of metal concentration and distribution in the affected area, through obtaining concentration maps at various field

---

[*] To whom correspondence should be addressed: miquel@nb.edu

L. Simeonov and V. Sargsyan (eds.),
*Soil Chemical Pollution, Risk Assessment, Remediation and Security.*
© Springer Science+Business Media B.V. 2008

scales, even aiming at the identification of hot spots. However, these preliminary studies are insufficient for a proper risk assessment, since the decision making about soil use and the evaluation of the impact of the heavy metal contamination in soils require to gather information also about metal interaction in soils and subsequent mobility of metals, which is directly related to their partitioning between the soil solid phase and soil solution (Evans, 1989). The knowledge on metal soil interaction may also be useful to forecast how changes in environmental conditions (e.g., changes in soil pH) may affect metal mobility. Risk assessment also would benefit from information on mobility changes with time, which in turn depends on the interaction dynamics.

Metal mobility can be examined and predicted by experiments carried out at field and/or at laboratory scale. While field experiments (e.g., soil-plant transfer or sampling soil profiles to examine metal vertical migration) permit obtaining in situ information at real time, they are usually high-cost, long-time studies, with results difficult to be extrapolated to other scenarios due to the absence of elucidated mechanisms controlling metal mobility. Instead, laboratory experiments represent lower-cost tests than field experiments or continuous soil monitoring campaigns, and they permit to acquire the knowledge of the interaction mechanisms. Data from laboratory tests are easily used as input data for prediction models in which the Environmental Decision Support Systems are based, in order to establish and characterize the soil quality and to perform the risk assessment exercise. However, laboratory tests are operational and data from the same soil-metal combination may vary in various orders of magnitude due to a number of factors. To date there is a significant lack of harmonized and standardized protocols, although recommended methods are available by several organizations (OECD, 2000; ASTM, 2001).

## 2.   Incorporation of Metals to Soils: Sorption and Sorption Reversibility

The chemical form and speciation of metals strongly affect their movement through environmental media and uptake by biota. Specifically, the way that a metal is bound to solids eventually controls the amount of metal in solution, which directly influences the fraction of metal that may be incorporated by organisms. As seen in Figure 1, dissolved metal ions in the soil solution can bind to solid surfaces by a number of processes often classified under the broad term of sorption. Models for the description of metal sorption are still mostly based on empirical solid-liquid distribution coefficient ($K_d$) values. This approach is the simplest sorption model available and is the ratio of the concentration of metal sorbed on a specified solid to the metal concentration in a specified liquid phase at equilibrium ($K_d$, L/kg).

While for a limited number of cases the $K_d$ can be successfully predicted from soil properties and as the weighted result of sorption on homogeneous surfaces, most common approaches derive $K_d$ values from field contaminated soils (Goody et al., 1995), and from sorption and mass transport experiments at a laboratory scale with initially non-contaminated soils (Sastre et al., 2007). Among the laboratory studies, the most common approach is to undertake sorption experiments using batch methods conducted at variable initial metal concentrations, thus leading to the construction of sorption isotherms (Jenne, 1998). Regarding batch experiments, many factors affect the quantification of the $K_d$, since it is not only extremely dependent on the soil-metal combination, but also on the operational conditions applied in the sorption tests, such as contact (shaking) time, volume/mass ratios, and filtration of the resulting solution, especially if the target metals exhibit an association with colloids. Moreover, there is a crucial effect of the composition of the contact solution, specifically the pH and the concentration of the major elements. Finally, the way as the source term is simulated in the batch experiments will also determine how representative the $K_d$ values are. Besides the initial metal concentration, the $K_d$ may be strongly dependent on whether monometallic or multimetallic scenarios are considered.

The simple $K_d$- based model relies on the hypothesis that the metal on the solid phase is in equilibrium with the metal in solution, and thus can exchange with it. However, the elapsed time since the incorporation of the metal is known to affect the quantification of $K_d$, since a fraction of the incorporated metal may become fixed by the solid phase (an aging effect related to sorption dynamics).

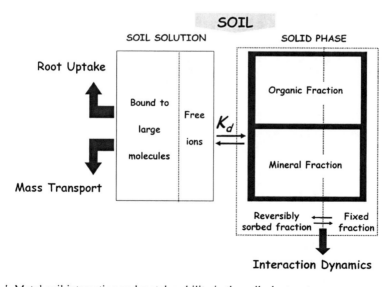

*Figure 1.* Metal-soil interaction and metal mobility in the soil-plant system.

As shown in Figure 1, as the metal speciation in the solid phase may change with time, an estimation of the changes in the reversibility of the sorption in the short and medium term is also required, through any experimental approach designed to estimate the reversibly sorbed fraction ($f_{rev}$), and how it varies with time. This permits to correct the $K_d$ derived from the sorption experiments and to define the adjusted $K_d$:

$$K_{d, \text{adjusted}} = K_d / f_{rev} \qquad (1)$$

The $K_{d,adjusted}$ is the $K_d$ value recommended for risk assessment exercises, since it includes the correction by the degree of sorption reversibility.

The reversibly sorbed fraction depends on the previous sorption step (e.g, sorption sites; sorption conditions, such as ionic status of the sorption solution and application of drying-wetting cycles after the sorption step; time elapsed after metal sorption), and it can rarely be predicted from soil and sorption data. Various are the laboratory tests to estimate the $f_{rev}$, such as the construction of desorption isotherms under the same conditions as in the sorption step (Comans et al., 1991) or the application of extraction tests to quantify the desorption yields (Kennedy et al., 1997). While the first approach throws some light on the mechanisms involved in sorption reversibility, the second represents a simpler, faster test to estimate the fraction of the metal that may be remobilized when the environmental conditions change (Ure, 1996). However, the mechanisms involved in metal sorption may also be elucidated through leaching tests at various pH (Dijkstra et al., 2004).

In this work, a number of examples are shown to illustrate the degree of dependency of the data derived from the sorption-desorption tests on the experimental design of the tests, and how it can affect the assessment of the risk derived from a contamination event in soils. Examples focus in purely experimental factors (e.g., contact time in sorption tests), and in factors related to the source term, such as the initial metal concentration and the presence of a competitive metal affecting the sorption of a target heavy metal.

## 3. Examples of Factors Affecting the Data Derived from Sorption-Desorption Tests

### 3.1. THE EFFECT OF THE CONTACT TIME IN SORPTION EXPERIMENTS

Although the contact time in the sorption step is an operational parameter whose importance is often underappreciated, it may have a significant influence on the estimation of the $K_d$ values. Figure 2 shows some examples of the effect of the contact time on the quantification of the $K_d$, by examining the changes in the $K_d$ values of Pb and Cu in two soils (clay and organic) at various contact times (5, 16, 24 and 48 h). The rest of the experimental conditions (1.5 g of soil sample; 30 ml of 2 meq/L metal solutions) were kept constant.

The pattern observed depended on the metal-soil combination. A $K_d$ stable over time was observed for Pb in the clay soil and for Zn in the organic soil. In all these cases, the $K_d$ reached a quasi-plateau value within 5 hours of contact time. On the contrary, for Pb in the organic soil and Zn in the clay soil, the $K_d$ showed an initial increase followed by a quasi-plateau for contact times between 16 and 24 hours. Additionally, in the case of Zn in the clay soil, a further increase was noticed when increasing the contact time from 24 to 48 hours.

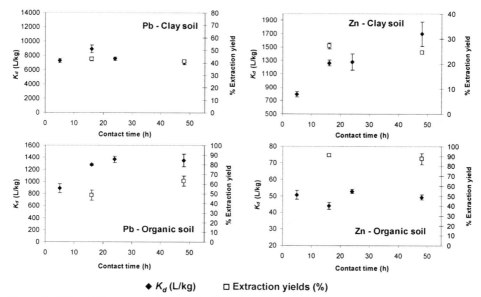

*Figure 2.* Effect of the contact time on the $K_d$ and extraction yields.

Experiments on the metal extractability in the resulting soil samples were also carried out to determine whether the increases in the $K_d$ values were associated with decreases in the metal sorption reversibility. The desorption test was conducted by adding 40 ml of 2 eq/L $CaCl_2$ to the solid residues originating from the sorption test, and shaking the suspension for 16 h at room temperature (Rigol et al., 1998). As shown in the Figure 2, the metal extractability was not influenced by the sorption contact time. Therefore, it can be difficult to set up a common contact time to examine metal sorption in soils of contrasting characteristics. Since the risk is basically overestimated by accepting lower $K_{d, adjusted}$ values than those obtained at longer contact times, the effect is (economically) significant if the decision making changes due to a risk overestimation (for instance, when deciding an intervention action instead of simply accepting to monitor the contaminated area).

## 3.2. THE EFFECT OF THE INITIAL METAL CONCENTRATION

The increase in the initial metal concentration usually provokes a decrease in the metal $K_d$ (Mesquita and Vieira e Silva, 1996). This is well illustrated by the Figure 3, which shows the sorption isotherm of Cd in a mineral soil, with a more than two-order of magnitude decrease in the $K_d$ values when increasing the initial Cd concentration. This fact is responsible for the limited validity of the linear models to describe the metal sorption, and it makes that the $K_d$ data are better fitted by non-linear models, such as Freundlich and Langmuir equations (Harter and Baker, 1997; Hooda and Alloway, 1998; Elzinga et al., 1999; Utermann et al., 2005).

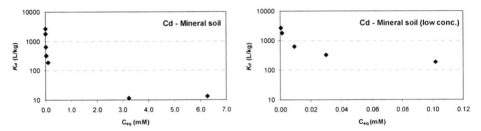

*Figure 3.* Effect of the initial metal concentration on the $K_d$.

Table 1 summarizes an example of a fitting exercise of the data of the Figure 3 to the Freundlich model. This model represents an ideal situation, which does not include the possible saturation of the sorption sites (Sparks, 1995). The linear form of the Freundlich isotherm equation is defined as follows:

$$\log C_{sorb} = \log K_F + (1/n_F) \log C_{eq} \qquad (2)$$

where $C_{sorb}$ is the sorbed metal concentration (meq/kg), $C_{eq}$ is the metal concentration in the final contact solution (meq/L), $1/n_F$ is the Freundlich sorption exponent, and $K_F$ is the Freundlich sorption coefficient. As seen in Table 1, the experimental results showed good agreement with the Freundlich mathematical model, with an r value of 0.98. The $n_F$ parameter, which relates to the selectivity of the sorption sites, was higher than 1, which is consistent with the lack of a linear dependence between $C_{sorb}$ and $C_{eq}$. The other fitting parameter, $K_F$, accounts for a weighted estimation of the $K_d$ value. The $K_d$ (min) and $K_d$ (max) corresponded to the scenarios with the high and low initial metal concentrations, respectively. If the $K_F$ value is compared with the range of the $K_d$ values obtained, it clearly approaches to the $K_d$ (min). This indicates that the $K_F$ value may have a limited use for risk assessment purposes, since it may lead to an overestimation of the metal mobility.

TABLE 1. Data from Cd sorption-desorption in a soil: Freundlich fitting and estimation of the $K_{d,adjusted}$

| $K_F$ | $n_F$ | r | \multicolumn{2}{c}{$K_d$ (L/kg)} | \multicolumn{2}{c}{Extraction yields (%)} | \multicolumn{2}{c}{$K_{d,adjusted}$ (L/kg)} |
|---|---|---|---|---|---|---|---|---|
|  |  |  | $K_d$(min) | $K_d$(max) | High conc. | Low conc. | High conc. | Low conc. |
| 53 | 2.4 | 0.98 | 12 | 2,720 | 90 | 95 | 13 | 2,860 |

In this specific example, the high sorption reversibility of the Cd in this soil, regardless the initial Cd concentration, makes that the $K_{d,adjusted}$ depends only on the $K_d$ value. It is thus obvious that due to the extreme dependency of the $K_d$ to the initial metal concentration, the risk assessment would require a laboratory test that permits to obtain a $K_d$ based on a metal concentration representative for the source term and field situation, to decrease the variability range of the assessment, as well as over and underpredictions of the impact derived from the metal incorporation in the soil.

## 3.3. THE EFFECT OF THE COMPOSITION OF THE CONTACT SOLUTION

Sorption isotherms are usually obtained under the same background electrolyte conditions in order to permit comparing results of the soil samples (Sparks, 1995). This approximation assumes that major cations from the sorption medium (Ca, Mg, Na and K) act only as a background electrolyte and their effects are disregarded in most cases (Jenne, 1998). However, the competitive effects for metal sorption of cations and protons originating from the sorption medium (Pardo, 1997; Voegelin et al., 2001) and desorbed from soil solid phases during the sorption process (Harter and Baker, 1997; Ponizovsky and Tsadillas, 2003) have been widely described.

Figure 4 shows the effect of the changes in the concentration of major cations (i.e., Ca, Mg, K and Na) in the contact solution on the $K_d$ determination by comparing the sorption isotherms of Cd in two soils (organic and clay saline soils), and in two ionic media (in 0.02 eq/L $CaCl_2$ and in a medium that simulates the cationic composition of the soil solution).

For the clay saline soil, the $K_d$ values obtained in the soil solution medium were from one to two orders of magnitude lower than those quantified in the 0.02 eq/L $CaCl_2$ solution, especially at the low initial metal concentration range. This was due to the higher ionic strength and higher concentrations of the major elements in the soil solution of this soil (around 75 meq/L in Ca; 180 meq/L in Mg; 470 meq/L in Na; and 6.5 meq/L in K) with respect to the $CaCl_2$ background solution. On the contrary, higher $K_d$ values were quantified in the organic soil in the soil solution medium, due to the lowest ionic strength of the solution simulating the cationic composition of the soil solution (1.3 meq/L in

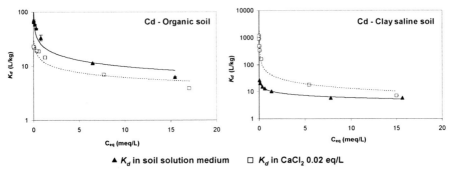

*Figure 4.* Effect of the composition of the contact solution on the $K_d$.

Ca; 10 meq/L in Mg; 6.5 meq/L in Na; 0.4 meq/L in K) with respect to the CaCl$_2$ solution, in which the higher Ca concentration reduced metal sorption by competition for the sorption sites.

A more detailed analysis of the data is shown in Table 2. Regarding the $K_d$ values, the effect of the contact solution was more significant at low concentrations, where the existence of specific sites were evidenced, leading to higher $K_d$ values, while at the high concentration range the effect of the composition of the contact solution was almost negligible, with low $K_d$ in all scenarios. The changes in the extraction yields due to the variations in the metal initial concentration and in the composition of the contact solution did not follow a clear pattern with respect to the initial metal concentration, although in general they increased in the soil solution medium when decreasing the initial metal concentration.

TABLE 2. Effect of the composition of the contact solution on the quantification of the sorption-desorption data of Cd in two soils

|  |  | $K_d$ (L/kg) | | Extraction yields (%) | | $K_{d,adjusted}$ (L/kg) | | Ratio of $K_{d,adjusted}$ CaCl$_2$ vs. soil solution | |
|---|---|---|---|---|---|---|---|---|---|
|  | $K_F$ | $K_d$ (min) | $K_d$ (max) | High conc. | Low conc. | High conc. | Low conc. | High conc. | Low conc. |
| *Clay saline* |  |  |  |  |  |  |  |  |  |
| Soil Solution | 11 | 6 | 27 | 49 | 60 | 12 | 45 | – | – |
| CaCl$_2$ | 48 | 7 | 1,150 | 76 | 51 | 9 | 2,255 | 0.8 | 50 |
| *Organic* |  |  |  |  |  |  |  |  |  |
| Soil Solution | 24 | 6 | 70 | 73 | 97 | 8 | 72 | – | – |
| CaCl$_2$ | 12 | 4 | 23 | 77 | 71 | 5 | 32 | 0.6 | 0.4 |

As the extraction yields were clearly lower than 100%, the use of the $K_{d,adjusted}$ was recommended. While for the scenarios with a high initial metal concentration (which are less representative of an environmental scenario) the effect of the contact solution in the risk assessment was minor, when dealing with scenarios with a low initial metal concentration the risk assessment depended on the choice of the contact solution in the sorption experiments.

When using data obtained in a common background solution like $CaCl_2$ the risk can be underestimated (more than 50 times in the clay saline soil in this specific example) or overestimated (around 2.5 in the organic soil), depending on the soil considered. It is thus suggested that sorption tests should be done that simulate real field conditions as much as possible (Murali and Aylmore, 1983), as is reproducing the cationic composition or the ionic strength of the soil solution in the sorption medium (Sauvé et al., 2000).

## 3.4. DETERMINATION OF THE $K_d$ IN MONOMETALLIC AND IN MULTIMETALLIC SCENARIOS

The determination of the $K_d$ of a target metal also depends on the presence of other metals that may be sorption competitive. This is due to the fact that the selective sites in the solid phase of a soil, which lead to the highest $K_d$ for a given metal, are not sufficiently metal specific for a given element, but metals with a similar chemical behaviour may compete for sorption at the same sites. Therefore, laboratory experiments based on monometallic scenarios may be of limited application for risk evaluation (Echeverría et al., 1998).

Figure 5 illustrates how the sorption of Zn, as a target metal, is affected by the presence of a competitive metal (Pb), at two concentrations (0.2 and 10 meq/L), in two soils of contrasting properties. When Pb was present at a low concentration, there was a minor, although significant, effect of the presence of Pb on the Zn sorption, especially in the mineral soil. However, when the concentration of the competitive metal increased, the sorption of Zn decreased various orders of magnitude in both soils and, as seen in the organic soil, the sorption isotherm even changed its pattern, since the increase in the Zn concentration led to an unexpected increase in the $K_d$ (Zn), which was not observed in the monometallic solution. This fact relates to the ability of Zn to compete with Pb for the sorption at selective sites, and thus increasing the $K_d$ (Zn), only when its concentration increased.

As shown in Table 3, in the mineral soil the inclusion of the Pb in the Zn sorption experiments shifted the $K_d$ (Zn) to lower values for all Pb concentrations, with outstandingly low values when examining the scenario with the highest Pb concentration. In the organic soils, the effect of including the Pb was

TABLE 3. Effect of a competitive metal on the sorption-desorption of a target metal

| | $K_d$(L/kg) | | Extraction yields (%) | | $K_{d,\ adjusted}$ (L/kg) | | Ratio of $K_{d,adjusted}$ Zn vs. (Zn+Pb) | |
|---|---|---|---|---|---|---|---|---|
| | $K_d$ (min) | $K_d$ (max) | High conc. | Low conc. | High conc. | Low conc. | High conc. | Low conc. |
| *Mineral soil* | | | | | | | | |
| Zn | 84 | 41,890 | 40 | 9 | 210 | 465,400 | – | – |
| Zn + Pb (0.2) | 65 | 17,210 | 27 | 6 | 240 | 286,800 | 0.9 | 1.6 |
| Zn + Pb (10) | 3 | 39 | 58 | 1 | 5 | 3,900 | 42 | 120 |
| *Organic soil* | | | | | | | | |
| Zn | 21 | 64 | 77 | 98 | 27 | 65 | – | – |
| Zn + Pb (0.2) | 14 | 70 | 53 | 52 | 20 | 135 | 1.4 | 0.5 |
| Zn + Pb (10) | 2 | 3 | 100 | 100 | 3 | 2 | 9 | 32 |

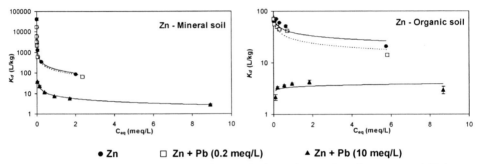

● Zn      □ Zn + Pb (0.2 meq/L)      ▲ Zn + Pb (10 meq/L)

*Figure 5.* Effect of the presence of a competitive metal (Pb) on the sorption of a target metal (Zn).

apparently less important, since the $K_d$ (Zn) were much lower than for the mineral soil. However, and as seen in the Figure 5, the Zn sorption pattern changed, probably due to the higher affinity of Pb to the organic sites (Dijkstra et al., 2004), and the $K_d$ (Zn) practically did not vary with the initial Zn concentration (in fact, for this case, the $K_d$ (min) and $K_d$ (max) obtained in the experiment with the highest Pb concentration did not correspond to the high and low Zn concentration scenarios, but the opposite was observed). As seen in the section above, the low and erratic values of the reversibly sorbed fraction, quantified by the extraction yields, justified the calculation and use of the $K_{d,adjusted}$. The examination of these values highlighted again the significant effect of the way the source term is simulated in the laboratory sorption-desorption tests on the risk assessment, since the decreases in the $K_{d,adjusted}$ when comparing mono with multimetallic scenarios were as high as two orders of magnitude, as seen here in the mineral soil with the highest Pb concentration. Therefore, data from laboratory tests with monometallic solution may underestimate the risk derived from a target metal when incorporates in the soil from a multimetallic source term.

## Conclusions and Recommendations

Sorption-desorption tests increase the understanding of the mechanisms of the metal-soil interaction, and they are suitable tools to predict metal mobility after their incorporation in soil. They must be applied not only when assessing the risk derived from a contamination event, but also to back up the design of remediation strategies (e.g., use of soil amendments to decrease metal mobility in contaminated soils). Regarding the data derived from the application of the sorption-desorption tests, the use of the $K_{d,adjusted}$ (the $K_d$ divided by the reversibly sorbed fraction) is suggested as the most appropriate input data for the risk assessment models.

The characteristics of the source term (e.g., concentration of the target metal; presence and concentration of potential competitive metals) and of the field scenarios (e.g., soil characteristics; cationic composition of the soil solution) must be considered in the design of the sorption-desorption tests.

Sorption-desorption tests must be harmonized and standardized, since to date there is still a lack of validated protocols and (certified) reference materials, especially dealing with sorption tests.

Finally, the combination of field and laboratory data is required for an optimum risk assessment, especially because the validity of the laboratory tests must be checked with field experiments.

## Acknowledgements

Part of the research presented in this work was funded by the Spanish Government (CICYT, contract CTM2005-03847, and by the Secretaría General para la Prevención de la Contaminación y el Cambio Climático del Ministerio de Medio Ambiente, contracts 1.2-193/2005/3-B and 240/2006/2-1.2).

## References

ASTM D4319-93. 2001. Standard test method for distribution ratios by the short-term batch method.

Comans RNJ, Haller M, De Preter P. 1991. Sorption of cesium on illite: non-equilibrium behaviour and reversibility. Geochim Cosmochim Acta 55: 433–440.

Dijkstra JJ, Meeussen JCL, Comans RNJ. 2004. Leaching of heavy metals from contaminated soils: an experimental and modeling study. Environ Sci Technol 38: 4390–4395.

Echeverría JC, Morera, MT, Mazkiarán C, Garrido JJ. 1998. Competitive sorption of heavy metal by soils. Isotherms and fractional factorial experiments. Environ Pollut 101: 275–284.

Elzinga EJ, Van Grisven JJM, Swartjes FA. 1999. General purpose Freundlich isotherms for cadmium, copper and zinc in soils. Eur J Soil Sci 50: 139–149.

Evans LJ. 1989. Chemistry of metal retention by soils. Environ Sci Technol 23: 1046–1056.

Goody DC, Shand P, Kinniburgh DG, Van Riemsdijk WH. 1995. Field-based partition coefficients for trace elements in soil solutions. Eur J Soil Sci 46: 265–285.

Harter RD, Baker DE. 1997. Applications and misapplications of the Langmuir equation to soil adsorption phenomena. Soil Sci Soc Am J 41: 1077–1080.

Hooda PS, Alloway BJ. 1998. Cadmium and lead sorption behaviour of selected English and Indian soils. Geoderma 84: 121–134.

Jenne EA. 1998. Adsorption of metals by Geomedia: data analysis, modelling, controlling factors and related issues. In: Jenne EA editor. Adsorption of metals by Geomedia. Variables, mechanisms and model applications. Academic, San Diego, CA; p. 1–73.

Kennedy VH, Sanchez AL, Oughton DH, Rowland AP. 1997. Use of single and sequential chemical extractants to assess radionuclide and heavy metal availability from soils for root uptake. Analyst 122: 89R-100R.

Mesquita ME, Vieira e Silva JM. 1996. Zinc adsorption by calcareous soil. Copper interaction. Geoderma 69: 137–146.

Murali V, Aylmore LAG. 1983. Competitive adsorption during solute transport in soils: 3. A review of experimental evidence of competitive adsorption and an evaluation of simple competition models. Soil Sci 136: 279–290.

OECD, 2000. OECD Guideline 106. Guideline for the testing of chemicals: Adsorption-desorption using a batch equilibrium method.

Pardo MT. 1997. Influence of electrolyte on cadmium interaction with selected Andisols and Alfisols. Soil Sci 162: 733–740.

Ponizovsky AA, Tsadilas CD. 2003. Lead (II) retention by Alfisol and clinoptilolite: cation balance and pH effect. Geoderma 115: 303–312.

Rigol A, Vidal M, Rauret G, Shand CA, Cheshire MV. 1998. Competition of organic and mineral phases in radiocesium partitioning in organic soils of Scotland and the area near Chernobyl. Environ Sci Technol 32: 663–669.

Sauvé S, Hendershot W, Allen HE. 2000. Solid-solution partitioning of metals in contaminated soils: dependence on pH, total metal burden and organic matter. Environ Sci Technol 34: 1125–1131.

Sastre J, Rauret G, Vidal M. 2007. Sorption-desorption tests to assess the risk derived from metal contamination in mineral and organic soils. Environ Int 33: 246–256.

Sparks DL. 1995. Environmental Soil Chemistry, Academic, San Diego, CA.

Ure AM. 1996. Single extraction schemes for soil analysis and related applications. Sci Total Environ 178: 3–10.

Utermann J, Heidkamp A, Meyenburg G, Gäbler HE, Altfelder S, Böttcher J. 2005. Pedotransfer functions for sorption of trace elements in agricultural and forest soils. In: Uhlmann O, Annokkée J, Arendt F editors. Con Soil pp. 258–265.

Voegelin A, Vulava VM, Kretzschmar R. 2001. Reaction-based model describing competitive sorption and transport of Cd, Zn and Ni in an acidic soil. Environ Sci Technol 35: 1651–1657.

# SOIL-PLANT TRANSFER OF ORGANIC CHEMICALS AND DERIVATION OF TRIGGER VALUES

KONSTANTIN TERYTZE[1,2*], WERNER KÖRDEL[3], MONIKA HERRCHEN[3], INES VOGEL[1] AND ANGELIKA NESTLER[2]
[1]*Federal Environment Agency of Germany,
Wörlitzer Platz 1, Dessau 06844, Germany*
[2]*Free University of Berlin,
Malteserstr. 74-100, Berlin 12249, Germany*
[3]*Fraunhofer Institute for Molecular Biology and Applied Ecology
Auf dem Aberg 1, Schmallenberg 57392, Germany*

**Abstract.** The Federal Soil Protection Act [Anonymus, 1998] was technically realized by the Federal Soil Protection and Contaminated Sites Ordinance (BBodSchV) of July 12th, 1999. Precautionary values and trigger values as laid down in the Law[1] are an important instrument to realise its requirements. Trigger values are related to various soil uses and objectives of protection, which are "human health", "quality of food and feed", and "leachate to groundwater". Any procedure to derive trigger values for the objective "quality of food and feed" has to consider the soil-plant transfer of chemicals for soils under agricultural use and household gardens. Thus, the production function of soil is taken into account.

The exact procedure to obtain trigger values for "quality of food and feed" is explained.

By this means, trigger values for "quality of food and feed" already were derived for some metal compounds, namely arsenic, lead, mercury and thallium. Beside metal compounds several organic substances are also of priority; for example, a trigger value for benzo(a)pyrene and an action value for PCBs has to be defined.

---

§ 8 Federal Soil Protection Act:

Precautionary values: "soil values which, if exceeded, shall normally mean there is reason that concern for a harmful soil change exists, taking geogenic or wide-spread, settlement-related pollutant concentrations into account."

Trigger values: "values which, if exceeded, shall mean that investigation with respect to the individual case in question is required, taking the relevant soil use into account, to determine whether a harmful soil change or site contamination exists."

---

*To whom correspondence should be addressed: konstantin.terytze@uba.de

L. Simeonov and V. Sargsyan (eds.),
*Soil Chemical Pollution, Risk Assessment, Remediation and Security.*
© Springer Science+Business Media B.V. 2008

The pathway soil-plant is a sensible exposure route due to potential accumulation in the food chain.

A summary on the methodology and results of trigger value derivation for some selected organic chemicals previously used as pesticides and of different data sizes with respect to the soil-plant transfer process is given.

**Keywords:** soil-plant transfer of organic substances, precaution levels, trigger levels, action levels

## 1. Introduction

The Federal Soil Protection Act (Anonymous, 1998) was put into practice by the Federal Soil Protection and Contaminated Sites Ordinance (BBodSchV) of July 12, 1999 (Anonymous, 1999a). Precautionary values and trigger values as laid down in the Law[1] are an important instrument to realise its requirements.

**"Precaution levels"** are indicating a certain chance of future oil problems which need to be addressed in order to avert upcoming damages.
**"Trigger values"** are triggering further investigations to ascertain (verify/ falsify) whether a contamination implies a danger. Furthermore there are **"Action levels"** generally indicating a danger which has to be warded off; further investigations to ascertain the danger are usually not necessary.

Trigger values are related to various soil uses and objectives of protection, which are "human health", "quality of food and feed", and "leachate to groundwater". Any procedure to derive trigger values for the objective "quality of food and feed" has to consider the soil-plant transfer of chemicals for soils under agricultural use and household gardens. Thus, the production function of soil is taken into account.

The exact procedure to obtain trigger values for "quality of food and feed" is laid down in the Federal Bulletin No. 161a (Anonymous, 1999b) and has to be followed when deriving justifiable values:

-   Determination of the highest permissible pollutant concentrations in plants (preparation of the plant-oriented assessment standard)
-   Description of substance transfer from the soil into plants, followed by calculatory derivation of the highest soil concentration that will still ensure compliance with the highest permissible plant concentration
-   Checking of the calculated soil values for plausibility, including estimation of the toxicological load from vegetables growing in gardens polluted with organic or mineral substances
-   Definition of trigger values

By this means, trigger values for "quality of food and feed" already were derived for some metal compounds, namely arsenic, lead, cadmium, mercury and thallium. Beside metal compounds several organic substances are also of priority. So far, a trigger value was defined for benzo(a)pyrene.

Furthermore soil trigger values for the following substances are necessary: HCB, DDT, HCH. They are considered to be added to the Federal Soil Protection and Contaminated Sites Ordinance at the first amendment of the ordinance taking place this time.

It is within the scope of the presentation to give a summary on the methodology and results of trigger value derivation for some selected organic chemicals previously used as pesticides and of different data sizes with respect to the soil-plant transfer process.

## 2. Database

Various research projects were conducted to derive trigger values for priority organic pollutants in the soil-plant pathway.

Due to a lack of data resources for the deriving of trigger values, a study framework first had to be developed to gather empirical data. The result is TRANSFER, a database using data from joint German Federal States research programmes and other data holdings at the Federal Environment Agency. The database allows soil concentrations of both inorganic and organic pollutants to be matched up with plant concentrations on the same site, and forms part of BIS, the national soil information system.

The Transfer database combines and integrates stocks of data that were managed separately in national and Federal States repositories in former times. Efficient data exchange at both Federal States and federal level became possible by new data compatibility standards. This aids the uniform enforcement of soil protection law. Data resources had also been set up to update and amend substance classifications and the precautionary, trigger and action values laid down in the Federal Soil Protection and Contaminated Sites Ordinance.

## 3. Methodology

### 3.1. OBJECTIVE OF PROTECTION: QUALITY OF FOOD

**1 Step: Consideration of maximum residue levels in/on plants**

Trigger values are calculated by inclusion of maximum residue levels (MRL-levels, "Rückstandshöchstmengenverordnung" of 21.10.1999, modified on 20.11.2000) and ADI-values, respectively. In case these official values have not been derived for the compound under consideration a preliminary MRL´-value is assessed using N(L)OAEL-values and additionally applying a safety factor (SF$_{\text{toxicological reference}}$). In case none of the toxicologically relevant data is available the soil trigger value cannot be derived.

## 2 Step: Quantitative description of soil-plant transfer and derivation of a maximum acceptable soil content

The soil-plant transfer coefficient is defined as the quotient of the substance content in the respective plant compartment (given in dry weight) and the soil content (also given in dry weight):

$$f_{\text{transfer (i)}} = \frac{C_{\text{plant (i)}} \,[\text{mg/kg dm}]}{C_{\text{soil}} \,[\text{mg/kg dm}]}$$

Since the soil-plant transfer depends on both, soil and plant properties, ideally each food item and all representative soils should be tested. However, such a broad variety of experimental studies is not achievable and thus, the following assumptions and definitions were applied:

-   The ideal data set is characterised by five soils and ten representative food items. As long as a heterogeneous data base is available with information, which is difficult to interpret, the five soils/ten food items data base is considered to be the optimum. In case systematic studies are published, a data set characterised by three soils and six food items is considered to be sufficient.
-   In case the optimal soil data set is not available a safety factor (SF$_{\text{soil}}$) is applied to the mean transfer coefficient for each food item.
-   The experimentally determined transfer coefficients for the individual food items each are combined and a mean is calculated.

The maximum tolerable soil content is, in a first step, calculated separately for each tested food item by using one of the alternative equations depending on the availability of MRL- and ADI-value, respectively.

In case MRL-values are available the equation is:

$$\text{max. tolerable soil content (i) [mg/kg dm]} = \frac{\text{HF x MRL (i) [mg/kg ww]}}{f_{\text{transfer (i)}} \,[1 - (\text{water content [\%] / 1oo})]}$$

with:

|   |   |   |
|---|---|---|
| i | = | vegetable food (i) |
| MRL | = | maximum residue level [mg/kg wet weight] |
| f$_{\text{transfer(i)}}$ | = | mean of transfer coefficients for food item (i) |
| HF | = | hazard factor according to approach "toxicological hazard assessment of chemicals" (Anonymous, 1999c) |

In case no MRL-value is available the equation is:

$$\text{max. tolerable soil content (i) [mg/kg dm ]} = \frac{\text{HF x MRL}' \text{ (i) [mg/kg ww]}}{f_{transfer\,(i)}\ [1-(\text{water content [\%] / 1oo})]}$$

where MRL′ is:

$$\text{MRL}'\text{ (i) [mg/kg]} = \frac{\text{ADI [mg/kg bw d] x 20 kg x portion (i) in food basket}}{\text{daily intake (i) [mg/kg]}}$$

with:

ADI   = acceptable daily intake
20 kg = reference for daily intake: girl, 4–6 years of age, sensitive sub-
        group, weight 20 kg

The finally proposed maximum tolerable soil content, which is the basis for a plausibility check and expert-judgment, is identical with the lowest value out of the ensemble of calculated maximum tolerable soil contents for the individual food items. In case not all of the ten food items have been tested, again safety factors ($SF_{food\,item}$) are applied.

The finally proposed maximum tolerable soil content is multiplied with a so called "hazard factor" (HF). Maximum residue levels are derived as precautionary values. Food with residues below these levels should not cause adverse health effects for all population groups including sensitive subgroups such as children. However, a maximum tolerable soil content – in the sense of a trigger value – does not reflect the precautionary principle but the avoidance of hazard to human health. Thus, a hazard factor is added. The use of such a "hazard factor" is in accordance with the derivation of soil trigger values for the objective of protection "human health, direct soil contact" as published in the Federal Bulletin No. 161a.

### 3 Step: Plausibility check

and

### 4 Step: Final stipulation of trigger values

The finally calculated maximum tolerable soil content has been multiplied with appropriate safety factors and the hazard factor is subjected to a plausibility check. The check comprises – among others – a comparison with background values and precautionary values in order to make the trigger values operable. Finally, the suggested triggers are stipulated in the course of a moderated round table discussion including expert-judgement.

## 3.2. OBJECTIVE OF PROTECTION: QUALITY OF FEED

For the derivation of maximum tolerable soil contents for the objective of protection "quality of feed" the procedure as laid down in the Federal Bulletin No. 161a is followed:

- Grassland and soil under agricultural use (for maize) are treated identically.
- The legal basis for maximum tolerable plant levels is Directive (29/99/EEC) as well as the German "Futtermittelverordnung".
- The Directive (29/99/EEC) gives maximum tolerable plant levels for Aldrin, DDT, HCB, HCHs and Dioxins without differentiating between feed items.
- For other than these compounds or compound groups the maximum tolerable soil content currently cannot be derived.

## 4. Calculation of the Maximum Tolerable Soil Content for Selected Compounds

### 4.1. HEXACHLOROCYCLOHEXANE

For hexachlorocyclohexane comprehensive and numerous data sets exist on soil-plant-transfer studies. However, these studies are heterogeneous with respect to intention and design which ranges from using soils from contaminated sites to spiked soils and from outdoor field studies to small scaled laboratory studies. Consequently, the studies yield contradictory results for the degree of soil-plant transfer, the contribution by different exposure pathways, target organs and differentiation between HCH-isomers.

It was concluded that there is principle need to derive trigger values for soil-plant uptake. The acquisition and evaluation of further homogeneous data sets in so far unpublished or "grey" literature is recommended. Additionally, data selection criteria should be developed and applied to the entire data set. In particular the latter needs the implementation of an ad-hoc working group for consensus finding.

### 4.2. HEXACHLOROBENZENE

Homogeneous results on small-scaled laboratory studies are available. Though data from field studies is preferred over laboratory studies maximum tolerable soil contents are derived exemplarily. The soil-plant transfer factors are given in Table 1.

TABLE 1. Soil-plant transfer factors for HCB derived from laboratory studies

| Food | Study design | Quality of study | Transfer factor[1] |
|------|-------------|------------------|-------------------|
| Oat, roots | desiccator, 1 soil | Low | 36 |
| Cress | desiccator, 1 soil | Low | 105 |
| Corn, roots | desiccator, 1 soil | Low | 53 |
| Rape, roots | desiccator, 1 soil | Low | 152 |
| Carrots | desiccator, 1 soil | Low | **2,250** |
| Salad, roots | desiccator, 1 soil | Low | 1230 |

[1] Safety factor ($SF_{soil}$) of 5 included because of information from one soil only available

In order to follow a worst-case approach the highest transfer of 2,250 is used to calculate the maximum tolerable soil content:

Maximum tolerable soil content (carrots) = 5.5 x 0.05/2250 [1 − 88.2/100] = 1.1 µg/kg

MRL-value          = 0.05 [mg/kg dm]
hazard factor (HF)  = 5.5
water content       = 88.2%

A comparison of the calculated maximum tolerable soil content with the stipulated trigger value for the path soil-human being (playground) as well as with background values shows comparability with background values (Table 2).

TABLE 2. Plausibility check: comparison with trigger and background values

| | |
|---|---|
| Background value (grassland) | 0.2 − 12 µg/kg |
| Background value (agricultural soil) | 0.5 − 13 µg/kg |
| Trigger value (path soil − human being, playground) | 4 mg/kg |
| Calculated maximum tolerable soil content (soil − plant uptake) | 1.1 µg/kg |

It was concluded that there is principle need to derive a trigger value for the path soil–plant uptake. Calculated maximum tolerable contents derived from laboratory studies are in the range of background values. Such a result can be explained by a principally observable systematic overestimation of the chemicals transfer in laboratory studies using desiccators. Thus, for the derivation of justiciable and operable trigger values results from field and lysimeter studies, respectively, are needed.

## 4.3. DDT

Because of the long-time use of the pesticide DDT comprehensive studies, in particular field studies, are available. Again, for the derivation of the maximum tolerable soil content safety factors are used depending on the number of soils and food items tested. Table 3 presents input information, measured transfer factors and calculated maximum tolerable soil contents.

TABLE 3. Soil-plant transfer factors and maximum tolerable soil contents for DDT derived from field studies

| Food | Study design | Quality of study | Transfer factor | Maximum tolerable soil content [mg/kg][4, 5] |
|------|-------------|------------------|-----------------|-------------------------------|
| Carrots | field, 5 soils[2] | Relatively high | 0.6 1 | 2.3 |
| Beets | field, 1 soil[3] | Relatively high | 0.05 | 6.7 |
| Potatoes | field, 1 soil | Relatively high | 0.2 | 0.7 |
| Radish | field, 1 soil | Relatively high | 0.35 | 1.3 |
| Rapeseed | field, > 5 soils | Very high | 0.3[1] | 0.7 |

[1] mean
[2] safety factor        = 1 (5 soils tested)
[3] safety factor $_{soil}$   = 5 (1 soil tested)
[4] hazard factor HF   = 10
[5] safety factor $_{food}$   = 3 (5 food items tested)

As for hexachlorobenzene, the calculated maximum tolerable soil content is compared with the stipulated trigger value for the path soil – human being (playground) as well as with several sets of monitoring data. The calculated tolerable contents for the path soil–plant are in the range of monitoring data but far below the trigger value for the path soil–human being (Table 4).

TABLE 4. Plausibility check: comparison with trigger values and monitoring data

| | |
|---|---|
| Monitoring data (industrial area) | 0.16–0.25 mg/kg |
| Monitoring data (agricultural soils) | 0.03–1.5 mg/kg |
| Trigger value (path soil – human being, playground) | 40 mg/kg |
| Calculated maximum tolerable soil content (soil – plant uptake) | 0.7 mg/kg |

It was concluded that there is principle need to derive a trigger value for the path soil – plant for DDT. As for hexachlorocyclohexane an ad-hoc task force will be implemented for the final value stipulation.

## 4.4. PENTACHLOROPHENOL

Due to the degradation potential of pentachlorophenol the transfer factors obtained on the basis of short-term laboratory studies are not comparable to long-term studies. At soil contents below toxicity to micro-organisms metabolism under aerobic conditions occurs. The transfer factor of PCP is lower than 0.01 in alkaline soils, and even lower in acid soils due to reduced bioavailability.

From these observations it was concluded that there is no need to derive trigger values.

**Discussion**

A comparison of soil trigger values derived for various objectives of protection (e.g. "human health – direct soil contact" versus "quality of food and feed" versus "groundwater protection") yields very low values for the soil-plant transfer path. From the discrepancies it can be concluded:

- Due to possible accumulation in the food chain the soil-plant transfer path is a sensible one and has to be considered carefully.
- The derivation of maximum tolerable soil contents using the suggested procedure and assumptions leads to trigger values which are lower as those obtained for the path "direct soil contact", and to values in the range of soil background contents.
- A uniform experimental design should be used as basis to obtain comparable transfer coefficients for all substances under discussion. This currently is not the case, and thus an overestimation of the soil-plant transfer might occur.
- Recent studies are heterogeneous in respect to design of the study (field testing – lysimeter – laboratory study...), mode of soil contamination, investigated and published soil data, extraction methods and analytical procedures and missing inclusion of wet and dry deposition.

**Conclusions**

The procedure to derive soil trigger values for the objective of protection "quality of food and feed" is conclusive and presents a pragmatic approach. The approach differs in respect to the procedure selected for metals and metal compounds. Though 10 representative food items are suggested for inclusion in the procedure, "special cases" in respect to a selective plant uptake (such as spinach) should be considered additionally. In case of an insufficient data base safety factors are applied. The suggested factors can be interpreted as "signals" for further discussions in ad-hoc working groups, and for optimisation or mutually agreed interpretation of the current data set. For organic chemicals the airborne impact, i.e. wet or dry deposition, additionally should be taken into account. It is a challenging task to integrate all exposure pathways and a jointly designed approach should be elaborated as soon as possible.

Several substances and groups of substances were identified for which need exists to derive soil trigger values for the objective of protection "quality of food and feed". These chemicals are: aldrin, PAHs, DDT, hexachlorobenzene, HCHs, and PCBs.

Furthermore the following activities in the field of soil protection are necessary:

- Harmonization and standardization efforts in soil protection in Central and Eastern Europe
- Assessment of soil quality, harmonization of strategies for developing of soil values for priority pollutants
- Generation of data on the current status of soil contamination as a function of the magnitude of contamination, the type of land use and the parent material involved in soil formation
- Improving knowledge of the contents and dynamics of substances in soils from background territories of different geographical regions
- Tracing trends in soil contamination by environmental chemicals and monitoring the effectiveness of initiated measurements
- Equitation of knowledge of the effect and behaviour of substances under different conditions (type of ecosystem, input patterns, pedological, geographical and climatic situation)

## References

Anonymous (1998): Federal Soil Protection Act of March 17th 1998 (Federal Law Gazette I, p. 502)

Anonymous (1999a): BBodSchV – Federal Soil Protection and Contaminated Sites Ordinance dated July, 12th, 1999 (BGBl. I 1999 S. 1554)

Anonymous (1999b): Promulgation of Methods and Standards for Derivation of Trigger Values and Action Values pursuant to the Federal Ordinance on Soil Protection and Contaminated Sites (Bundes-Bodenschutz- und Altlastenverordnung (BBodSchV)), Federal Bulletin No. 161a, August 28

Anonymous (1999c): Gefährdungsabschätzung von Umweltschadstoffen. Toxikologische Basisdaten und ihre Bewertung. T. Eikmann, U. Heinrich, B. Heinzow und R. Konietzka (eds.), Loseblattsammlung im Erich Schmidt Verlag

# A SINGLE CHEMICAL EXTRACTION SCHEME FOR THE SIMULTANEOUS EVALUATION OF THE POTENTIAL MOBILITY OF A METALLOID (As) AND METALLIC ELEMENTS (Cd, Cu, Ni, Pb, Zn) IN CONTAMINATED SOILS AND MINE WASTES

MARIA ANGÉLICA WASSERMAN[2], MONIKA
KEDZIOREK[1*], VINCENZO CANNIZZARO[3], AGNÈS
WOLLER[1], HANS-RÜDI PFEIFER[3] AND ALAIN BOURG[1]
[1]Environmental HydroGeochemistry Laboratory (LHGE),
Department of Earth Sciences, University of Pau, BP 1155, 64013
Pau Cedex, France
[2]Instituto de Radioproteção e Dosimetria/Comissão Nacional de
Energia Nuclear (CNEN/Brazil). Av. Salvador Allende s/n°,
Recreio - Rio de Janeiro - RJ- Brazil, CEP: 22780-160, Brazil
[3] Mineral Analysis Center, Earth Sciences Department, University
of Lausanne, BFSH2, CH-1015 Lausanne, Switzerland

**Abstract.** The total concentration of toxic chemical elements present in contaminated soils and mine wastes does not provide reliable information about the risk they present to Man and the Environment. Realistic assessments imply determining the fraction of total pollutant load susceptible to be solubilized and thus potentially transferred to the hydrocycle and the biosphere. Chemical extraction is the most commonly used method to reach this objective. Here we investigate two of the classical methods of sequential extraction, the first developed for cationic metallic elements (Tessier et al., 1979), the second for the oxyanionic phosphorus, a chemical analogue of arsenic (Woolson et al., 1983) and we propose a single protocol capable of simultaneous giving useful information about the reactivity of both cationic and anionic pollutants (often present together in contamination due to mining and smelting activities or heavy chemical industry). If the method of Tessier et al. (1979) is used for arsenic, it underestimates its potential mobility. The method we propose does

* To whom correspondence should be addressed: monika.kedziorek@univ-pau.fr

L. Simeonov and V. Sargsyan (eds.),
*Soil Chemical Pollution, Risk Assessment, Remediation and Security.*
© Springer Science+Business Media B.V. 2008

not provide direct clues about the geochemical associations of the toxic elements in the contaminated solids but it gives instead information about (a) the potentially mobile fraction of these elements and (b) their geochemical behavior in the presence of a large range of physical-chemical conditions (acidic, alkaline, oxidizing, reducing), data of interest for remediation measures.

**Keywords:** arsenic, chemical extraction, contaminated soils, metals, mine waste, potential mobility

## 1. Introduction

Soils may be contaminated by metallic elements (heavy metals and metalloids, such as As). Major polluting sources are mining operations, ore processing and associated industrial activities. In Western Europe alone more than 1 million sites have been identified (Dudka et al., 1997). Questions are raised about the fate of these pollutants. How much and how fast any pollutant entering a soil is susceptible to be transferred to plants or to groundwater? Excessive accumulation of metallic pollutants can have deleterious effects on soil ecosystem functions, and might consequently constitute a risk to Man and associated ecosystems.

In contaminated soils and mine wastes, heavy metals and metalloids can be found under a diversity of geochemical and mineralogical forms, each one having its own potential for dispersion in the hydrocycle and, possibly, its own toxicity. A chemical element can be adsorbed on surfaces, complexed with organic matter, present as individualized mineral phases (as precipitates) or incorporated in mineral amorphous or crystalline structures. Depending on the form under which it is present, the element is more or less subject to solubilization. For example, if adsorbed on a solid surface it can be easily remobilized, especially if environmental conditions change (changes in acidity, in redox conditions, presence of complexing agents, ...). In contrast, if present within a difficultly solubilisable structure, its potential for re-entering biogeochemical cycles will be low.

It is now widely accepted that determining only the total concentration of toxic elements in solids is insufficient to assess their environmental impact (e.g., Ma et al., 1997; Mester et al., 1998). For a realistic environmental risk assessment, it is necessary to determine the geochemical form or association of each toxic element in the solid of concern. Knowledge of the processes controlling the solubility of the element (i.e., adsorption/desorption, precipitation/dissolution, redox reactions) should then permit evaluation of the risk of solubilization connected to each geochemical form.

The classical approach to determine the geochemical fractionation of elements in solids is by using multiple chemical extraction procedures. The current protocols can be conveniently divided into two main groups: (1) a sequence of chemical extracting solutions applied on the same sample (e.g., Tessier et al., 1979; Woolson et al., 1983; Berti et al., 1997; Rauret, 1998; Simeonova and Simeonov, 2006) or (2) a series of parallel extracting solutions on different aliquot samples (Heron et al., 1994; Crouzet et al., 2000). Parallel extractions were developed only for a few elements (e.g., Fe, S). Sequential chemical extractions have been the object of many criticisms mainly because they are proposed as selective of given geochemical forms (Davidson et al., 1994; Biester et al., 1997).

Pollutants like heavy metals and metalloids can be divided into two categories according to their chemical properties and subsequent reactivities, elements present in dissolved state usually as cations (most transition elements) or as anions (chromate and oxyanions formed with metalloids, such arseniate and arsenite). Sequential extraction procedures were developed mostly to investigate cationic forms while only few are proposed for anionic elements such as P, As and Se (Woolson et al., 1983; Macleod et al., 1998).

For many pyrite-related activities, arsenic is found in polluted soils and waste material associated with heavy metals. This is especially the case within and in the vicinity of gold mining operations. In such situations it would be advantageous to apply a single extraction procedure for the fractionation of both types of contaminants (heavy metals and arsenic).

In this paper we are addressing two problems:

1. The feasibility of using methods developed for heavy metals for fractionating soil As or other metalloids. Considering that As is present under dissolved state in soil as anionic oxyanions, an alternative approach has been to use extraction schemes designed to investigate the forms of P in soils. This method is based on the principle that As should have chemical properties close to that of P. Both elements are known to form oxyanions in the +V oxidation state and they are located in the same column of the periodic table.
2. The definition of what is geochemical fractionation. The identification of definite geochemical associations of metallic elements in polluted solids has been seriously questioned (see above). Chemical extractions are not geochemical association-specific (e.g., Howe et al., 1999).

The objective of this study is to propose a single sequential chemical extraction protocol capable of simultaneously determining the potential mobility of cationic chemical elements and arsenic (and other elements present as oxyanions) in solids. This situation is representative of contamination of solids by mining, smelting and related industrial operations. We used as references the most widely accepted chemical sequential extraction procedure for cations, the protocol of Tessier et al. (1979) and a procedure used for phosphorus, a chemical element expected to present reactive behavior similar to arsenic, the protocol of Woolson et al. (1983) Four samples were investigated: a certified sediment reference material, a contaminated soil, a sample from an industrial site and a mining waste.

In this work we focus on two aspects of the potential mobility of chemical elements from contaminated solids. What is the fraction of the total element concentration potentially mobile? Under which geochemical aggression conditions are they mobile? The first question is relevant to risk assessment investigations, the second to developing decontamination strategies.

## 2. Materials and Methods

### 2.1. MATERIALS

The solids investigated are from different origins and offer a wide range of concentrations of arsenic and metallic pollutants (Table 1). The first sample (IS) is an arsenic rich sample obtained from an industrial site in France. It is a basic, sandy sample (pH = 9.9 in a 0.01 M $NaNO_3$ electrolyte and 32% of grain size <250 μm), with low organic carbon (0.5%). Its arsenic content is quite high (about 12%). It is also rich in Cu, Ni and Pb. The second sample (MS) is a sandy soil collected close to (and downwind from) the gold mine of Astano (Switzerland). Its contaminant content is rather low except for As. The third sample (MW) was collected in a waste pile of the Astano gold mine. It contains high concentrations of As, Pb and Zn. The exploitation area is rich in pyrite, arseno-pyrite, galena, sphalerite, native gold and silver. A reference material, CRM-320 (CRM), was used to validate the extraction protocols and analytical methods, it is a sandy river sediment collected close to industrial areas. A sediment sample had to be chosen since no soil reference material was at the time certified for As. This solid has low contents in heavy metals and As (Table 1).

TABLE 1. Total elemental concentrations, in μg/g*, and standard deviations (± sd[a])

| Sample | As | Cd | Cu | Ni | Pb | Zn |
|---|---|---|---|---|---|---|
| IS | 123,070 | 16.22 | 1,366 | 1,390 | 570 | 173 |
| (± sd) | (± 3,970) | ± 3.65) | (± 63) | (± 60) | (± 38) | (±11) |
| MS | 175 | 0.13 | 1.9 | 5.3 | 9 | 36 |
| (± sd) | (± 5) | (± 0.02) | (± 0.4) | (± 0.1) | (± 1) | (± 2) |
| MW | 1,770 | 4.13 | 39 | 45 | 286 | 442 |
| (± sd) | (± 71) | (± 0.15) | (± 2) | (± 1) | (± 13) | (± 5) |
| CRM-320 | 78 | 0.43 | 43 | 67 | 26 | 142 |
| (± sd) | (± 1) | (± 0.01) | (± 2) | (± 2) | (± 1) | (± 14) |

*Sum of the five fractions using method 3 described below.
[a] Triplicates.

The CRM sample was used as delivered, the other samples were dried at 40°C for one week and sieved at 250 μm prior to the chemical treatments.

## 2.2. METHODS

The potential solubility of As and of heavy metals (Cd, Cu, Ni, Pb and Zn) was evaluated using three different sequential extraction schemes (Table 2): (a) method 1: the protocol of Tessier et al. (1979), developed for cationic

TABLE 2. Description of the methods tested

| Extracted fraction | Method 1[1] (1g of sample) | Method 2[2] (2.5 g of sample) | Method 3[3] (1g of sample) |
|---|---|---|---|
| Water soluble | not considered | 50 ml of DIW | not considered |
| Exchangeable weakly bound to major constituents | 16 ml $MgCl$ (1M); pH 7. | 50 ml $NH_4F$ (0.5N) | extracted in the slightly acid step (below) |
| Slightly Acid Method 3: easily bioavailable Method 1: bound to carbonates | 35 ml $CH_3COONa$ (0.6 M); pH 5. | not considered | 40 ml $CH_3COOH$ (2M) + $CH_3COONa$ (2 M) 1:1; pH 4.7. Shake for 16 hrs at RT* |
| Reducible Method 3: bound to Mn oxides. Method 1: bound to Fe and Mn oxides | 35 ml $NH_2OH.HCl$ (0.1 M) in $CH_3COOH$ 25% (V/V). 1h at 96°C | not considered | 40 ml $NH_2OH.HCl$ (0.1 M); pH 2. Shake for 16 hrs at RT* |
| Oxidizable Method 3: bound to organic matter Method 1: bound to organic matter or sulfur | 10 ml $H_2O_2$ (30%)+ 6 ml $HNO_3$ (0.02M); pH 2. 1h at 98°C | not considered | 20 ml $H_2O_2$ (30%) + 50ml of $CH_3COONH_4$ (1M); pH 2. Shake for 16 hrs at RT* |
| Alkaline Bound to resistant organic matter or iron compounds | not considered | 50 ml NaOH (0.1 M); | 50 ml NaOH (0.1 M); pH 12. Shake for 18 hrs at RT* |
| Acid Adsorbed to soil compounds and/or carbonate bound | not considered | 50 ml $H_2SO_4$ (0.5N) | not considered |
| Occluded Al-associated Incorporated to clay minerals | not considered | 50 ml $NH_4F$ (0.5N) | not considered |
| Residual Method 3: not mobilized in previous phases Methods 1 and 2: incorporated to crystal structure | 8 ml of aqua regia + HF. Heat to 60°C/30min. | 3.2 ml aqua regia + 4.8 ml HF until dryness. Add 20 ml $HNO_3$ | 8 ml of aqua regia. Heat to 50°C/30min. |

[1] Protocol of Tessier et al. (1979); [2] Protocol of Woolson et al. (1983); [3] Proposed protocol; *Room temperature.

elements; (b) method 2: the protocol developed for phosphorus by Woolson et al. (1983) and applied to arsenic, an element expected to present a similar chemical behavior; and (c) method 3: a procedure proposed here which involves a sequence of physical-chemical conditions (acidic, reducing, oxidizing and alkaline, followed by a total attack of the residual solid). To better simulate expectable environmental conditions and to provide easier determinations, all extractions in method 3 were performed at room temperature, except for the residual fraction which was obtained using *aqua regia* at only 50°C, following recommendations of Förstner and Wittman. (1983) for samples with elevated concentrations of As or Cd. The decision to exclude fluorhydric acid from the extraction of the residual fraction, was based on previous experience, where recoveries for a reference material were worst for some elements in the presence of this acid, when the extracts were analyzed by ICP-MS. The weak extraction used for the residual fraction of our proposed protocol is not able to destroy all refractory minerals, lowering recovery for lithogenic elements in natural concentrations (in this case, Pb), but no important information for risk assessment is lost.

The concentrations of the elements in the extracts were analyzed by ICP-MS (Perkin Elmer Elan 6000). Triplicate extractions were carried out for each sample and overall uncertainties were usually below 5%.

## 3. Results and Discussion

### 3.1. VALIDATION OF THE THREE EXTRACTION METHODS USING THE CRM REFERENCE MATERIAL

Total elemental concentrations (in µg/g) of the reference material CRM-320 obtained by the sequential extraction methods and their recovery (in %) based on the certified values are presented in Table 3.

Only about 59% of the As was extracted by method 1, developed for cationic elements, while the two other methods provide total recovery (within experimental uncertainty of one standard deviation). Method 3 was, on an overall basis, more efficient than the two classical methods as it extracted between 81% and 100% of most of the elements (including As), with nonetheless slightly lower recoveries for Pb (about 60%). This low Pb recovery observed for method 3 is certainly due to a high lithogenic contribution of this element in this sample, where the use of *aqua regia* at relatively low temperature (50°C) in the last extraction was not strong enough to solubilize the remaining metals (residual) in the CRM reference material, as compared to the two other methods.

TABLE 3. Total elemental concentrations* (in µg/g) of the reference material CRM-320 obtained by the sequential extraction methods and recovery (in %) based on the certified values

|  | As | Cd | Cu | Ni | Pb | Zn |
|---|---|---|---|---|---|---|
| Certified values | 77 ± 3 | 0.53 ± 0.03 | 44 ± 1 | 75 ± 1 | 42 ± 2 | 142 ± 3 |
| Method 1 (µg/g ± sd[a]) | 46 ± 8 | 0.42 ± 0.12 | 30 ± 1 | 53 ± 3 | 26 ± 2 | 92 ± 5 |
| Recovery (%) ± sd[a]) | 59 ± 10 | 79 ± 23 | 68 ± 2 | 71 ± 4 | 62 ± 5 | 65 ± 4 |
| Method 2 (µg/g ± sd[a]) | 108 ± 38 | 0.32 ± 0.03 | 37 ± 5 | 54 ± 3 | 32 ± 2 | 94 ± 8 |
| Recovery (%) ± sd[a]) | 140 ± 49 | 60 ± 6 | 84 ± 11 | 72 ± 4 | 76 ± 5 | 66 ± 6 |
| Method 3 (µg/g ± sd[a]) | 78 ± 1 | 0.43 ± 0.01 | 43 ± 2 | 67 ± 2 | 26 ± 1 | 142 ± 14 |
| Recovery (%) ± sd[a]) | 101 ± 1 | 81 ± 2 | 98 ± 5 | 89 ± 3 | 62 ± 2 | 100 ± 10 |

## 3.2. EXTRACTION EFFICIENCY OF THE 3 PROTOCOLS FOR SAMPLES IS, MS AND MW

For the three contaminated solids (Table 4) method 3 was by far the most efficient for solubilizing all elements in all samples, with the exception of

TABLE 4. Total elemental concentrations (in µg/g)[a] for the non-certified samples (IS, MS and MW) and relative extraction efficiency (in %) of method 1 and 2 [b]

| Sample | Method | As | Cd | Cu | Ni | Pb | Zn |
|---|---|---|---|---|---|---|---|
| IS | Method 1 | 135,410 | 3.51 | 1,113 | 1,064 | 325 | 157 |
|  | (± sd[c]) | (± 59,540) | (± 0.78) | (± 360) | (± 257) | (± 81) | (± 37) |
|  | Recovery[b](%) | 110 | 22 | 81 | 77 | 57 | 91 |
|  | Method 2 | 141,000 | 1.4 | 968 | 1,272 | 223 | 104 |
|  | (± sd[c]) | (± 3,000) | (± 0.1) | (± 13) | (± 23) | (± 25) | (± 5) |
|  | Recovery[b] (%) | 115 | 9 | 71 | 92 | 39 | 60 |
|  | Method 3 | 123,070 | 16.22 | 1,366 | 1,388 | 570 | 173 |
|  | (± sd[c]) | (± 3,970) | (± 3.65) | (± 63) | (± 60) | (± 38) | (± 11) |
| MS | Method 1 | 83 | 0.06 | 4.12 | 4.3 | 15 | 31 |
|  | (± sd[c]) | (± 29) | (± 0.01) | (± 1.09) | (± 1.2) | (± 5) | (± 15) |
|  | Recovery[b] (%) | 47 | 50 | 221 | 80 | 168 | 86 |
|  | Method 2 | 148 | 0.19 | 3.9 | 3.1 | 12.7 | 21 |
|  | (± sd[c]) | (± 3) | (± 0.02) | (± 0.1) | (± 0.3) | (± 0.1) | (± 3) |
|  | Recovery[b] (%) | 85 | 146 | 210 | 58 | 141 | 58 |
|  | Method 3 | 175 | 0.13 | 1.86 | 5.3 | 9 | 36 |
|  | (± sd[c]) | (± 5) | (± 0.02) | (± 0.40) | (± 0.1) | (± 1) | (± 2) |
| MW | Method 1 | 1,110 | 3.21 | 35 | 45 | 181 | 283 |
|  | (± sd[c]) | (± 461) | (± 0.51) | (± 12) | (± 14) | (± 157) | (± 49) |
|  | Recovery[b] (%) | 63 | 78 | 91 | 99 | 63 | 64 |
|  | Method 2 | 1,350 | 1.74 | 32 | 38 | 143 | 264 |
|  | (± sd[c]) | (± 23) | (± 0.19) | (± 3) | (± 1) | (± 3) | (± 26) |
|  | Recovery[b](%) | 76 | 42 | 82 | 84 | 50 | 60 |
|  | Method 3 | 1,767 | 4.13 | 39 | 45 | 286 | 442 |
|  | (± sd[c]) | (± 71) | (± 0.15) | (± 2) | (± 1) | (± 13) | (± 5) |

[a] Sum of five fractions of sequential extraction.
[b] Relative to the amount extracted by method 3.
[c] Average of triplicates.

Cd, Cu and Pb in sample MS. Amongst these three metals, as mentioned above, Pb also presented low recoveries with method 3 for the certified sample. This confirms that when Pb has a natural origin, its total extraction using method 3 is not efficient.

The total elemental content of a sample will be divided for the discussion below into two fractions, the *extractable* (total content minus the residual) and the *residual* fractions. The first is the only one of interest for risk assessment studies. Details of the extractable content will be discussed on specific cases, in relation with the elemental geochemical reactivity. All figures (except Figure 2) give relative fractions extracted, the black part of the bars represent the residual, the others are the potentially extractable fractions.

## 3.3. POTENTIAL MOBILITY OF ARSENIC

According to method 1 (developed for cationic elements) from 80% to 99% of the As, depending on the solid, is present as residual fraction and therefore assumed to be incorporated within resistant minerals (Figure 1). Arsenic would therefore be considered as not available for environmental transfer processes. For the IS and CRM samples the residual As is higher than 95%. For the two other samples it amounts to about 80% and, amongst the extractable fractions, the most efficient is with hydroxylamine, which suggests, in agreement with observations of many authors (e.g., Belzile et al., 1990; Bombach et al., 1994; Jones et al., 1997; Ribeiro et al., 1997; Mclaren et al., 1998), a connection with oxidized (i.e., reducible) Fe. As we will see below method 1, sometimes used for As (e.g., Matera and Le Hecho, 2001), provides inappropriate results as it overestimates the residual As.

The residual fraction obtained by method 2 (developed for anionic elements) is considerably smaller than method 1 for all samples (15–65%). A large fraction of As (30–40%) is solubilized by sodium hydroxide at pH 12 (alkaline extraction), suggesting either an association with iron(III) solids or the occurrence as calcium arseniate (Mclaren et al., 1998). The acid attack in this protocol achieved also significant extraction for samples CRM and MW. The geochemical meaning of the acid extraction is uncertain, since the 0.5 N sulfuric acid extracting solution is strong enough to solubilize not only elements adsorbed to various solids but also to destroy components such as Ca arseniate. In the MS solid more than 40% of the As was extracted by ammonium fluoride as exchangeable.

Compared to the first two methods, the residual fraction obtained with our method 3 is the smallest for As (25–35%). Results for the IS sample obtained by this procedure are in agreement with batch experiments where the solubility of As was determined as a function of pH in the extracting sample suspension,

60% of the total As was solubilized at pH 12 (Figure 2). Method 3 provided higher or similar values for extractable As as compared to the two other methods.

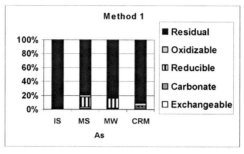

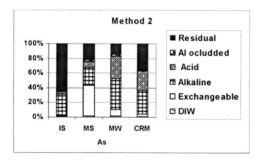

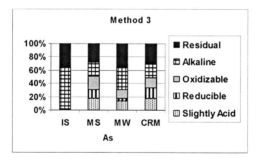

*Figure 1.* Solubilization of As by each step of the three protocols investigated (in %) (Based on the relative fraction of the total amount extracted by the given protocol for each sample).

In conclusion, method 3 is globally the most effective to quantify the As potentially available for environmental transfer in all samples. Its sequence of extractions provides more detailed information for the assessment of the potential solubility of this element in response to changes in the environmental conditions. Alkaline conditions can mobilize a large fraction of As, but all other physical-chemical conditions (acidic, oxidizing and reducing) can solubilize significant quantities of As in all samples as well.

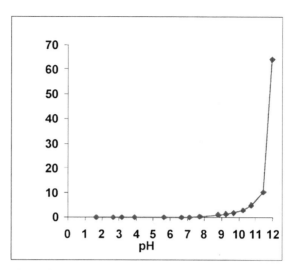

*Figure 2.* Fraction (%) of arsenic solubilized as a function of pH in the IS solid.

## 3.4. POTENTIAL MOBILITY OF HEAVY METALS USING METHOD 2

Two trends are observed in Figures 3–7. The potential mobility is underestimated as the residual pool is usually higher than with both other methods (or equivalent for Cu and Ni with method 1, Figures 4 and 5, and Ni and Zn with method 3, Figures 5 and 7, respectively) and the extractable pool is due to a large extent to the strongly acidic attack (0.5 N $H_2SO_4$). Such very aggressive conditions not only destroy several components of the solid (including reducible phases and carbonates) but they also desorb elements from various solid surfaces, giving therefore little geochemical information.

## 3.5. POTENTIAL MOBILITY OF CADMIUM

According to Methods 1 and 3, a large fraction of the Cd present in the solids is potentially available for environmental transfer processes: from 65% to 100% for method 1 and from 25% to 80% for method 3 (Figure 3). The residual fraction from method 3 is greater than for method 1, which is explained by the less drastic extraction conditions for this fraction (see Methods section).

In method 3, slightly acidic conditions can dissolve a large fraction of Cd in all samples. The fraction of Cd from this extraction is very close to the sum of exchangeable and carbonate extractions obtained using method 1.

Indeed at pH 5 not only is the Cd bound to carbonates efficiently extracted, but the Cd adsorbed on many other surfaces can be desorbed (e.g., Kim et al., 1991).

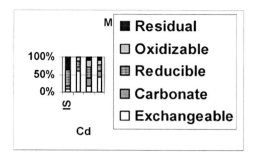

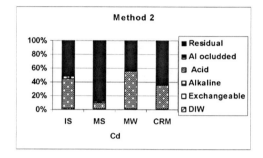

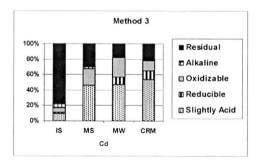

*Figure 3.* Solubilization of Cd by each step of the three protocols investigated (in %) (Based for each one on the relative fraction of the total amount extracted by the given protocol).

Very little Cd is solubilized in the reducing extraction of method 3 compared to that of method 1. The heat used in the reducible phase of method 1 enhances the metal recovery, as observed by Kim et al. (1991) and Lopez-Sanchez et al. (1993) for Cd associated with organic matter or crystalline Fe oxides.

According to methods 1 and 3, acid, and to a lesser extent, oxidizing and reducing conditions favor the solubility of Cd, in agreement with other observations (e.g., Reimann and De Caritat., 1998; Bourg and Loch, 1995).

## 3.6. POTENTIAL MOBILITY OF COPPER

According to method 3, extractable Cu is mainly associated with organic compounds or carbonates (Figure 4), in agreement with observations by Lopez-
Sanchez et al. (1993). As described above and below, other elements such as Cd
and Pb, often associated with organic material, did not show this association
when method 1 was used, which indicates that the organic solids can be
significantly affected in the reducible phase extraction (step 3) using method 1.

For method 3 the residual fraction is smaller than with method 1 or similar
(for IS and MW). The difference for CRM and MS is certainly due to the
contribution of the oxidizable (CRM) or alkaline (MS) extractions. Method 3

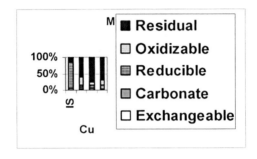

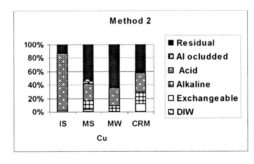

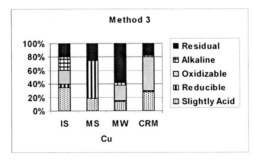

*Figure 4.* Solubilization of Cu by each step of the three protocols investigated (based for each one
on the relative fraction of the total amount extracted by the given protocol).

can be effective to mobilize Cu potentially available for the transfer processes, giving at the same time interesting geochemical information. No general trend is observed, the potential dispersion is sample specific.

## 3.7.  POTENTIAL MOBILITY OF NICKEL

Nickel is considered to be highly mobile under acid conditions and moderately mobile under oxidizing conditions (e.g., Reimann and De Caritat, 1998). This behavior agrees best with the results of method 3, where Ni was mostly residual (60–70%) with for the remaining content relative mobility under all physico-chemical conditions (Figure 5).

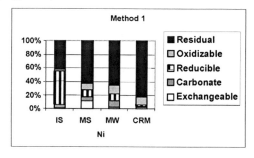

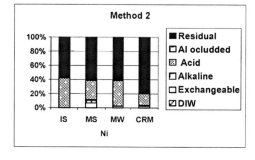

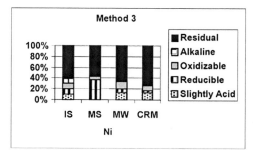

*Figure 5.* Solubilization of Ni by each step of the three protocols investigated (based for each one on the relative fraction of the total amount extracted by the given protocol).

Ni total recovery for the reference material (CRM-320) was about 89% with method 3 and 71% with method 1 (Table 3). Better recovery of Ni with method 3 was possible due to the efficient attack of the mobiles phases. Considering the non-certified samples, method 3 provides the highest values for Ni, giving also reliable geochemical information.

## 3.8. POTENTIAL MOBILITY OF LEAD

Lead is poorly mobile in the environment as it is a lithogenic element and moreover also strongly immobilized in the humic fraction of soils, with sulfides and carbonates acting as additional geochemical barriers (e.g., Reimann and

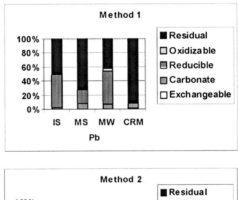

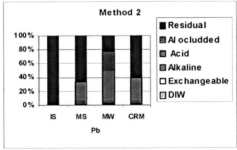

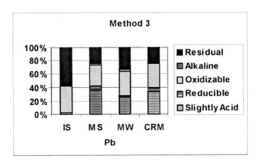

*Figure 6.* Solubilization of Pb by each step of the three protocols investigated (based for each one on the relative fraction of the total amount extracted by the given protocol).

De Caritat, 1998). In agreement with results obtained by Lopez-Sanchez et al. (1993) using a three-step procedure, extraction of Pb by method 3 does indicate that this element is mainly associated with organic compounds and/or carbonates (Figure 6). In contrast, method 1 and a protocol similar to method 1 used by Lopez-Sanchez et al. (1993) showed that Pb is mainly in the reducible phase, associated with Fe and Mn oxides. This provides additional evidence that the organic phase can be attacked in the reducible extraction conditions as observed by Kim et al. (1991).

For the reference material (CRM-320) total Pb recoveries were about 62% with method 3, 61% with method 1 and 76% with the method 2 (Table 3). In the residual phase, Pb concentrations obtained with method 3 was 6 µg/g, while with method 1 it was 22 µg/g and 19 µg/g with method 2. Since the fraction of Pb linked with crystalline mineral structures is probably important, this can explain poor recoveries for Pb with our protocol. In contrast, our protocol was able to give more reliable geochemical information than provided by methods 1 and 2.

### 3.9. POTENTIAL MOBILITY OF ZINK

Zinc is reputedly highly mobile in the environment especially under acidic and oxidizing conditions. Adsorption on clays, Fe-Mn oxides and organic matter as well as co-precipitation with iron and manganese are main geochemical barriers for this element (e.g., Reimann and De Caritat, 1998). This agrees with the results of method 3, where Zn was observed mainly associated with the mineral structures in natural sample and with relative mobility under all physico-chemical conditions for contaminated samples (Figure 7). According to the results of our protocol the order of Zn mobility decreases as follow: acid ~ oxidizable ~ reducible > alkaline.

For the reference material (CRM-320) Zn recovery was about 100% with method 3, 66% with method 2 and 65% with method 1 (Table 3). Better recovery of Zn with method 3 was possible due to its better efficiency to attack mobiles phases. Considering the polluted samples, method 3 obtained the highest potentially mobile concentration for Zn, giving therefore potentially interesting geochemical information.

### Conclusions

Recovery for the CRM-320 reference material depended on the method used and the element investigated. The best results were obtained with our method (method 3) for As, Cd, Cu, Ni. and Zn. These results were associated with the better efficiency to attack mobile phases, when a large range of physico-chemical

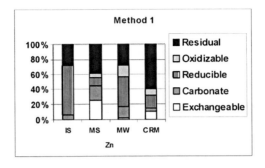

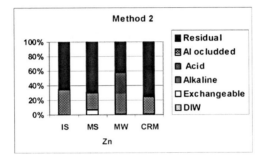

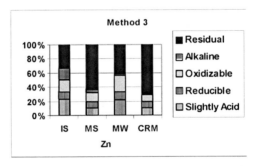

*Figure 7.* Solubilization of Zn by each step of the three protocols investigated (based for each one on the relative fraction of the total amount extracted by the given protocol).

conditions is considered (low and high pH, redox conditions). Also considering the polluted samples, method 3 achieves the best extraction for all investigated elements and samples, with the exception of Cu, Pb and Cd in the mine soil (MS) sample where these elements are present in very low concentration. Method 3 can mobilize the same amount or more As than the As specific method (method 2), giving in addition reliable geochemical information.

Method 2 presents better recovery for Pb but geochemical information is lost due to the use of sulfuric acid as extracting reagent in the first step, solubilizing not only elements adsorbed to various solids but also major solid components. The use of *aqua regia* at 50°C for the residual phase of Method 3 was not able to destroy some minerals, diminishing recovery for lithophilic elements such as Pb, linked to refractory minerals. Results obtained using the classical method 1 can be applied for cations, yet some interpretational problems have been pointed out, and clearly method 1 does not provide reliable geochemical information for As.

In summary the protocol proposed here (method 3) gives reliable geochemical information for both heavy metals (cationic) and metalloids (arsenic-anionic). The results of method 3 suggest that under reducing conditions the risk of mobility of both cationic and anionic forms of pollutants is minimized for the contaminated samples studied (IS and SW). It is a simple method adapted and tested to be analyzed by ICP-MS.

## Acknowledgements

This work was funded by a grant from Rhodia, with additional funds from the CNEN-Brazil (Brazilian Agency for Nuclear Energy), the Swiss FNRS (Swiss National Foundation for Research) and the European Union (ERDF Program). We thank Olivier Donard for access to ICP/MS.

## References

Belzile, N. and Tessier, A. 1990. Interactions between Arsenic and Iron Oxyhydroxides in Lacustrine Sediments. *Geochimica et Cosmochimica Acta*, **54**, 103–109.
Berti, W.R., Cunningham, S.D. and Jacobs, L.W. 1997. Sequential Chemical Extraction of Trace Elements: Development and Use in Remediating Contaminated Soils. In: R. Prost (Ed.), *Biogeochemistry of Trace Elements*. Paris: INRA Editions, 121–131.
Biester, H. and Scholz, C. 1997. Determination of Mercury Binding Forms in Contaminated Soils: Mercury Pyrolysis versus Sequential Extractions. *Environmental Science & Technology*, **31**, 233–239.
Bombach, G., Pierra, A. and Klemm, W. 1994. Arsenic in Contaminated Soil and River Sediment. *Fresenius Journal of Analytical Chemistry*, **350**, 49–53.
Bourg, A.C.M. and Loch, J.P.G. 1995. Mobilization of Heavy Metals as Affected by pH and Redox Conditions. In: W. Salomons and W.M. Stigliani (Eds.) *Biogeodynamics of Pollutants in Soils and Sediments: Risk Assessment of Delayed and Non-Linear Responses*. Berlin: Springer, 87–102.
Crouzet, C., Kedziorek, M.A.M., Altmann, R.S. and Bourg, A.C.M. 2000. Speciation of Sulfur in Aquifer Sediments Contaminated by Landfill Leachate Using Chemical Extraction Techniques. *Environmental Technology*, **21**, 285–296.
Davidson, C.M., Thomas, R.P., Mcvey, S.E., Perala, R.D.L. and Ure, A.M. 1994. Evaluation of a Sequential Extraction Procedure for the Speciation of Heavy Metals in Sediments. *Analytica Chimica Acta*, **291**, 277–286.

Dudka, S. and Adriano, D.C. 1997. Environmental Impacts of Metal Ore Mining and Processing: A Review. *Journal of Environmental Quality*, **26**, 590–602.

Förstner, U. and Wittmann, G.T.W. 1983 *Metal Pollution in the Aquatic Environment.* Heidelberg: Springer. 486 p.

Heron, G., Crouzet, C., Bourg, A.C.M. and Christensen, T.H. 1994. Speciation of Fe(II) and Fe(III) in Contaminated Aquifer Sediments using Chemical Extraction Techniques. *Environmental Science & Technology*, **28**, 1698–1705.

Howe, S.E., Davidson, C.M. and Mccartney, M. 1999. Operational Speciation of Uranium in Inter-Tidal Sediments from the Vicinity of a Phosphoric Acid Plant by Means of the BCR Sequential Extraction Procedure and ICP-MS. *Journal of Analytical Atomic Spectrometry*, **14**, 163–168.

Jones, C.A., Inskeep, W.P. and Neuman, D.R. 1997. Arsenic Transport in Contaminated Mine Tailings Following Liming. *Journal of Environmental Quality*, **26**, 433–439.

Kim, N.D. and Fergusson, J.E. 1991. Effectiveness of a Commonly Used Sequential Extraction Technique in Determining the Speciation of Cadmium in Soils. *The Science of the Total Environment*, **105**, 191–209.

Lopez-Sanchez, J.F., Rubio, R. and Rauret, G. 1993. Comparison of Two Sequential Extraction Procedures for Trace Metal Partioning in Sediments. *International Journal of Environmental Analytical Chemistry*, **51**, 113–121.

Ma, L.Q. and Rao, G.N. 1997. Chemical Fractionation of Cadmium, Copper, Nickel, and Zinc in Contaminated Soils. *Journal of Environmental Quality*, **26**, 259–264.

Macleod, F., Macgaw, B.A. and Shand, C.A. 1998. Sequential Extraction of Selenium from Four Scottish Soils and a Sewage Sludge. *Communications in Soil Science and Plant Analysis*, **29**, 523–534.

Matera, V. and Le Hécho, I. 2001. Arsenic Behavior in Contaminated Soils: Mobility and Speciation, In, Selim, H.M. and Sparks, D.L. Eds., Heavy Metals Release in Soils, Lewis Pub., Chelsea, 207–235.

Mclaren, R.G., Naidu, R., Smith, J. and Tiller, K.G. 1998. Fractionation and Distribution of Arsenic in Soils Contaminated by Cattle Dip. *Journal of Environmental Quality*, **27**, 348-354.

Mester, Z., Cremisini, C., Ghiara, E. and Morabito, R. 1998. Comparison of Two Sequential Extraction Procedures for Metal Fractionation in Sediment Samples. *Analytica Chimica Acta*, **359**, 133–142.

Rauret, G. 1998. Extraction Procedures for the Determination of Heavy Metals in Contaminated Soil and Sediment. *Talanta*, **46**, 449–455.

Reimann, C. and De Caritat, P. 1998. *Chemical Elements in the Environment: Fact sheets for the Geochemist and Environmental Scientist.* Heidelberg, Germany: Springer. 398 p.

Ribeiro, A.B. and Nielsen, A.A. 1997. An Application of Discriminant Analysis to Pattern Recognition of Selected Contaminated Soil Features in Thin Sections. *Geoderma*, **76**, 253–262.

Simeonova, B. and L. Simeonov. 2006. An application of a phytoremediation technology in Bulgaria. The Kremikovtzi Steel Works experiment Remediation Journal, Spring edition 2006 Wiley Periodicals, New York, pp. 113–123.

Tessier, A., Campbell, P.G.C. and Bisson, M. 1979. Sequential Extraction Procedure for the Speciation of Particulate Trace Metals. *Analytical Chemistry*, **51**, 844–851.

Woolson, E.A., Axley, J.H. and Kearney, P.C. 1983. The Chemistry and Phytotoxicity of Arsenic in Soils: II. Effects of Time and Phosphorus. *Soil Science Society's America Proceedings*, **37**, 254–259.

# NUMERICAL MODELS FOR PREDICTION OF FLOW AND TRANSPORT IN SOIL AT THE FIELD SCALE

MATHIEU JAVAUX[1,2] AND MARNIK VANCLOOSTER[1]
[1]*Department of Environmental Sciences, Université Catholique de Louvain,*
*2 Croix de Sud, box 2, Louvain-la-Neuve, Belgium*
[2]*Agrosphere, Forschungszentrum Juelich, Juelich, Germany*

**Abstract.** Numerical models are essential tools for predicting water flow and solute transport under potential future scenarios, performing risk assessment studies, resource management or to develop remediation strategies. Thorough understanding and characterization of the relevant mechanisms governing the water flow and fate and transport of solutes in soils are important steps before a suitable modelling approach could be selected. In this article we briefly review the basics of solute transport and water flow in soils and point out important challenges for modelling these processes at the field scale. Finally, we conclude by giving some opportunities which currently exist for improving modelling tools.

Keywords: numerical models, solute transport, water flow model, model uncertainty

## 1. Introduction

Soil and groundwater are important natural resources that should be protected from human induced pollution. Dissolved substances entering the soil from diffuse and point pollution sources, further referred to as solutes, are carried with the water phase through the soil zone, where they are subject to a range of phase exchange and transformation processes. Along with the boundary conditions at the soil surface, these transport, phase exchange and transformation processes will determine the ultimate concentrations and fluxes of dissolved substances in soil and groundwater. As numerical models can deal with these

---

* To whom correspondence should be addressed: mathieu.javaux@uclouvain.be

L. Simeonov and V. Sargsyan (eds.),
*Soil Chemical Pollution, Risk Assessment, Remediation and Security.*
© Springer Science+Business Media B.V. 2008

complex and interacting mechanisms, they are excellent candidates to predict water flow and solute transport under potential future (climatic, land-use, etc.) scenarios, and can therefore be used for performing risk assessment, resource management or to develop remediation strategies. However, a thorough understanding and characterization of the relevant mechanisms governing the water flow and fate and transport of solutes in soils are prerequisites for designing appropriate numerical models.

The most challenging issue when dealing with flow and transport modelling in soil is definitely the huge space and time variability (Feyen et al., 1998). Variability of soil and environmental boundary conditions influences considerably the flow and transport at the different scales. Whereas in the past, theoretical approaches were well developed and validated to model flow and transport at the local scale (e.g. the laboratory column or the small field plot), there is nowadays a general consensus that this knowledge cannot directly be transferred to the larger scale where the management problems occur (Pachepsky et al., 2003). This scale paradigm forces hydrologist, soil scientists and environmental engineers to seek for effective approaches to model the flow and transport problem at different spatial scales (Feyen and Wiyo, 1999; Vanclooster et al., 2002, 2005).

Although no consensus currently exist about the way how flow and transport problems in soil should be addressed consistently at the different spatial scales, all approaches are underlined by implicit or explicit assumptions regarding the system hierarchy (Vogel and Roth, 2003; Lin and Rathbun, 2003). The type of hierarchy followed by the system model is the way the soil property or behaviour is related with scale. Multiscale system models may present a structural or a functional hierarchy. Structural hierarchy characterizes systems which can be decomposed into interacting physical subunits. On the other hand, functional hierarchy deals with processes evolving with scale. Obviously both may be interconnected.

A well known example is the distributed modelling approach, particularly designed to model flow and transport at the regional scale. Distributed modelling was typically developed in hydrology (Jensen and Mantoglou, 1992) and other disciplines of environmental sciences as it is conceived to incorporate the variability of soil and land surface properties that can be characterized from classical soil surveys or remote sensing surveys. Such an approach assumes a discrete structural hierarchy. Landscapes are decomposed in discrete homogeneous blocks (referred to as representative elementary areas) for which a process based numerical model can be implemented. The model performance at the larger scale, i.e. the scale at which most management problems occurs, will therefore depend on the performance (and limitations) of the model at the local scale.

Uncertainty, bias and error of the model at the local scale will be propagated through the distributed model at the larger scale. As such, controlling model performance at the local scale remains a prerequisite for larger scale modelling application, and needs continuous scientific effort.

The objective of this chapter is to describe some concepts and challenges for modelling flow and transport at the small field plot scale. More detailed reviews on this subject are given by Feyen and Wiyo (1999), Addiscott (2003), Feddes et al. (2004), Alvarez-Benedí et al. (2005). It is our understanding that improving and consolidating the numerical models also at this scale is significant as rather local field scale models are the vehicle for modelling approaches at the larger scale of the management unit.

## 2.   Concepts for Modelling Flow and Transport in Soils at the Field Scale

Most of water flow models rely on the mass balance equation, which writes for a 1-D soil profile of length $L_r$ (L):

$$\frac{\partial}{\partial t}\left(\int_0^{L_r} \theta(z)dz\right) = R(t) - I(t) - E(t,\theta) - T(t,\theta) - Q(t,\theta) \tag{1}$$

where $\theta(z)$ is the volumetric soil water content profile ($L^3 L^{-3}$), $t$ is the time (T), $R(t)$ the rainfall ($LT^{-1}$), $I(t)$ the canopy interception ($LT^{-1}$), $E(t, \theta)$ the soil evaporation ($LT^{-1}$), $T(t, \theta)$ the plant transpiration ($LT^{-1}$) and $Q(t,\theta)$ the net water loss computed as the sum of runoff, drainage and leaving interflow ($LT^{-1}$).

Assuming incompressibility and isotropy of the porous matrix, and isothermal single phase flow, the temporal change of soil water potential through space can be written with the 3 dimensional Richard equation:

$$\frac{\partial \theta}{\partial t} = \nabla \cdot \left[\mathbf{K}\nabla(h-z)\right] - S \tag{2}$$

where $K$ the hydraulic conductivity ($LT^{-1}$), $h$ the water potential on weight basis (L), $S$ the sink term representing root water uptake ($T^{-1}$) and $z$ the vertical coordinate (L). Analytical solutions for Eq. (1) exist in 1-D but are limited to very simple boundary and initial conditions (for instance uniform and constant rainfall on a uniform soil) and therefore numerical models are needed to solve it under more realistic boundary conditions in 1, 2 or 3 dimensions.

Numerous other models based on mass balance Eq. (1) can be used to predict soil water dynamics. A typical example are the so-called 'capacity models' which considers only gravity driven flow in the soil profile. In contrast to the fully mechanistic Richards equation based model, capacity models only need some points of the retention and hydraulic conductivity curve to model flow in soils (typically saturated hydraulic conductivity, soil moisture field capacity and soil moisture wilting point). Further, in contrast to the labor intensive laboratory experiments which are often needed to parameterize the full mechanistic Richards equation model, simple proxy and in-situ measurement techniques are available for estimating the capacity model parameters. In addition, parameters of capacity models can easily be linked to data available on the regional scale and pedotransfer functions for capacity models are often more robust as compared to those of the Richards equation model. Hence, capacity models are easier to parameterize as compared to the full mechanistic Richards model. Capacity models are therefore also more popular for modeling flow and transport at the larger regional scales (see e.g. references in Seyfried, 2003). However, the simplified capacity models are less versatile because it cannot model upward water flow in the soil profile, nor the capillary rise from shallow water tables. These soil hydrological phenomena become particularly important in areas with perched water tables, or with shallow groundwater systems such as alluvial planes. Therefore, the validation level of capacity models in those conditions will generally be lower than those based on the Richards equation (see e.g. Diekkrüger et al., 1995). The simpler parameter identification problem of capacity models does not necessarily outweigh this lower validation status of these models. Therefore recently, also Richards equation based models are implemented to model flow and transport at the larger scales such as illustrated by Birkinshaw and Ewen (2000) and Tiktak et al. (2004) amongst others.

The solute transport can be described with the Convection-Dispersion Equation (CDE):

$$\frac{\partial C_l^r}{\partial t} + \frac{\rho_b}{\theta}\frac{\partial S}{\partial t} + \vec{v}.\nabla C_l^r - \nabla.\left(\mathbf{D}.\nabla C_l^r\right) = 0 \tag{3}$$

where $C_l^r = C_t^r/\theta$ is the resident fluid concentration of the chemical substance (M L$^{-3}$) and $C_t^r$ is the total resident volume averaged concentration (M L$^{-3}$), $\rho_b$ (M L$^{-3}$) is the soil bulk density, $S$ (M M$^{-1}$) is the sorbed solute mass per unit mass of dry soil, $\mathbf{D}$ (L$^2$ T$^{-1}$) is the hydrodynamic dispersion tensor and $\overline{v}$ is the velocity vector. Equation (3) can also be solved numerically for complex, realistic boundary conditions and is linked to Eq. (2) through $\overline{v}$ and $\theta$ distributions.

Numerous analytical and numerical solutions exist for Eqs. (2) and (3) (van Genuchten and Alves, 1982; van Genuchten et al., 1999). In contrast to analytical models, numerical models allow the implementation of the deterministic variability of soil properties and the dynamic boundary conditions in an elegant way which makes numerical models powerful tools for supporting soil and water management in variable environments. The increased versatility and flexibility of numerical models however is at the expense of more intensive computing power, more complex parameterization schemes, and often decreased accuracy. The choice of the proper model and numerical approach depend primarily on the conceptual framework, dataset available and objectives of the project.

## 3. Challenges for Modelling Flow and Transport at the Field Scale

When numerical modelling is used for management purpose or decision support, the choice of the appropriate model is a critical point (Vanclooster et al., 2004). The goal is to choose a model for which the validation status is appropriate and which will keep the model uncertainty as low as possible, given the uncertainty in the available data. Model uncertainty can arise from (i) a bad characterization of the relevant processes, leading to a bad conceptual model/parameterisation; (ii) a poor model accuracy in terms of numerical code; (iii) a high uncertainty in the experimental data supporting the estimation of the model input and model parameters. In addition to the model uncertainty, there is also the user subjectivity which is incorporated in the modelling process. Modelling uncertainty should therefore be distinguished from the model uncertainty, the former including the subjectivity of the model user in the process of generating numerical results.

   In the following sections, we will illustrate the numerous challenges which must be overcome in order to reduce the uncertainty in modelling.

### 3.1.   IMPROVING PROCESS KNOWLEDGE AT THE FIELD SCALE

Much to the chagrin of soil scientists, convective flow through and diffusion within field soils do not behave in the uniform and isotropic way that their models such as in Eqs. (2) and (3) often demand. Rather, soil structure in the form of either aggregates, cracks or bio-pores, serves to create an apparently chaotic flow regime that is rapid and far-reaching. The incorporation of the heterogeneous flow and transport determined by the soil structure rather than the soil texture is definitely one of the most important conceptual challenges in modelling of flow and transport in soil (Flühler et al., 2001; Jarvis, 2007).

   Water is an essential factor for plant. Typically, crop plants transpire 200 to 1,000 kg of water for every kg of dry matter produced (Martin, 1976) and this water is principally extracted from the soil by roots. Therefore, modeling of soil water uptake by plant roots can be of importance for a series of environmental and agronomy purposes. Root water uptake is an important mechanism which dramatically affects the spatio-temporal water content distribution in the upper layers of vegetated soils. However, the root water uptake processes and its interactions with soil are still poorly understood nowadays. One reason, which explain this lack of understanding is the intrinsic difficulty of observing below-ground processes and assess soil and root properties. Another one is the knowledge gap in understanding biological processes governing the water extraction by roots. Today new advances in plant biology and the extended use of non invasive techniques open new avenues for investigating more deeply root water uptake in relation with 3-D root architecture and soil variability. New models have been developed (Javaux et al., 2007).

## 3.2.  IMPROVING MODEL ACCURACY

The choice of the numerical solution for Eqs. (2) and (3) may also lead to a certain degree of uncertainty. A benchmarking study was performed to compare the performance of several 1-D numerical models with analytical solutions of flow and transport equations that exist for given simplified initial and boundary conditions (Vanderborght et al., 2005).  Because the models are based on the same flow and transport concepts and use the same functions to describe the unsaturated hydraulic soil properties, differences between model simulations can be attributed to differences in the numerical solution. For the water flow scenarios, the largest deviations between numerical models and analytical solutions were observed for the case of soil limited evaporation while for solute transport, accurate modeling of solute dispersion posed most problems.

## 3.3.  MODEL PARAMETERIZATION

### 3.3.1.  *Solving the Scale Issue*

A classical problem in modelling field scale flow and transport comes up from the scale gap between usual experimental methods and the spatial resolution of the numerical model, referred to as the scale issue. Spatial variability of flow and transport properties at the field scale is in general very large (Biggar and Nielsen, 1976; Jury et al., 1987; Mallants et al., 1996a, b; Javaux and Vanclooster, 2006) and the classical support scale of soil parameterization techniques (e.g., the classical laboratory scale soil physical parameterization techniques) is generally much more smaller than the field extent for which the numerical model is implemented. This pertinent scale gap which exists when applying the numerical model at the field scale is often ignored. For instance, in large scale hierarchical distributed modelling studies (e.g., Tiktak et al., 2006), soil mapping units or fields are parameterized using pedotransfer functions which on their turn are developed at the support scale which is much lower than the local field scale or the scale of the soil mapping unit. Ignoring this scale gap at the field scale may result in considerable bias when large scale assessments are made using field scale model (Leterme et al., 2007).

An *upscaling* approach is therefore needed to assess the input parameters corresponding to the model resolution from point scale measurement of basic soil properties. The optimal upscaling approach should be selected following the spatial variability of the point scale measurements/the scale dependency of local properties. This will also affect the dimensionality of the model. Indeed, numerical models distinguish also by the degree of explicitness with which the structural variability of the medium is considered in the model. In 1-D models, the spatial variability of soil properties is embedded in a completely implicit way in the transport and flow models. This assumption is usually justified by the narrow lateral variation of the boundary conditions and the soil structure as compared to the vertical variability in structure trend due to soil layering. On the other hand, for 3-D models, the complex structure of the porous medium and of the boundary conditions can be explicitly considered and 3-D flow and transport processes are explicitly solved in a medium with known structure. However, upscaling approach is still needed following the grid resolution of these models (Zhu and Mohanty, 2003).

*Example: Modeling reactive transport in heterogeneous medium with 1-D model*

Dye tracer techniques can be used to visualize flow patterns with high spatial resolution (Kasteel et al., 2005). Using image analysis techniques, images of dye patterns can be transformed to concentration images. Dyes tracers sorb to soil particles and provide additional information about the sorption process in the soil. Inert and dye tracer experiments in a sandy soil monolith taken from a tertiary sandy sediment illustrated that the effect of the soil structure on transport was more pronounced for sorbing than an inert tracer (Javaux et al., 2006). An analysis of the dye concentrations further revealed that dye sorption in the unsaturated sandy sediment was smaller than in batch experiments. This demonstrates that this 1-D information, which is commonly obtained in tracer experiments, is not sufficient to gain relevant information about the transport mechanisms in a soil profile. Therefore, there is a need for experimental techniques which deliver information about the 3-D structure of flow and transport processes.

### 3.3.2. *Incorporating Data Uncertainty*

It is important to characterize the degree of uncertainty in the dataset which is used for calibrating models or assessing model properties. In this example, we investigated a long term data set of chloride concentrations in an artificial leaking lake and in the underlying vadose zone was to assess in-situ vertical transport through a tertiary sandy deposit (Javaux and Vanclooster, 2004a, b). A vertical profile of porous cups was installed in the vadose zone below the lake, which monitored major solute ions for 17 years. The lake water was also regularly sampled. Uncertainty resulted from two sources: (i) the low temporal resolution in the chloride concentration time series and (ii) the low spatial resolution of the porous cups. We demonstrated this latter effect by comparing a layered convection-dispersion (CD) model with the observations. It was shown that the dispersivity profile can not be explained with the model, due to non-representative sampling using porous cup solution samplers (PCS). We hypothesize that fingering flow or convergence phenomena below sand-clay interfaces, leads to non-representative artificially high dispersivity values.

### 3.3.3.   *Reducing the User Subjectivity*

Although numerical modeling relies very much on computer technology and artificial intelligence, numerical modeling remains a technical skill that is operated by model users which are individuals, each having their own background, their educational and technical skills, and their perception of the functioning of the soil system. Numerical models for flow and transport in soil are in addition based on highly non-linear processes and suffer therefore, by definition, very often from low robustness, i.e. small changes in model input and model parameters may influence considerably the modeling result. As such, small subtle changes induced by model users may be subjective and reduce the power of the so-called objective modeling approach. This problem is well illustrated for modeling pesticide behavior in soil (Jarvis et al., 2000).

Boesten (2000), e.g., compared the estimation of the pesticide half life at 10°C of ethoprophos and bentazone as estimated by 20 model users using a common laboratory degradation study at reference temperature. The coefficients of variation of estimated half life were 29% and 46% for ethoprophos and bentazone, respectively. The principal cause of this important user-subjective variability was the lack of guidance on the transformation rate dependency on soil temperature. Brown et al. (1996) compared the outputs from three models operated by five modelers. Differences between the output data from the five modelers using the same model were of a similar magnitude to the variation associated with field measurements. They concluded that model development should seek to reduce subjectivity in the selection of input parameters and improve the guidance available to users where subjectivity cannot be eliminated. Therefore strategies of good modeling practice should be generalized.

### Conclusions

Numerous challenges exist in using numerical modelling for predicting solute and transport in soils at the field scale. First, the model type and model dimensionality must be chosen in function of the type of structure in the medium or boundary conditions, the level of details within the dataset and the dimension of the relevant processes. The proper parameterization is a critical point especially how the subscale spatial variability is taken into account in the parameters. Second, the accuracy of the model must be assessed or known. The scale at which parameters are obtained should match with the resolution of the model grid. If not, an appropriate upscaling approach for parameterization is needed. Third, uncertainty in the dataset should be characterized and taken into account when model calibration is performed.

Multidimensional simulation of flow and transport processes is computationally quite expensive. Distribution of the computational load over different processors by parallelized simulation models is a promising way to reduce the simulation time (Hardelauf et al., 2007) and parallel computing on PC clusters becomes accessible to more and more users. Another important progress is the improvement of experimental methods with which multi-dimensional simulations can be validated. Geophysical tools are promising in that regard, since they allow a non-invasive assessment of the spatial variability of the soil properties.

## References

Addiscott, T.M. 2003. Modeling: Potential and limitations. In: *Handbook of processes and modeling in the soil-plant systems* (eds D.K. Benbi and R. Neider), The Haworth Press, New York.

Alvarez-Benedí, J., Muñoz-Carpena, R. & Vanclooster, M. 2005. Modeling as a tool for characterization of soil water and chemical fate and transport. In: *Soil - Water - Solute Process Characterization: An Integrated Approach* (eds J. Alvarez-Benedí & R. Muñoz-Carpena), CRC, Boca Raton, London, New York, Washington.

Biggar, J.W. & Nielsen, D.R. 1976. Spatial variability of the leaching characteristics of a field soil. *Water Resources Research*, **12**(1), 78–84.

Birkinshaw, S. & Ewen, J. 2000. Nitrogen transformation component for SHETRAN catchment nitrate transport modelling. *Journal of Hydrology*, **230**, 1–17.

Boesten, J.J.T.I. 2000. Modeller subjectivity in estimating pesticide parameters for leaching models using the same laboratory data set. *Agricultural Water Management*, **44**, 389–409.

Brown, C.D., Baer, U., Günther, P., Trevisan, M. & Walker A. 1996. Ring test with models LEACHP, PRZM-2, VARLEACH: Variability between model users in prediction of pesticide leaching using a standard data set. *Pesticide Science*, **47**, 249–258.

Diekkrüger, B., Sondgerath, D. & Kersebaum, C. 1995. Validity of agro-ecosystem models: a comparison of the results of different models applied to the same data set. *Ecological Modelling*, **81**, 3–29.

Feddes, R.A., de Rooij, G. & van Dam, J. 2004. *Unsaturated Zone Modeling: Progress, Challenges and Applications*, Wageningen, The Netherlands, Kluwer Academic.

Feyen, J., Jacques, D., Timmerman, A. & Vanderborght, J. 1998. Modelling water flow and solute transport in heterogeneous soils: A review of recent approaches. *Journal of Agricultural Engineering Research*, **70**(3), 231–256.

Feyen, J. & Wiyo. 1999. *Modeling of Transport Processes in Soils*, Wageningen, The Netherlands, Wageningen Pers.

Flühler, H., Ursino, N., Bundt, M., Zimmerman, U. & Stamm, C. 2001. The preferential flow syndrome a buzzword or a scientific problem? In: *Soil erosion research for the 21st century symposium and 2nd international symposium on preferential flow* (eds D.C. Flanagan and Ascough, J. C), Hawai, USA, ASAE.

Hardelauf, H., Javaux, M., Herbst, M., Gottschalk, S., Kasteel., R., Vanderborght J. & Vereecken, H. 2007. PARSWMS: a parallelized model for simulating 3-D water flow and solute transport in variably saturated soils. *Vadose Zone Journal*, 5, 255–259.

Jarvis, N., Brown, C.D. & Granitza, E. 2000. Sources of error in model predictions of pesticide leaching: a case study using the MACRO model. *Agricultural Water Management*, 44, 247–262.

Jarvis, N.J. 2007. A Review of Non-Equilibrium Water Flow and Solute Transport in Soil Macropores: Principles, Controlling Factors and Consequences for Water Quality. *European Journal of Soil Science*, 58, 523–546.

Javaux, M., Kasteel, R., Vanderborght, J. & Vanclooster M. 2006. Interpretation of dye transport in a macroscopically heterogeneous unsaturated subsoil with a 1-D model. *Vadose Zone Journal*, 5, 529–538.

Javaux, M., Schroeder, T., Vanderborght, M. & Vereecken, H. 2007. Use of a three-dimensional detailed modelling approach for predicting root water uptake. *Vadose Zone Journal* (accepted).

Javaux, M. & Vanclooster, M. 2004a. In situ long-term chloride transport through a layered, non saturated subsoil. 1. Dataset, interpolation methodology and results. *Vadose Zone Journal*, 3, 1322–1330.

Javaux, M. & Vanclooster, M. 2004b. In situ long-term chloride transport through a layered, non saturated subsoil. 2. Effect of layering on solute transport processes. *Vadose Zone Journal*, 3, 1331–1339.

Javaux, M. & Vanclooster, M. 2006. Scale-dependency of the hydraulic properties of a variably saturated heterogeneous sandy subsoil. *Journal of Hydrology*, 327, 376–388.

Jensen, K.H. & Mantoglou, A. 1992. Future of Distributed Modeling. *Hydrological Processes*, 6, 255–264.

Jury, W.A., Russo, D., Sposito, G. & Elabd, H. 1987. The spatial variability of water and solute transport properties in unsaturated soil. I. Analysis of property variations and spatial structure with statistical models. *Hilgardia*, 55, 1–32.

Kasteel, R., Burkhardt, M., Giesa, S. & Vereecken, H. 2005. Characterization of field tracer transport using high-resolution images. *Vadose Zone Journal*, 4, 101–111.

Leterme, B., Vanclooster, M., Van der Linden, A.M.A., Tiktak, A. & Rounsevell, M. 2007. Including spatial variability in Monte Carlo simulations of pesticide leaching. *Environmental Science and Technology*, 41(21), 7444–7450.

Lin, H. & Rathbun, S. 2003. Hierarchical frameworks for multiscale bridging in hydropedology. In: *Scaling Methods in Soil Physics* (eds Y. Pachepsky, et al.), Boca Raton, Florida, CRC.

Mallants, D., Mohanty, B.P., Jacques, D. & Feyen, J. 1996a. Spatial Variability of Hydraulic Properties in a Multi-Layered Soil Profile. *Soil Science*, 161, 167–181.

Mallants, D., Vanclooster, M. & Feyen, J. 1996. Transect study on solute transport in a macroporous soil. *Hydrological-Processes*, 10, 55–70.

Martin, J., Leonard, W. & Stamp, D. 1976. *Principles of Field Crop Production* (Third Edition), New York, Macmillan.

Pachepsky, Y., Radcliffe, D.E. & Selim, H.M. 2003. *Scaling Methods in Soil Physics*, Boca Raton, Florida, CRC.

Seyfried, M.S. 2003. Incorporating of remote sensing data in upscaled soil water model. In: *Scaling Methods in Soil Physics* (eds Y. Pachepsky et al.), Boca Raton, Florida, CRC, pp. 309–346.

Tiktak, A., Boesten, J.J.T.I., Van der Linden, A.M.A. & Vanclooster, M. 2006. Mapping the vulnerability of European groundwater to leaching of pesticides with a process based meta-model of EuroPEARL. *Journal of Environmental Quality*, **35**, 1213–1226.

Tiktak, A., De Nie, D.S., Pineros-Garcet, J.D., Jones, A. & Vanclooster, M. 2004. Assessing the pesticide leaching risk at the pan European level: the EuroPEARL approach. *Journal of Hydrology*, **289**, 222–238.

van Genuchten, M.T. & Alves, W.J. 1982. *Analytical Solutions of the One-Dimensional Convective-Dispersive Solute Transport Equation*. U.S. Department of Agriculture, Technical Bulletin No. 1661, 151p.

van Genuchten, M.T., Schaap, M.G., Mohanty, B.P., Simunek, J. & Leij, F. 1999. Modeling flow and transport processes at the local scale. In: *Modeling of Transport Processes in Soils* (eds J. Feyen & Wiyo),Wageningen, The Netherlands, Wageningen Pers, pp. 23–45.

Vanclooster, M., Boesten, J., Tiktak, A., Jarvis, N., Kroes, J., Clothier, B.E. & Green, S. 2004. On the use of unsaturated flow and transport models in nutrient and pesticide management. In: *Unsaturated Zone Modelling: Progress, Challenges and Applications* (eds R. Feddes et al.), Kluwer Academic, pp. 331–361.

Vanclooster, M., Javaux, M., Hupet, F., Lambot, S., Rochdi, A., Piñeros-Garcet, J.D. & Bielders, C.L. 2002. Effective approaches for modelling chemical transport in soils supporting soil management at the larger scale. In: *Sustainable Land Management – Environmental Protection. A Soil Physical Approach. Advances in Geoecology 35* (eds M. Pagliai & Jones R.), International Union of Soil Science (IUSS), pp. 171–184.

Vanclooster, M., Javaux, M. & Vanderborght, J. 2005. Chap. 69. Solute transport in soil at the core and field scale. In: *Encyclopaedia of Hydrological Sciences*, M.G. Anderson edn, Wiley, pp. 1041–1054.

Vanderborght, J., Kasteel, R., Herbst, M., Javaux, M., Thiery, D., Vanclooster, M., Mouvet, C. & Vereecken, H. 2005. A set of analytical benchmarks to test numerical models of flow and transport in soils. *Vadose zone Journal*, **4**, 206–221.

Vogel, H.J. & Roth, K. 2003. Moving Through Scales of Flow and Transport in Soil. *Journal of Hydrology*, **272**, 95–106.

Zhu, J. & Mohanty, B.P. 2003. Upscaling of hydraulic properties of heterogeneous soils. In: *Scaling Methods in Soil Physics* (eds Y. Pachepsky et al.), Boca Raton, Florida, CRC, pp. 97–118.

# DIMINISHING OF HUMAN EXPOSURE FROM ACTIVE LEAD AND ZINC MINING DUMPS

RAFAL KUCHARSKI, ALEKSANDRA SAS-NOWOSIELSKA[*] AND MARTA POGRZEBA
*Institute for Ecology of Industrial Areas*
*6 Kossutha Str., 40-832 Katowice, Poland*

**Abstract.** Lead and zinc mining dumps are still an active sources of living being exposure to metals, especially when their surface is not covered with plants. Numerous soil cleaning technologies are available, but most of them due to high cost performance and unwanted side effects have a limited use. The only feasible approach appear to be the biotechnologies, which are relatively cost-effective and environmentally friendly. The most realistic is phytostabilization, i.e., immobilization of pollutants in soil by plants. In the process of phyto-stabilization, heavy metals are immobilized in soil as a consequence of root action and the introduction of chemicals to the soil. Actually, metals are not removed but their adverse environmental effects are reduced. Techniques which allow stabilizing of pollutants in soil using chemical or biological methods, could be very practical, considering their technical simplicity and low cost performance. The combination of both approaches could be particularly useful as (i) the chemicals bind the excess of metals, help to maintain an appropriate pH and required plant nutrition and (ii) the plants built up a dense root mat, which prevents wind erosion of contaminated material, reduces the leaching of contaminated water and accumulates metals in roots. This philosophy was applied for an 80-year old and still active dump, originated from mechanical enriching of Zn/Pb/Cd ore. As the technologies of ore processing used those days were rather primitive, the material deposited on the dumps still contains elevated quantities of metals), amounting totally to 7% Zn, 1% of Pb and above 0,03% of Cd in kg dried soil. The material becomes compacted after geo-chemical modifications and wetting, and as a consequence of mentioned processes its mechanical structure does not support the plant growth. Due to wind and water erosion, the population dwelling in the neighborhood as well as the sur-rounding agriculture mainly vegetables and potatoes, is suffering from elevated

---

[*]To whom correspondence should be addressed: sas@ietu.katowice.pl

L. Simeonov and V. Sargsyan (eds.),
*Soil Chemical Pollution, Risk Assessment, Remediation and Security.*
© Springer Science+Business Media B.V. 2008

metal concentrations. Over the last 20 years, several attempts of revegetation of the dump were performed, using routine methods and commonly applied plant species. However, they were unsuccessful, probably due to phytotoxic concentrations of contaminants and adverse texture of soil. The approaches as described towards diminishing the human exposure, i.e., immobilizing the noxious substances in contaminated ground using either local species colonizing spontaneously the area, or applying the evapotranspiration cover, seems the only feasible way for the time being, in case of developing economies.

**Keywords:** remediation, reclamation, dumps, cadmium, lead, zinc

## 1. Introduction

Soil pollution became a worldwide problem due to various human activities over the centuries. The seriousness of the problem was recognized relatively late, when huge areas of land were already heavily contaminated with persistent pollutants of various origin and different chemical composition. Figure 1 shows the degree of contamination of the foodstuff originated from industrialized region of Poland.

The activities towards soil cleaning started on the second half of the last century. It was quickly found, that the resulting expenditures are prohibitive high, even for wealthy countries. At present, only when the land is of particular value and contaminants are easy to be removed, traditional technical methods of soil clean-up are being applied.

Instead, the other approaches as stabilization of pollutants in the upper layer of soil slightly conditioned with chemicals, or covering the surface of soil with active biological barriers are considered. The above are cost-effective, low-tech and environmentally friendly.

The most recommended, however, for developing economies are those approaches which apply the adequate policy of the land use and require minimal technical involvement. In the case of agricultural land this kind of remediation is also very efficient in reduction of human exposure to pollutants coming from the contaminated crop.

In this paper, an example of such approach is given, where pollution of soil and plant with lead, cadmium and zinc from dumps plays a crucial role in human environmental exposure.

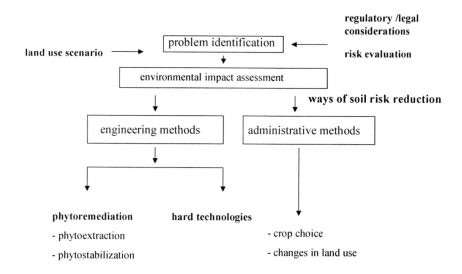

*Figure 1.* Possible scenarios for arable land reclamation.

## 2. Problem Description

Lead and zinc mining and smelting begun at southern part of Poland in the XIIIth century. In the beginning, the ancient explorers were mainly interested in zinc for tableware production, and in silver as an accompanying material, being used for the manufacture of jewelry. In very early period of development, the enterprises made small family businesses, digging the mining shafts up to 6 meters deep, which was limited by the level of watertable (Majorczyk, 1985). A regular industrial activity had begun in the XIXth century, when mine galleries were built and ore was enriched for further smelting. The enrichment technologies, which were used those times were rather uncomplicated, which resulted in the deposition of considerable content of metals in wastes on the heaps. The most common technology was that of ore washing (Girczys and Sobik-Szoltysek, 2002). The heaps were difficult to revegetate due to high concentration of metals, the material texture unfavorable for plant growth and adverse air/water ratio in deposited material.

A typical example of the problem is a spoil pile of heavy metal containing material, situated at Upper Silesia Industrial Region, southern part of Poland, located at the area of a former Pb/Zn/Cd ore mine. The pile is located in the neighborhood of arable land, which is highly contaminated as a result of dump activity. The problem becomes complex, as the remediation processes should be completed after the dump is not active. A professional and definite solution of the problem, as removal to a safe place, or permanent stabilization of the harmful material, is beyond of financial capacities of the present land owner. Therefore, an attempt has been made towards the use of phytostabilization, which is a temporary but cost-effective method of instant reduction of human exposure to pollutants. The method rather isolate than solve the problem, but proves good as a provisional countermeasure (Salt et al., 1995; Vangronsveld, 1998; Vangronsveld and Cuninigham, 1998; Li and Chaney, 1998; Berti et al., 1998; Berti and Cuningham, 2000; Knox et al., 2000).

## 3. Description of the Site

The heap was formed in years 1915–1930 from the waste material after process of ore washing. The ore was dug from a neighboring shaft, non-existing any longer. By virtue of high concentration of metals, a part of the heap was sold in early sixties for direct thermal reprocessing, but the buyer is no longer interested in this kind of activities. At present, the heap makes about one hectare area, sparsely covered with poorly growing plants, which are the remnants of unsuccessful attempts of former reclamation activities. There are also small spots of land, covered with plants of local metal-tolerant species which form a xerothermic grassland, which appear irregularly on the shoulder of the heap. The heap is elevated about five meter above surrounding area, two farm houses and the farming land are located in close distance. Water and air erosion processes create an elevated health risk to population dwelling in the vicinity, due to dust resuspension and metal-contaminated washout polluting the neighboring fields and penetrating to the lower ground levels.

Metal concentrations in heap material and in neighboring soils are shown in Table 1.

TABLE 1. Metal concentrations in soil samples

| Location | Pb (mg/kg) | Cd (mg/kg) | Zn (mg/kg) |
|---|---|---|---|
| Heap | 10,320 | 340 | 68,220 |
| Eastern part of heap | 12,810 | 500 | 51,820 |
| Field close to heap | 1,530 | 440 | 3,680 |
| Field about 500 m from heap | 1,290 | 750 | 5,280 |

Actually there are two problems to be solved on the site:

- Still active heap, polluting the air with resuspended and falling metal-bearing dust. The dust in turn, is inhaled by humans and contaminating local soils and groundwater, also the wash-outs and the leakage from the heap may have a considerable adverse effect on the quality of the local water.
- Local farmland, which is highly polluted due to 80-years activity of the heap.

## 4. Possible Approaches to Reclamation of the Site Considering the State of the Art

The following options of site reclamations are to be taken into account: Each of them has its pros and cons and should be carefully evaluated taking into account technical, environmental and economical considerations.

The best scenario for contaminated arable land are administrative methods which include a right choice of crops non cumulating pollutants, or in extreme cases, elimination of very heavily polluted areas from agricultural production. The other choice might be phytoremediation, particularly phytostabilization (Figure 2), which was successfully implemented on lead and zinc contaminated

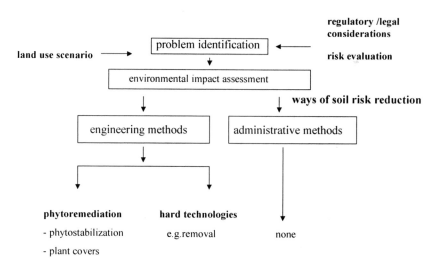

*Figure 2.* Possible scenarios for dump reclamation.

area in Poland (Kucharski et al., 2005). Using local ecotype of *Deschampsia caespitosa*, the phosphate fertilizer-amended soil, a very effective cover was created, whose dense root mat prevented from wind and water erosion. The other biological option was phytoextraction, which was rejected because of unacceptable time consumption.

## 4.1. REMOVAL OF THE HEAP FOR FURTHER PROCESSING

It is a final solution. If a party interested in buying the material can be found, the removal of the heap for the environmentally friendly use would be the best approach to the problem. As a temporary increase of dust emission can be expected when dismantling the heap, the necessary countermeasures should be undertaken to prevent the surroundings against an additional exposure.

## 4.2. STABILIZATION

The commonly used method of revegetation, i.e., covering the heap with a layer of biosolid and planting a ready-made mixture of plant seeds is one of the options. The technology as described, however, is not effective here for a long run. It has already been proven that after relatively short time, the nutrients from the seed and those from the biosolid are getting exhausted by plants and washed out by the run-off water. Moreover, the biodiversity of plants on the area can be easily disturbed, as the newly introduced, well fed species would eliminate quickly the vulnerable native plants.

## 4.3. EVAPOTRANSPIRATION COVER

The evapotranspiration cover (ET) consist of the subsequent layers of material and grass, which isolate the surface from the contaminated deposits underneath.

Typically the ET consists of compacted, locally easily available material which prevents in the same time the leakage of rainfall water. On the surface, the engineered vegetative cover is planted to facilitate water storage and evaporation of moisture into the atmosphere through the process of plats transpiration.

Construction of xerothermic grassland is an option of phytostabilization in which top soil is remodeled towards the grassland needs. This approach would be feasible in case of natural presence of the right species and limited human intervention into process of the plant growth.

## Summary

Considering the existence of numerous unfavorable factors, remediation method used here has to be custom-tailored in terms of artificial soil composition, soil conditioners and water regime, to achieve satisfactory results.

In course of laboratory experiments, factors impairing plant growth are to be identified, specific soil conditioning scheme will be worked out, which would support selectively the growth of required plant species. Finally, plot scale field applications should be conducted to confirm the laboratory findings.

## References

Berti W. R., Cunningham S. D. and Cooper E. M. 1998. Case studies in the field – in-place inactivation and phytorestoration of Pb-contaminated sites. *In* Metal-contaminated soils *in situ* inactivation and phytorestoration. Eds. J. Vangronsveld and S. D. Cunningham. pp. 235–248. Springer, Heidelberg, Berlin, and R.G. Landes Company, Georgetown, TX, USA.

Berti W. R. and Cunningham S. D. 2000 Phytostabilization of metals. *In* Phytoremediation of toxic metals: using plants to clean up the environment. Eds. I. Raskin and B. D. Ensley. pp. 71–88. Wiley, New York, USA.

Girczys J. and Sobik-Szołtysek J. 2002. Odpady przemysłu cynkowo-ołowiowego (Non-ferrous industry waste – in Polish) Monografie. Wydawnictwo Politechniki Częstochowskiej.

Knox A. S., Seaman J., Adriano D. C. and Pierzynski, G. 2000. Chemophytostabilization of metals in contaminated soils. *In* Bioremediation of contaminated soils. Eds. D. L. Wise, D. J. Trantolo, E. J. Cichon, H. I. Inyang and U. Stottmeister. pp. 811–836. Marcel Dekker, New York, Basel.

Kucharski R., Sas-Nowosielska A., Małkowski E., Japenga J., Kuperberg J. M., Pogrzeba M. and Krzyżak J., 2005. The use of indigenous plant species and calcium phosphate for stabilization of metal-polluted sites. Plant and Soil, 273: 291–305.

Li Y. M. and Chaney R. 1998. Case studies in the field – industrial sites: phytostabilization of zinc smelter-contaminated sites: the Palmerton case. *In* Metal-contaminated soils: *in situ* inactivation and phytorestoration. Eds. J. Vangronsveld and S. D. Cunningham. pp. 197–210. Springer, Berlin, Heidelberg and R.G. Landes Company, Georgetown, TX, USA.

Majorczyk R. 1985. Historia gornictwa kruszcowego w rejonie Bytomia. (Non-ferrous ore mining in Bytom area – in Polish). Ed. ZUP Piekary Sl.

Salt D. E., Blaylock M., Kumar N. P. B. A., Dushenkov V., Ensley B. D., Chet I. and Raskin I. 1995. Phytoremediation: a novel strategy for removal of toxic metals from the environment using plants. Biotechnol., 13, 468–474.

Vangronsveld J. 1998. Case studies in the field – industrial sites: phytostabilization of zinc-smelter contaminated site: the Lommel-Maatheid Case. *In* Metal-contaminated soils: *in situ* inactivation and phytorestoration. Eds. J Vangronsveld and S. D. Cunningham. pp. 211–216. Springer, Berlin, Heidelberg and R.G. Landes Company, Georgetown, TX, USA.

Vangronsveld J and Cunningham S. D. 1998. Introduction to the concept. *In* Metal-contaminated soils: *in situ* inactivation and phytorestoration. Eds. J Vangronsveld and S D Cunningham. pp. 1–15. Springer, Berlin, Heidelberg and R.G. Landes Company, Georgetown, TX, USA.

# SOIL REMEDIATION SCENARIOS FOR HEAVY METAL CONTAMINATED SOIL

ALEKSANDRA SAS-NOWOSIELSKA[1*], RAFAŁ KUCHARSKI[1], MARTA POGRZEBA[1] AND EUGENIUSZ MALKOWSKI[2]
[1] *Institute for Ecology of Industrial Areas*
*Kossutha Str. 6, Katowice, Poland*
[2] *Department of Plant Physiology, Faculty of Biology and Environmental Protection, University of Silesia*
*Jagiellonska Str. 28, Katowice, Poland*

**Abstract.** The countries of the former Eastern European Block continue to suffer from environmental contamination resulting from former economic practices, where environmental issues were not given a priority. Considerable environmental progress has already been made over the past decade, especially concerning air quality. Nevertheless, persistent pollutants in soils, particularly heavy metals still remain deposited. Current remedial technologies for soils contaminated with heavy metals involve ex situ physical and chemical methods such as, solidification, electrokinetics, soil washing, pyrometallurgical separation, excavation etc. are generally very costly and unaffordable to developing countries. For such sites introduction of different kinds of phytoremediation technologies seems to be an acceptable solution.

**Keywords:** phytoextraction, phytostabilization, metal uptake, lead, cadmium, zinc

## 1. Introduction

Heavy metals are highly persistent in soils with residence times in the order of hundreds or event thousands of years. Specially, the bioavailable fractions of heavy metals are an issue of particular concern from an ecological, toxicological

---

[*] To whom correspondence should be addressed: sas@ietu.katowice.pl

L. Simeonov and V. Sargsyan (eds.),
*Soil Chemical Pollution, Risk Assessment, Remediation and Security.*
© Springer Science+Business Media B.V. 2008

and health standpoint, as these may easily penetrate most environmental components, including the food chain (Kabata-Pendias and Pendias, 1999; Knox et al., 2001).

Most industrial activities in Poland is located in southern part of the country. Hard coal and polymetalic ores are the geological treasures. As a heritage of long years working industry, completely deteriorated areas, the brownfields and the land surrounding the former or still operating industrial premises had appeared. Anthropogenic pressure to the environment can cause significant degradation of all its compartments especially for soils. Soil protection and rehabilitation of contaminated sites provides the necessary framework for rescuing the integrity of soil and groundwater. The selection of the most appropriate soil remediation methods are based on the site characteristics, concentration and form of existing elements as well as a decision of the end use. Theoretically there is a choice between immobilization, extraction, separation or pollutant isolation. The techniques, which allow stabilizing contaminants in soil using chemical or biological methods could be very practical, considering their technical simplicity and low cost performance (Berti et al., 1998; Salt et al., 1995; Li and Chaney, 1998; Vangronsveld, 1998). The chemicals bind the excess of metals, help to maintain an appropriate pH and required plant nutrition. The well-chosen plants built up a dense root mat, which prevents wind erosion of contaminated material, reduces the leaking of contaminated water and cumulate metals in roots (Vangronsveld and Cuningham, 1998; Berti and Cuningham, 2000; Knox et al., 2000). Phytoremediation offers large-scale, cost-effective, on-site treatment of contaminated land areas (Li and Chaney, 1998; Vangronsveld, 1998; Simeonov et al., 1999; Simeonova and Simeonov, 2006, 2006a).

In the presented article, soil remediation scenarios for heavy metal contaminated soil will be discussed. A special intention will be put on biological methods, specially, phytoextraction and stabilization.

One of the major obstacles of the efficiency of phytoremediation process is a high concentration of pollutants in the soil, which impairs the plant growth essential for producing the high biomass. The surface layer of the soil can be even contaminated to the extent, which makes the germination very difficult.

The described experiment was aimed at investigating the feasibility of use the ready seedlings prepared in a clean place, to receive the high plant biomass on the contaminated site.

## 2. Methods

The soil for experiment was collected from the vicinity of former non-ferrous metal smelter, located in Upper Silesia Region, southern part of Poland. Pots were located in growth chambers with supplemental high-intensity discharge sodium vapor lamps (from 12,500 to 15,000 Lx). The average temperature was 22°C ±1, humidity 58–62%, and photoperiod 16h/8h day/night. The initial soil moisture content was 60% of total water capacity. During experiment water was added if necessary. The experiment was performed in five replications. Pot positions were changed periodically to equalize light exposure. The growth, health, and vitality of investigated plants were observed.

The following plants species were tested: *Helianthus annuus* cv. Phyto and Albena, *Zea mays* cv. Hermes and Electra, *Ricinus communis* and *Canna x generalis*. In case of high soil metal concentrations the growth of plant seedlings might be restricted. To avoid this problem transplanting procedure was used according to Wu et al. (1999) and Cooper et al. (1999).

The optimal soil amendment application protocols was applied (5 mmol/kg EDTA and 5 mmol/kg acetic acid). Two methods of amendments application were used:

- Plants were treated directly after transplantation.
- Plants were treated one week after transplantation.

Soil was analyzed for metals and nutrients. Soil pH, electroconductivity (EC), organic matter (OM), and content of nutrients were determined according to the ISO methodology. Bioavailable metals fraction was determined after extraction of air-dried soil ground to <0.25 mm with 0.01 M $CaCl_2$ for 5 h; total metal concentrations by extraction with concentrated $HNO_3$ and HF.

Plant material was washed with tap water in an ultrasonic washer to remove soil particles and then dried at 70°C for three days. Approximately 1 g of dried ground material was wet-ashed using concentrated nitric acid (Merck) in a microwave system (MDS 2000, CEM, USA).

Concentration of metals in plant material and soil extracts was analyzed by flame atomic absorption spectrophotometer (Varian Spectra AA300) or by inductive coupled plasma spectroscopy (ICP-AES) (Varian, USA). Data reported in this paper were subjected to one way ANOVA using the computer software Statistica for Windows (Statistica'99). Where significant differences were found, differences among means were calculated using the LSD test. A probability of 0.05 or less was considered to be statistically significant.

## 3. Results and Discussion

Soil cleaning technologies are generally costly and unaffordable for developing economies. In case of different soil properties, contaminant removal should be done individually using the most appropriate remediation scenario. For contaminated sites, the integration of different kind of phytoremediation technologies seems to be acceptable solution. It establishes priorities and offers various ways to assess and manage the contaminated sites. In addition to using specific criteria, the policy provides guidelines for assessment and rehabilitation through risk analysis and management.

A highly contaminated area located in the vicinity of former lead and zinc smelter, served as a pilot site to facilitate a right choice of cleaning-up technology. The study was carried out in two steps, i.e., soil characterization and treatability study. Treatability study evaluate the growth potential of selected plant species and suggest soil amendment to optimize plant growth and metals uptake.

Based on analytical results, the soil was classified as loamy, with neutral pH, and medium OM content. The soil was very highly polluted with metals (total Pb – 8,992 mg/kg, Zn – 11,280 mg/kg, Cd – 530 mg/kg). About 73% of Pb, 69% of Zn – and 68% of Cd – existed in potentially bioavailable form (0.43 N $HNO_3$ extraction), whereas 0.06%, 3.3% and 8% in soil solution (0.01M $Ca(NO_3)_2$ extraction) respectively. The accompanying elements included Mn (378 ppm), Fe (2,528 ppm), Cr (2.3 ppm), Mg (713 ppm), Ca (3,589 ppm), Cu (37 ppm); As (187 ppm) and Sb (3.7 ppm).

In spite of very good care taken (fertilizing, watering) plants used for experiment have shown a symptom of very poor growth. The soil did not support the plant growth. There was a vague chance of recovering for some plants planting from seeds. Condition of transplanted plants during experiment was acceptable. The most valuable results was presented on Figures 1, 2 and 3.

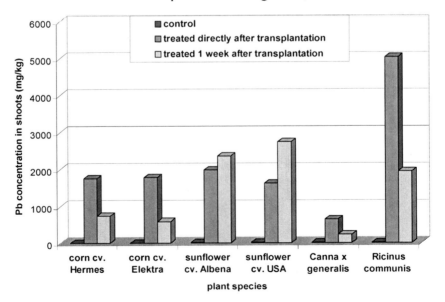

*Figure 1.* Comparison of lead concentration (mg/kg DW) in plants in relation to terms of amendment application to transplanted plants.

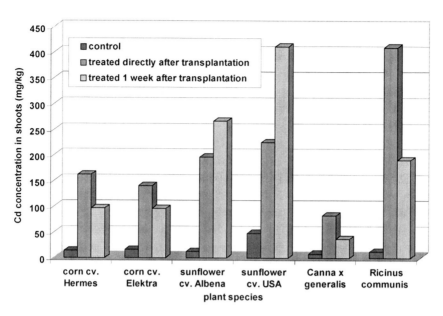

*Figure 2.* Comparison of cadmium concentration (mg/kg DW) in plants in relation to terms of amendment application to transplanted plants.

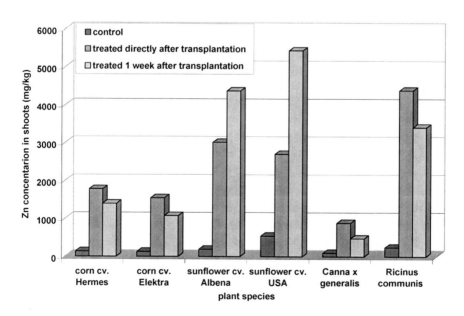

*Figure 3.* Comparison of zinc concentration (mg/kg DW) in plants in relation to terms of amendment application to transplanted plants.

Concentrations of Pb, Cd and Zn in shoots of transplanted plants were evaluated. The highest concentrations of the heavy metals were observed in shoots of both sunflowers cultivars and *Ricinus communis*. In sunflower cultivars the content of Pb, Cd and Zn was higher when plants were treated one week after transplantation. In contrary in case of *Ricinus communis*, higher concentrations of metals were observed in plants treated directly after transplantation. Presented results states that fitoextraction should be suggested only for low or medium contaminated sites.

For highly contaminated soil, stabilization seems to be the most appropriate method for remediation. Locally easily available soil additives which are known as heavy metal stabilizers (sewage sludge, mixture of dolomite and zeolite (Biodecol), zeolites, lignite, ammonium polyphosphate or calcium phosphate) may help a lot. Additives as Biodecol and sewage sludge may slightly change the pH and EC in soil solution but the rest of additives reduce concentration of bioavailable form of metals.

Plant species play a crucial role in plant cover creation. The species as *Agrostis capillaris*, *Salix viminalis*, *Festuca rubra*, *Armoracia lapathifolia* and *Helianthus tuberosus* are not suggested for this purpose (PhytoDec Project). Especially in extremely polluted sites, the use of local vegetation may give an opportunity to create a soil cover. Polluted site related species have already proven its ability to survive under pollution stress. Looking for more suitable species of plants, screening should be focused on local vegetation. A very good growth was observed of indigenous plant species *Deschampsia caespitosa L.*, *Silene inflata* and *Melandrium album* on the soil close to the foundry.

The techniques which allow for stabilizing pollutants in soil using chemical or biological methods could be very practical, considering their technical simplicity and low cost performance. The chemicals bind the excess of metals and help to maintain an appropriate pH and required plant nutrition. The well-chosen plants built up a dense root mat, which prevents wind erosion of contaminated material, reduces the leaking of contaminated water and cumulate metals in roots. Phytoextraction offers large-scale, cost-effective, on-site treatment of contaminated areas. This technology is suggested for remediation of medium or low heavy metal contaminated soils.

The combination of both approaches was successfully implemented in case of lead, cadmium and zinc contaminated megasite in southern part of Poland.

# References

Berti W. R., Cunningham S. D. and Cooper E. M. 1998. Case studies in the field – in-place inactivation and phytorestoration of Pb-contaminated sites. *In* Metal-contaminated soils *in situ* inactivation and phytorestoration. Eds. J. Vangronsveld and S. D. Cunningham. pp. 235–248. Springer, Berlin, Heidelberg and R.G. Landes Company, Georgetown, TX, USA.

Berti W. R. and Cunningham S. D. 2000. Phytostabilization of metals. *In* Phytoremediation of toxic metals: using plants to clean up the environment. Eds. I. Raskin and B. D. Ensley. pp. 71–88. Wiley, New York, USA.

Cooper E. M., Sims J. T., Cunningham S. D., Huang J. W. and Berti W. R. 1999. Chelate-assisted phytoextraction of lead from contaminated soils. J. Environ. Qual., 28: 1709–1719.

Kabata-Pendias, A. and Pendias, H., 1999. Biogeochemistry of Trace Elements. PWN, Warsaw, Poland (in Polish).

Knox A. S., Seaman J., Adriano D. C. and Pierzynski G. 2000. Chemophytostabilization of metals in contaminated soils. *In* Bioremediation of contaminated soils. Eds. D. L. Wise, D. J. Trantolo, E. J. Cichon, H. I. Inyang, and U. Stottmeister. pp. 811–836 Marcel Dekker, New York, Basel.

Knox A. S., Seaman J. C., Mench M. J. and Vangronsveld J. 2001. Remediation of metal- and radionuclides-contaminated soils by *in situ* stabilization techniques. In Environmental Restoration of Metal-contaminated Soils. Ed. I. K. Iskandar. pp. 21–60. Lewis Publishers, Boca Raton, London, New York, Washington, DC.

Li Y. M. and Chaney R. 1998. Case studies in the field – industrial sites: phytostabilization of zinc smelter-contaminated sites: the Palmerton case. *In* Metal-contaminated soils: *in situ* inactivation and phytorestoration. Eds. J. Vangronsveld and S. D. Cunningham. pp. 197–210. Springer, Berlin, Heidelberg and R.G. Landes Company, Georgetown, TX, USA.

Salt D. E., Blaylock M., Kumar N. P. B. A., Dushenkov V., Ensley B. D., Chet I. and Raskin I. 1995. Phytoremediation: a novel strategy for removal of toxic metals from the environment using plants. Biotechnol., 13, 468–474.

Simeonov L. I., Simeonova B. G. and Nikolova J. 1999. Phytoremediation of Industrially Polluted with Heavy Metals Lands in Bulgaria (First trials), Proc. of 1999 Contaminated Site Remediation Conf.: Challenges Posed by Urban and Industrial Contaminants, Freemantle, Australia, 551–557.

Simeonova B. and Simeonov L. 2006. An application of a phytoremediation technology in Bulgaria. The Kremikovtzi Steel Works experiment. Remediation Journal, Spring edition 2006, Wiley Periodicals, New York, pp. 113–123.

Simeonova B. and Simeonov, L. 2006a. Planning and execution a phytoremediation pilot experiment. In Chemicals as intentional and accidental global environmental threats, Borovetz, 2006, L. Simeonov, E. Chirila (eds). NATO Science Series C, Environmental Security, Springer, Dordrecht, pp. 297–303.

Cooper E. M., Sims J. T., Cunningham S. D., Huang J. W. and Berti W. R. 1999. Chelate-assisted phytoextraction of lead from contaminated soils. J. Environ. Qual., 28: 1709–1719.

Vangronsveld J. 1998. Case studies in the field – industrial sites: phytostabilization of zinc-smelter contaminated site: the Lommel-Maatheid Case. *In* Metal-contaminated soils: *in situ* inactivation and phytorestoration. Eds. J. Vangronsveld and S. D. Cunningham. pp. 211–216. Springer, Berlin, Heidelberg and R.G. Landes Company, Georgetown, TX, USA.

Vangronsveld J. and Cunningham S. D. 1998. Introduction to the concept. *In* Metal-contaminated soils: *in situ* inactivation and phytorestoration. Eds. J. Vangronsveld and S. D. Cunningham. pp. 1–15. Springer, Berlin, Heidelberg and R.G. Landes Company, Georgetown, TX, USA.

# DEGRADATION OF SOIL IN KAZAKHSTAN: PROBLEMS AND CHALLENGES

NURLAN ALMAGANBETOV[*] AND VLADIMIR GRIGORUK
*Land relation Department*
*Kazakh Scientific Institute of Economy, Agro-Industrial Complex and Rural Area Development,*
*30-B, Satpaeva Str., 050057 Almaty, Republic of Kazakhstan*

**Abstract.** Despite, agriculture in Kazakhstan has substantial potential with total agricultural land in 2005 accounting for 84.5 million hectares of which 20.5 hectares are arable Kazakhstan inherited to some environmental problems, affecting to health and living conditions of rural habitants. Unsustainable land practices, non-rational use of natural resources and environmental pollutions have led to varying degrees of land degradation and desertification in all regions of Kazakhstan. The process of land degradation and desertification occurs in high extend in regions with unfavourable ecological conditions such as the Aral and Caspian Sea, Lake Balhash regions. The deterioration of land quality also is caused by the process of soils contamination with oil and oil products, particles of heavy metals, radio nuclides and other pollutants. The national strategy for prevention and reduction of scales of land degradation and desertification is to be aimed to address a reversal of environmental degradation and loss of biodiversity, development of sustainable land use in marginal areas, which would contribute to increase of income and living conditions of rural habitants.

**Keywords:** land resources, ecosystems, degradation of lands; soil erosion, land contaminations and desertification

## 1. Introduction

Kazakhstan has a territory of 2,724, 900 sq. km, which makes it the ninth largest country in the world, roughly the same size as western Europe.

---

[*] To whom correspondence should be addressed: almag@nursat.kz

L. Simeonov and V. Sargsyan (eds.),
*Soil Chemical Pollution, Risk Assessment, Remediation and Security.*
© Springer Science+Business Media B.V. 2008

Kazakhstan is totally landlocked (UNDP, 2002). The country is one huge plain, sloping from north east to south west. The plain is bordered by extensive maintain ranges on the east and south-eastern borders. Its neighbours are Russia in the north and west, China in the east, and Kyrgyzstan, Turkmenistan and Uzbekistan in the south (ADB, 2004). Kazakhstan lies between the Siberian Taiga in the north and the Central Asia deserts in the south, the Caspian Sea in the west and the mountain range of the Tien-Shan and Altay in the east (UNDP, 2002).

Kazakhstan has a continental climate; temperatures in the north average – 18°C in January, while temperatures in the south are an average of –3°C in January. Summers are generally dry and the average temperature in July increases gradually from 19°C in the north to 28–30°C in the south. Precipitation in plain areas is generally low from 400 mm annually in the north to 100 mm annually in the southwest. In the mountainous regions, the precipitation ranges from 400–1,600 mm (ADB, 2003).

The rural Kazakhstan accounts for over 95% of Kazakhstan's territory, or approximately 260 million hectares. Approximately 43.7% of the population (total 14.9 million) live in rural area and 30.9% of the rural population live on less than the minimum subsistence income (approximately 35 USD per month) (ADB, 2004).

## 2. Land Resources

Kazakhstan has three principal ecosystems: desert areas, accounting for approximately 55% of rural areas; the steppe (grassland), accounting for approximately 30%; and mountains and foothills, accounting for the remaining 15% (ADB, 2003). These ecosystems provide a varied base for rural economic development encompassing agriculture, fishing, hunting, recreation/tourism, extractive industries and the further processing of natural raw materials (UNDP, 2002).

The Kazakhstan's climate makes possible to cultivate wheat, barley, oats and rye in most regions. Irrigated land in the southern regions provides the growing temperature sensitive crops such as cotton, tobacco, rice, sugar beet, grapes and other fruits. Natural pastures accounts for 187. 9 million hectares of land, which is enough to feed 70.5 million head of sheep or 7.05 million cattle. Kazakhstan's climate is favourable for live-stock farming. It is traditional occupation of Kazakh people and most pastures have been utilised during the year as a forage base (UNDP, 2002).

In 1990 the total area of agricultural land in Kazakhstan was about 220 million hectares. Out of this figure 35 million hectares of land was cultivated and over 180 million hectares of steppe was utilised as grazing pasture for cattle breeding (AS, 2002). Following land and agrarian reforms the agricultural economy suffered a deep financial crisis that lead to a decrease of the total area of cultivated land. The agricultural lands decreased by 2.5 times in comparison with 1990 year and the total area of agricultural lands is currently about 84.5 million hectares with arable land accounting for 20.5 million hectares and the remaining land are used for pasture and hayfields, amounting to 59 million and 2 million hectares respectively. All these abovementioned agricultural lands have been distributed to land users, including 20.5 million hectares of arable land (AS, 2002). The agricultural system of Kazakhstan consists of three main natural category resources such as rain-fed land, irrigated land and pasture.

The actual area of sustained competitive rainfed agriculture remains an unanswered question. The area of good quality soils is about 12 million hectares, mostly in the north of the country, but the area with quality soils and normally adequate annual rainfall for arable farming is less. The area suited to rainfed agriculture is mainly concentrated in the north and east. Even in these areas the expectation is that one year in three or four will be one of inadequate rainfall. Thus, rainfed arable agricultural activity in Kazakhstan is not only restricted, geographically, but faces high risks from climatic uncertainties, which impact on the productive capacity of farming entities (ADB, 2003).

The natural resource base provides the opportunity to irrigate substantial areas of reasonable quality soils, which is one way of reducing production risks (ADB, 2003). In the early 1990s Kazakhstan had 2.3 million hectares of irrigated land, which accounted for 6% of total sown area, yielding up to 30% of crop production. Subsequently, the area of irrigated agriculture has been reduced to 1.2 million hectares due to water shortages and deterioration of the infrastructure systems and yields have fallen 1.5–2 times. Most of the irrigated area is concentrated in the southern part of the country and has also undergone a contraction. The land reclamation qualities of soil have been deteriorating, while the technical condition of water stations has also worsened (UNDP, 2002).

The third natural resource category is permanent pasture (mostly steppe), which is the dominant natural vegetation of Kazakhstan. This category covers 90% of the country, but the quality of the pasture varies considerably depending on soil quality, temperature and rainfall regimes. The foothills of the mountains and parts of northern Kazakhstan support quality pasture but large areas are arid and support only scrub vegetation. The scale and range of quality of the natural grasslands determines that livestock systems must be the dominant form of agricultural system over much of the country (ADB, 2003).

The natural environment that supports agricultural production systems is fragile due to past inadequate attention to sustainable production practices which meant that substantial areas of lands suffered from saline soils, water logging, soil erosion and desertification. The loss of quality topsoil over the past 30 years is well recognised and parts of the black soil area have lost up to

30% of their humus content (ADB, 2003). All these abovementioned changes of soil qualities are determined as land degradation.

## 3. Degradation of Lands in Kazakhstan

The term of "land degradation" is broadly defined as "any form of deterioration of natural potentiality of land, which influences on integrity of ecosystems in relation to decreasing its stable ecological productivity or natural biological welfares and maintenance of sustainability". It is widely common phenomenon, basically influencing on productivity more than in 80 countries. The land degradation negatively influences on ecological integrity and yields productivity of 2 billions hectares or 23% of a landscape used by a man. Negative influence of land degradation is ecological and social-economic processes and has already jeopardize means of subsistence, economic welfare and situation with nutrition of more than 1 billion people in less developed countries by assessment of the Global Ecology Fund (GEF, 2003).

The main causes of soil degradation are unsustainable agricultural practices, excessive use of pastures for animal feeding, i.e. overgrazing of pastures and deforestation and forest degradation (GEF, 2003). These practices prevail in places, where lands, water and other natural resources are underestimated.

One example of inadequate use of agricultural lands is the Virgin Land Scheme in time of the Soviet Union. Under this scheme, during 1950s and 1960s cereal cultivation was extended to central part of Kazakhstan, with the objective to of increasing wheat production in the Former Soviet Union. Approximately, 35 million hectares mainly in steppe area were put under cereal production although the yield on these arable lands is unpredictable and unstable as a result of changeable climatic conditions and vulnerability of arable lands to wind erosion (WB, 2003). Thus, this extensive development of agricultural production has caused degradation of agricultural land and impoverishment of landscapes such ploughing the steppes and lay land.

Previously used system of land use and allocation of production capacities, without considering relating ecological limitation have led to forming regions and zones with increased technological influences and risks and ecological disaster for living people there. Thus, the land degradation is extended in these key regions with unfavourable ecologic conditions such as Aral and Caspian seas, Balhash lake regions.

Currently, the main cause of land degradation is a lack of attention to rational use of lands. It is relevant in cases, when people, which are not having land possession rights do not show the initiative to contributing to sustainable land management. Instead of it, they tend to concentrate on short-term economic needs to the detriment of an environment.

## 3.1. THE EROSION OF SOILS

The total area of agricultural lands, incurred to water erosion is about 5 million hectares including 1 million hectares of arable lands according to data on qualitative characteristics of lands. The biggest areas of washed off soils are located in the southern and western part of Kazakhstan and total area of agricultural lands with washed off soils is about 2.1 millions hectares and 1.27 millions hectares, relatively.

The total area of agricultural land with soils, incurred to wind erosion is about 25.5 million hectares. The wind erosion prevails on whole territory of the Republic of Kazakhstan, but agricultural lands, presented with carbonate soils and soils with easy mechanical structure are suffered significantly. The main areas, incurred to wind erosion are the southern, western part of Kazakhstan. The total area of eroded lands is about 13.1 million and 12.4 million hectares, relatively. As regards of the northern and eastern part of Kazakhstan the total area of eroded agricultural lands amounts of 3.87 million hectares and 1.28 million hectares, relatively.

Some agricultural lands incurred to wind and water erosions. The total area of such agricultural lands is about 0.2 million hectares and these lands are located, mainly in West Kazakhstan region. Thus, the total area of eroded agricultural lands is about 30.7 million hectares and the arable land incurred to all types of erosion amounts for 1.7 millions hectares.

The soils of 81.1% of arable lands or 1.4 million hectares are eroded inconsiderably and these arable lands require simple anti-erosion activities. 18.1% of arable lands or 309,000 hectares need intensive anti-erosion land improvements, because soils of these lands are eroded in medium extent. 13,300 hectares or 0.8% of arable land require conducting complicated anti-erosion land improvements such as sowing of these arable lands with perennial grasses (ALRM, 2006).

## 3.2. CONTAMINATIONS OF LANDS

The negative influence on qualitative condition of soils in great extent is determined by the process of soils contamination. The contamination of land soils revokes deterioration of living conditions of local people, changes of conditions of soils, diminishing quality of agricultural products. The basic polluting substances are radionuclide, heavy metals, oil and oils products.

The enterprises of oil and gas sectors take a first places in soils contamination with various chemical compounds. Regions of oil and gas extraction are concentrated on the western and south-western part of Kazakhstan and in the south (Kizil-Orda region). In spite of oil and gas industry takes a first place among other industries of Republic of Kazakhstan an exploitation of oil and gas fields is conducted by using obsolete technologies and outdated equipments, which leads to pouring of oils.

The total area of oil contaminations in West Kazakhstan is 194,000 hectares and volume of out flowed oils is more than 5 million tons according to official data. But the UNDP program in its review "Environment and sustainable development in Kazakhstan" has reported that total area of soils, contaminated by oil products is more 1.3 million hectares and thickness of soil contamination reaches 10 meters in some oil fields. Two hundred and sixty-seven sites of radio-active pollution with various levels of radioactive contamination are revealed in 22 largest oil fields. The total area of radioactive contamination amounts 650 hectares and volume of radioactive waste is 1.3 million cubic meters.

The main source of soil contamination is overflow of oil to land barns, pouring of oils and water-oil mixes at break of oil pipelines, outflow of oils on surface of soils in repairing of well. The destruction of vegetation on 70–80% in radius of 500–800 m happens around each boring machine on exploratory and extracting drilling. Thus, the size of soil contaminations might be diminished by the introducing rigid environmental control over activities of oil companies particular in contaminations of environment.

Kazakhstan has suffered from activity of a nuclear testing range in Semi-palatinsk region. The Semipalatinsk nuclear testing range has been created on August 21 of 1947 and was situated in territory of three regions – East Kazakhstan, Karaganda, Pavlodar and occupied the area of 18,500 sq. km. Four hundred and seventy nuclear explosions were made on Semipalatinsk nuclear testing centre over period of 1949 and 1991 years and about 2 million hectares of agricultural lands in this region were incurred to radioactive contamination mainly by radionuclide.

Seventeen nuclear explosions were made on the nuclear testing range "Azgir" located in Atyray region and as a result of which 9 underground cavities and one artificial lake in diameter of 600 m were formed. Cavities were filled by a salt solution which is polluted by radio nuclides.

The range "Kapustin Iar" is located in West Kazakhstan and Aryray regions and occupies territory from above 3 million hectares. Some regions of Atyray and West Kazakhstan have incurred to radioactive contamination.

The tight ecological conditions in central part of Kazakhstan are connected to activity on a complex of "Baikonur", where environmental services are power-less, because of deficiencies of the legislative bases. The contamination of lands occurs in territory of the complex and also in areas of falling separated parts of space rockets during flight of space rockets. The area of falling pace rockets parts takes significant territories, located in Karaganda, Akmola, Pavlodar and East Kazakhstan regions. The total area of lands, contaminated with products of rocket fuel combustion and separated rocket parts is about 9.6 million hectares according to estimates of experts.

The complex of "Baikonur" makes a negative influence on a soil cover in its surrounding territories. This occurs by contaminations of soils with pouring of fuels and waste products. The greatest contamination of soils happens in emergency situations on start up of rockets.

Development of the mining industry has strengthened process of soil contamination by toxic substances. In particular, according to the information of the Ministry of Environment protection of the Republic Kazakhstan, as a result of mining work it was formed 4 billion tons of waste heaps, including 1.1 billion tons of wastes of ores enrichment and extracted minerals, 105 million tons of metallurgical enterprises waste.

The significant waste heaps from mining and processing have been formed by enterprises of Karaganda, Aktubinsk, East Kazakhstan, Pavlodar, Zambul, West Kazakhstan and Atyray regions, involved in extraction of coals, ferrous metals and phosphorites. The land of East Kazakhstan oblast is contaminated by combination of copper, zinc, cadmium, lead, arsenic. The soils contamination in the Karaganda region is caused by waste of coal, mining and steel industry. The basic source of land contamination in Kizil-Orda region is an activity of oil companies, making pollution of environment with heavy metals and oil products. Thus, an activity of mining, oil and metallurgical industries enterprises are main cause of heavy land contamination in central and eastern and western part of Kazakhstan (ALRM, 2006).

## 3.3. DESERTIFICATTION OF LANDS

The desertification of land is a degradation of lands in droughty, semi-droughty and dry areas in result of various factors action. The problem of desertification is crucial for Kazakhstan with its significant part of territory, occupied with sand (up to 30 million hectares) and saline lands (about 34 million hectares). The natural features of Kazakhstan stipulate weak sustainability to anthropogenic influences. The desert and steppe ecosystems are very sensible to anthropogenic impact and can be easily destroyed as they have low renewal capability (ALRM, 2006).

Currently, all administrative regions of Kazakhstan are involved in the process of desertification and there is tendency for their acceleration. 179.9 million hectares out of 272.5 millions hectares or 66% of total Kazakhstan's area are incurred to desertification (CSD, 2002). According to data of National Program for prevention of desertification in the Republic of Kazakhstan on 2005–2015 years, elaborated by the Ministry of Environmental Protection about 75% of Kazakhstan's territory is a subject of high risk of ecological disaster (NPSD, 2005).

Out of 188.9 million hectares of pasture 26.6 million hectares have incurred to strong desertification and this data emphasizes a high level of desertification. The pastures, locating to villages and well have incurred to high level of desertification.

The soils of all arable lands incurred to desertification have lost humus. The total area of arable lands incurred to losses of humus is 11.2 millions hectares, including 5.2 millions hectares to moderate extent and 1.5 millions hectares to high extent of humus losses. The losses of humus on irrigated lands are revealed on 0.7 million hectares. The salinity of irrigated lands causes growth of saline lands and secondary salinity of irrigated lands. The share of saline soils amounts 31.3% of total irrigated lands.

To process of desertification takes a place in a great extend in the regions with unfavourable ecological conditions such regions as Aral Sea, Caspian Sea and Lake Balhash regions. The desertification of Caspian Sea region is connected mainly with intensive research and extracting natural resources and exploitation of oil and gas fields.

The pastures of this region are located in desert areas, which characterised with extreme natural and climatic conditions such as law precipitation, high temperatures and continental climate with sharp change of daily and seasonal temperatures and strong winds as well. The soils of this area are less sustainable to such factors and the degradation of soil and vegetative covers together with soils erosion are common in this area.

The situation is worsened with intensive exploitation of oil and gas fields and transportation of these hydrocarbons and these activities are an ecology dangerous type of business activities. Mechanical infringements soil – vegetative ecosystems, contaminations of soils by toxic chemical substances such as oil and oil products, metals and compound of sulphur negatively influence on soil of pastures in this region.

The area of ecology disaster of Aral Sea region occupies 59.6 million hectares and agricultural lands in this region amount for 43.4 million hectares. The agricultural lands consist of 42.4 million hectares of pasture, 0.6 millions of arable lands and 0.4 millions of hay fields. Almost all lands are saline in the delta of the Syrdariay river. The area of arable saline lands in Kazakhstan part of Aral Sea has increased in two times and content of humus has decreased on 30–50% for last 15 years. As results of diminishing of Aral Sea size and unevenly growing of vegetation on dried place of this sea the lands with sands and saline near Aral Sea have increased significantly.

The degradation of natural hay lands and pastures continues. As result the pastures of spring and summer seasons of use began to disappear in this region, because ecology situation is unfavourable. The area of hay lands in this region to compare before starting crisis period has decreased on 140,000 hectares and about 47,000 hectares are littered by poisonous plants.

The area of Lake Balhash embraces considerable territory and includes northern part of Almaty oblast and south-eastern part of Karaganda region and northern part of Zhambul region. The agricultural lands basically are presented with deserted pastures (91.6%). The significant areas of this region lands are incurred to erosion, salinity and chemical pollutions. Non-ferrous industries such production of copper and molybdenum ores near Balhash city negatively influence on soils in this region.

The intensive use of water of from Ili, Karatal, Aksu and Lepsi rivers in Almaty oblast had led to decreasing a level of water in Lake Balhash. The ecology infringements has happened in result of regulation of the river Ili flowing and creation of a water reservoir in Kapshagai. This had led to salinity of considerable part of lands and degradation of soils near Balhash Lake (ALRM, 2006).

The problem of soils degradation as consequence of desertification is one serious problem in land issues in Kazakhstan. Desertification in Kazakhstan is real threat for sustainable development of a society, loss and depletion of soil, water, forest and pasture resources. This concerns increase of poverty, unemployment, decrease of environmental security and economic capacity of farming entities. Poverty, pollution and environment degradation jeopardise people health, and leads to establishment of unstable economy and over-consumption of the natural resources (CSD, 2002).

Desertification is a problem that has environmental and social-economic consequences such as empty barns and granaries, famished cattle, and seeded fields covered by coming sands. Therefore, desertification leads to great losses of economic resources (CSD, 2002).

Economic consequences of lands desertification and degradation are decrease in yields of cereals, reduction of a number of sheep, cattle, horses and camels, decrease of volume of agricultural products export, stagnation in development food and other relating industries and sharp decrease in paid proceeds from agricultural and processing sectors.

According to prior calculation the loss from pasture degradation in Kazakhstan is 963.2 million USD annually. The missed profit from soil erosion is 779 millions USD annually. The missed profit from re-salinity, swamping, etc., is 375 millions USD. Annual damage from humus loss is 2.5 billions USD (CSD, 2002).

The desertification also negatively impact on several local human generation, i.e., birth and natality decrease, death and country people's migration increase. The country people's migration is confirmed by the decrease of the number of people working in the farming sector. Nature degradation reduced the environmental and productivity security of the country, economic capacity of incomes and consumption, increased sickness and death rate, migration of population (CSD, 2002).

An intensive contamination of soils, waters and airs, an exhaustion of natural resources have led to destruction of ecosystems, to desertification of lands and significant losses of biodiversity. On preliminary estimates the damages from degradation of pastures, the missed benefit of erosion and re-salinity of farm-lands makes about 300 billion tenge by official data (CES, 2003).

## 4.  Review of Some International Project Aimed for Development of Sustainable Land Management System in Deserted Lands

A number of projects related to sustainable land management have been implemented in Kazakhstan by international donors and agencies during last decade. The systematic appraisal of these projects implementation is not pos-sible due to involvement of various agencies and donor organizations. In many cases, projects operated locally, both duplicating and large-scale applications of results of projects were not observed. The reasons is in specific technical nature of these projects aimed on solving local problems, insufficient publications of learnt lessons and results, which can be applied, or difficulties  to duplicate the project approaches based on external financing for the same projects without such financial sources.

The pilot project on sustainable land management was implemented by GTZ CCD jointly with the Institute on Ecology and Sustainable Development, which acted as a national partner. The deserted pasture and damp pastures prevail in a pilot zone, which is located in the southern part of Balhash region. The project had paid significant attention in development of sustainable management of pastures and recommendations to local farmers, involved in cattle breeding.

The cooperation of scientists with local farmers has a potentiality to make scientific knowledge practically applicable for needs of farmers. But it is very difficult to develop approaches and methods without further external financing. The elaborated activities for sustainable pasture management required financial resources and made impossible to apply by local farmers. Nevertheless, the Project team tried to adapt new approaches and intensify involvement of local farmers in project activities. Thus, project had difficulties to duplicate project approaches without financial support as it was done sustainable pasture management activities by a Project team during implementation of this project.

The World Bank implements Drylands Management Project by request of the GEF in Kazakhstan. The development objective of this project is to demon-strate and promote sustainable land uses in the marginal dryland ecosystem of a pilot area in the Shetsky rayon of Karaganda region. This pilot project intends to test the environmental, social and economic viability of shifting from the current unsustainable cereal-based production system to the traditional livestock-based production system.

In order to achieve the above-mentioned development objectives the Project aimed to develop sustainable land use systems, provide an initial service support to producer groups and improve national capacity to quantify carbon sequestration. Moreover, the project intended to develop a strategy so that project interventions could be replicated in similar areas of Kazakhstan and other Central Asian countries. By promoting sustainable land use practices, the Project emphasizes an integrated ecosystem management approach to achieving ecological, economic and social goals that are expected to yield benefits at a local, regional and global level (WB, 2003).

The activities of this Project are supposed to re-cultivate abandoned arable lands in dry steppe area and to achieve global improvements of environment. But this Project is more agricultural investment project rather than environmental project by essence. Declaring about regeneration of natural vegetation of steppe and preservation of a biodiversity, significant carbon sequestration the Project actually implements sowing of fodder vegetation, but not for natural flora of steppe. Thus, value of a biodiversity of artificial fodder vegetation is not higher than fallow vegetation. The carbon sequestration of fodder plants with prevailing lands biomass is limited in comparison with natural vegetation of steppe.

This project presents itself as a technology for improvement of pastures and cultivation of fodder crops for animal breeding and it is doubtful that farmers, involved in animal breeding would be able to implement the activities under this technology without external assistance and financing. Besides, the Project had shown that insufficient consideration and utilization of ecological knowledge in elaborating Project aimed to reverse environmental degradation may lead to improper use of funds, directed for achieving regional and economic goals or rather than for benefit of individual enterprises and certain groups of farmers.

## Conclusions and Recommendations

Although the Kazakh officials are aware about problems, connected with land desertification and contaminations of soils and relevant state program aimed for preventing land desertification had been already adopted the lack of efficient strategy for environment protecting, including preventing land desertification and degradation is evident.

In order to elaborate the efficient strategy for environment protection and restoration of degraded lands the relevant analytical ecological knowledge should be considered and applied in elaboration of this strategy by the policy-makers and officials.

This strategy should be addressed by applying efficient policy instruments for environment protection that promote sustainable management of natural resources and overall environmental sustainability.

This national program for struggle against land desertification should be in line with this strategy and include restoration of degraded lands and fertility of soils, development and introduction of economic mechanisms for sustainable land management and restoration of natural resource base, ensuring of strengthening ecological securities.

The goals of this Program should be prevention and reduction of lands desertification processes and scales of lands degradation, development of conservation and sustainable use of land resources, building institutional capacity for addressing lands degradation issues and increase of agricultural production efficiency.

## References

Agency for Statistics of the Republic of Kazakhstan, Statistical Yearbook of Kazakhstan, 2002.

Agency for Land Resources Management of the Republic of Kazakhstan, Consolidated.

Analytical Report on Land Conditions and Use in the Republic of Kazakhstan, 2006.

Asian Development Bank, TA No 3898-KAZ "Participatory Rural Sector Planning and Development", February, 2003.

Asian Development Bank, TA No 6078- REG "Rural Finance in Central Asia", 2004.

Commission on Sustainable Development, UNDP, Country Profile Report, World Summit on Sustainable Development, Johannesburg, 2002.

Concept on Environmental safety for the period 2004–2015, Government Resolution # of Kazakhstan, President Decree # 1241, 3 December, 2003.

Global Environment Facility, Operating Program on Sustainable Land Management # 15, 2003.

National Program for struggle against land desertification in the Republic of Kazakhstan for 2005–2015 years, Government Resolution # 49, 24 January, 2005.

UNDP, Human Development Report, Kazakhstan, 2002.

WB, Report No 25929, Project appraisal document on a proposed environment facility grant in the amount of US5.27 million to the Government of Kazakhstan for Drylands management project, May, 2003.

# ECOLOGY SCREENING AND MAPPING OF HEAVY METAL POLLUTION OF SOILS BY LASER MASS SPECTROMETRY

LUBOMIR SIMEONOV[1*] AND BIANA SIMEONOVA[2]
[1] Solar-Terrestrial Influences Laboratory,
Bulgarian Academy of Sciences
Acad. G. Bonchev Str. Bl.3, 1113 Sofia, Bulgaria
[2] Institute of Electronics
Bulgarian Academy of Sciences
72 Tzarigradsko Shossee Bul, 1784 Sofia, Bulgaria

**Abstract.** The present paper discusses an approach for performing an express ecology in-field screening with the help of transportable laser mass spectrometric system of easy-accessible small and medium soil areas, polluted with heavy metals. Presented are results from a demonstration experiment in which the heavy metals elemental and isotopic data, gathered by the laser time-of-flight mass analyzer are arranged according to the map of the places of soil sample-taking, thus providing a satisfactory evidence of the ecological status of the area in respect to further remediation activities.

**Keywords:** heavy metals, soil pollution, ecological screening, ecological mapping, laser mass spectrometry

## 1. Introduction

The presence of heavy metals in cultural soils is a result mainly of an anthropogenic activity. The magnitude of contemporary production and the introduction of new technologies in different branches of industry and in the living sphere at present time exceeds the potential of the biosphere ecosystem and creates a real hazard for the human being. The gradual contamination of the biosphere with heavy metals is explained by their stability in outer conditions, the solubility in atmospheric water and the ability to be absorbed in soil. For example, the soil

---

* To whom correspondence should be addressed: simeonov2006@abv.bg

L. Simeonov and V. Sargsyan (eds.),
*Soil Chemical Pollution, Risk Assessment, Remediation and Security.*
© Springer Science+Business Media B.V. 2008

retention time of Pb in soil is between 150 and 5000 years (McBride, 1989). According to some risk assessment evaluations, the overall contamination with heavy metals will inevitably take the first place in the near future, exceeding the pollution with organic substances and pesticides. In this context, the scientific research for the development of reliable and up-to-date methods and techniques for qualitative and quantitative environmental analysis, combined with effective clean-up technologies is quite actual.

The group of heavy metals consists of more than 40 elements, which have a specific gravity greater than 6 g/cm$^3$. From an ecological, toxicological and hygienist point of view, not all the elements from this group are equally important. The metals which have a wide scope of production and implementation in industry and everyday life, usually have high levels of accumulation in the environment and are considered hazardous because of their biological activity and toxicological properties. Especially hazardous are considered Cd, Pb, Hg and As in the first order and also Zn, Ni, Cu, Co, Sn, Sb, Mo, V and Cr$^{VI}$ (Ilin, 1992; Trachtenberg et al., 1994). Metals in soil environment exist as components of several different fractions: (1) free metal ions and soluble metal complexes; (2) metal ions, occupying ion-exchangeable sites and absorbed in inorganic soil constituents; (3) organically-bound metals; (4) precipitated or insoluble compounds, oxides, carbonates and hydroxides; (5) metals in the structures of silicate minerals. It is clear, that only metals in fraction (1) and partially from fraction (2) form the hazardous bio-available part of the total presence of metals in soil. Therefore, it is important to evaluate the magnitude of this particular part and to keep this in mind when analysing methods and techniques as conventional AAS (Atomic Absorption Spectroscopy) for example are used in chemical analysis of ecology samples.

In this article we present an alternative possibility to apply laser mass spectrometry for express qualitative and semi-quantitative elemental and isotopic analysis of the presence of heavy metals in polluted soils. The investigation is directed for the purposes of ecological screening and mapping in respect to future choice and application of the most appropriate soil remediation approaches and practices for every particular case. The developed approach can be successfully applied also for the evaluation and monitoring of long-term metal deposition in soil in cases of pollution of cultural lands near to heavy industrial complexes as well, as in cases of intentional or criminal releases of pollutants. The method is developed on the basis of a new design concept of traditional laser time-of-flight mass spectrometer. The instrument is created as a result of a technological transfer of a space research technique, originally constructed for the purposes of in-field determination of the elemental and isotopic composition of the surface layer of small space bodies without atmosphere, like asteroids and planetary satellites (Managadze et al., 1986; Simeonov, 1987). This is probably the smallest laser mass-analyser in terms of its size and whose design features several new elements like a two-stage time-of-flight reflector compartment in a combination with a chevron assembly MCP (micro-channel plate) particle detector with ring configuration. The specially designed construction of the sample lock chamber makes it possible to replace a holder with solid-state (soil) samples during less than one minute without gas leakage and venting the whole analyser vacuum system. The workout of the referred instrumental methodology was focused on several specific aspects: estimation of the instrumental analytical parameters as determination rate, reproducibility of measurement and quantification levels. A special attention was paid to the workout of the sampling strategy (complementary with AAS), clearing aspects as sample-taking, preparation of laser targets from environmental samples and solving of the questions in connection to the reference material.

## 2. Results and Discussion

On Figures 1 and 2 are shown two representative mass spectra, taken with the help of the portable laser time-of-flight mass analyser.

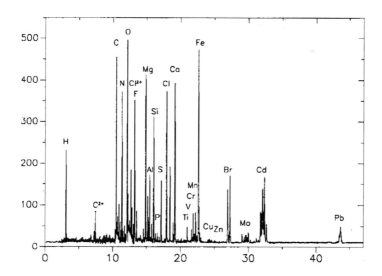

*Figure 1.* Representative spectrum of a soil sample with the presence of Cd and Pb, polluted by industrial waste waters taken with the help of the portable laser mass analyser. The abscissa shows the arrival time in µsec of the elements from H to Pb on the detector of the instrument.

On Figure 2 is shown a part of a mass spectrum in the mass range from 1 to 70 a.m.u. The elemental and isotopic identification of the time-of-flight peaks is achieved with a specially computer programme, which performs a 5-fold estimation based on the natural isotopic abundances of the Periodic System of Elements. An option of the computer programme allows the calculation of the isotopic abundances in percentage in relation to the detected intensities in every separate laser shot – mass spectrum.

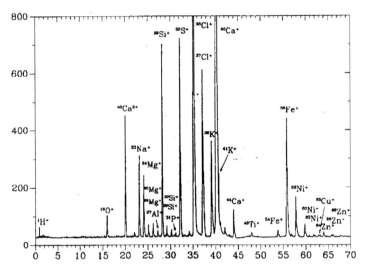

*Figure 2*. Representative mass spectrum in the mass range from 1 to 70 a.m.u. of a soil sample polluted by industrial waste waters with the presence of Mg, Al, Fe, Ni and Cu. The mass spectrum is taken with the help of the portable laser mass analyser.

The special application of the portable laser mass spectrometer for the purposes of the express ecologic screening and mapping of heavy metal pollution of soil is illustrated on Figures 3 and 4.

As it is shown on Figure 3 the investigated soil area is divided in by a grid into separate equal parcels, where the crossing points of the screen are the measuring points, or the places of sample-taking. Each crossing point gets a laser shots number, ten laser shots/mass spectra per point, so these laser shot numbers are related to the measuring location.

On Figure 4 is presented a plot of the detected elements in isotopic resolution in relation to the places where the samples are taken. This simple operation allows in a fast way conclusions about the degree of heavy metal contamination

|  |  |  |  |  |  |  |  |  |  |
|---|---|---|---|---|---|---|---|---|---|
| 101/110 | 111/120 | 121/130 | 131/140 |  |  |  |  |  |  |
| 1/10 | 11/20 | 21/30 | 31/40 | 41/50 | 51/60 | 61/70 | 71/80 | 81/90 | 91/100 |

*Figure 3.* An example of a grid or a screen over contaminated soil area under investigation with numbers of laser shots on points of sample-taking.

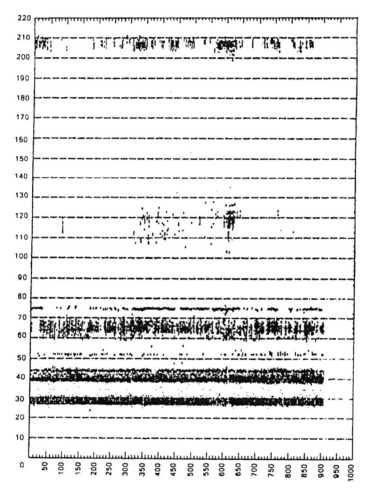

*Figure 4.* A plot of the estimated distribution of detected elements in isotopic resolution (y-axis) over the laser shot number or place of sample-taking (x-axis).

of the soil area to be made. The results for particular metals as Cu, Zn, Cd and Pb were verified independently by AAS and the check showed a satisfactory agreement.

A look on the distribution plot notes, that the mass numbers 27–30 and 39–40 are evidently present in nearly all laser shots/mass spectra. They originate from Al (mass 27), Si (masses 28–30), K (masses 39, 41) and Ca (masses 40, 42, 43, 44). These elements belong to the soil matrix.

In great number of mass spectra there are signals of mass number 52 and 75. They indicate contamination with Cr and As.

Another evidence for contamination is by mass numbers 63–70, which originates from the presence of Cu and Zn.

Separate signals in the mass region 110–140 demonstrate some local distribution of Cd and Sn.

Finally, Pb is found by mass numbers 206–208 in many places, but is distinguished in local distributions.

With this simple demonstration, which can be accomplished with the portable modification of the above described laser mass analyzer even directly in fields, a fast conclusion about the heavy contamination is possible, when needed. This would help to separate the harmful soil areas with metal contamination in respect to the country local Maximum Permissible Levels. The approach saves money and time in comparison to conventional and analytically more precise laboratory methods and gives a satisfactory and fast information for the decision-makers how to proceed with the remediation practices.

## References

McBride M.B. Advances of Soil Science, 10, 1989, 1–56.

Trachtenberg, I.M., V.G. Kolesnikov, V.P. Lukovenko. Heavy metals in the environment, Science and Technics Publ., Minsk, 1994, 11–23.

Ilin V.V. Heavy metals in the system soil-plant. Science Publ. Novosibirsk, 1992, 105–124.

Managadze, G.G., R.Z. Sagdeev, L. Simeonov. Proc. Of the Phobos Study, 1988, 163–186.

Simeonov, L. Comp. Rend. Acad. Bulg. Sci., 40, 1987, 67–70.

# ANALYTICAL TECHNIQUES AND BIOINDICATORS IN ENVIRONMENTAL CONTROL: HONEYBEES, MUSSELS, BIOLUMINESCENT BACTERIA: RAPID IMMUNOASSAYS FOR PESTICIDE DETECTION

STEFANO GIROTTI[1*], ELISABETTA MAIOLINI[1],
LUCA BOLELLI[1], SEVERINO GHINI[1], ELIDA FERRI[1],
NADIA BARILE[2] AND SVETLANA MEDVEDEVA[3]
[1]*Dipartimento di Scienza dei Metalli, Elettrochimica e Tecniche Chimiche (SMETEC), Università di Bologna, Via S. Donato 15, 40127 Bologna, Italy*
[2]*Center of Water Biology, Institute of Experimental Zooprophylaxis of Abruzzo and Molise "G. Caporale", Viale Marinai d'Italia 20, 86039 Termoli, Italy*
[3]*Institute of Biophysics, Siberian Branch of Russian Academy of Sciences, Academgorodok, 660036 Krasnoyarsk, Russia*

**Abstract.** The toxic pollutants are widely distributed on terrestrial and marine environment and their early identification is fundamental to prevent or to control the damages to humans and environment. The environmental monitoring of toxic pollutants by bioindicators like bioluminescent bacteria (BLB), mussels and honeybees, both in terrestrial and marine environment is reported. BLB and mussels have been employed to assess the toxicity of heavy metals in lagoon slime samples and seawater, respectively. Honeybees were applied, coupled to suitable immunoassays, to the determination of azinphos-methyl and thiram pesticides.

**Keywords:** bioindicators, bioluminescent bacteria, heavy metals, honeybees, mussels, immunoassay, azinphos-methyl, thiram

---

* To whom correspondence should be addressed: stefano.girotti@unibo.it

## 1. Introduction

The use of living organisms as indicators to determine the environment quality has long been widely recognised. Over the past few decades plants, animals, fungi, and bacteria have been employed as bioindicators and biomonitors in air, soil and water pollution surveys (Alfani et al., 1996; Gerhardt, 1999; Nimis et al., 1990; Henderson, 1996; Wolterbeek, 2002).

Biological monitoring may be defined as the measurement of the response of living organisms to changes in their environment and it can be divided into "passive" and "active". Passive monitoring is performed through the observation and analysis of organisms which are usual inhabitants of an ecosystem; active monitoring includes all methods which insert organisms, under controlled conditions, into the site to be monitored. In general, (Ceburnis and Valiulis, 1999) these techniques provide information on the response of living organisms to the integrated effects of environmental contaminants, which cannot be determined by direct analytical measurements. Biomonitoring may allow a detailed and reliable coverage of the territory with relatively low costs (Girotti et al., 2006a) and, in general, is based on sensitive or accumulative organisms, i.e. bioindicators or bioaccumulators (Batzias and Siontorou, 2007).

Bioindicators display a very high sensitivity to pollutants, providing information that is function of the quality of the environmental sector under examination (Butterworth et al., 1995). They can give information on: (1) decrease of the biotic diversity with a decrease of the species constituting the community and an increase of individuals belonging to few, resistant species; (2) presence or absence of a particular species; for example, *Plecoptera* can live only in cold water rich in oxygen; (3) appearance of a structural damage, reversible or permanent.

An organism, to be employed as a bioindicator, has to have the following characteristics: to be easily identified; to be easily sampled; to be widely spread in the studied area; to have a low mobility; to have a long-life cycle; to have a good genetic uniformity on all the considered area; to be present all-year long. Bioindicators permit to establish a network of monitoring sites with a consequent satisfactory coverage of the territory without the requirement of sophisticated and expensive equipments and power. In this view, the availability of a large number of reliable bioindicator species is an important step towards a wider application of the methodology (Iriti et al., 2006). In addition, data obtained by biomonitoring are not related to a single sampling, but they are the summa of very high number of sampling.

An organism is defined "bioaccumulator", when the modality of distribution of pollutants can be inferred thanks to their concentrations in the organism (Batzias and Siontorou, 2007). The accumulative species can store contaminants in their tissues, since bioaccumulation is an equilibrium process of biota compounds intake/discharge from and into the surrounding environment. The disadvantage to employ bioaccumulator organisms is the need for sometimes complicated analytical techniques and equipments to determine the compound of interest.

Since several years the Analytical Chemistry Section of SMETEC focuses its research activity on the development of assays based on luminescent re-agents, as well as on the environmental monitoring by bioluminescent bacteria (BLB), mussels and honeybees, employed as bioindicators (Girotti et al., 2002a, 2005a, 2006a, b).

The bioassays based on luminescence have been used for several decades, and test species include *Vibrio fischeri, Vibrio harveyi* and *Pseudomonas fluorescens* (Girotti et al., 2001). Between them, the most employed one is the naturally bioluminescent marine bacterium *Vibrio fischeri*, the use of which was developed in the 1970s (Trott et al., 2007). The bacterial assays are usually rapid, reproducible, cheap and in some case can replace tests using animals, resulting very useful to measure the toxicity of different kinds of environmental samples (Parvez et al., 2006). The toxicity test is usually based on the bioluminescence inhibition assay: the luminescent bacteria emit light when they find themselves in optimal conditions, whereas in presence of noxious substances their lumine-scence decreases. Thus, the presence of toxic molecules, as pesticides, heavy metals or organic compounds can be evaluated (Girotti et al., 2002b). The luminescent bacteria system can be applied both as short- and long-term test (El-Alawi, et al., 2002). Even if short test has some obvious advantages, it has also some not negligible drawbacks, i.e., it can reveal only molecules acting immediately and underestimate the toxicity of chemicals which show mainly long term effects. The bioluminescent bacterial tests can be used both for water and solid samples. In the latter case the tests are applied to elutriates and extractable components, being suitable also for the evaluation of soil bio-remediation (Plaza et al., 2005), as reported by Girotti et al. (2007). Recently, Trott et al. (2007) demonstrated as the response of bioluminescent bacterial assays for soils are comparable to those of higher organism and they can represent a valid complementary tool to traditional methodologies, when an accurate validation and an optimised extraction protocol have been obtained.

One of our more recent applications of the BLB tests is to the analysis of sludge and soil contamination and bioremediation (Girotti et al., 2005c). During sludge remediation treatments, for example on a proper boat, the biotoxicity of the materials and the effects of the treatment must be frequently evaluated, and such a control can be easily performed even in these conditions by BLB toxicity assays. This kind of assay is already widely employed and it has been standardized as an ISO procedure (BS EN ISO 11348-3, 1999).

Mussels are recognized as pollution bioindicator organisms because they accumulate pollutants in their tissues at elevated levels in relation to pollutant availability in the marine environment. Moreover, this ability has led to the adoption of the international "Mussel Watch Project" and several national programs on Mussel Watch in the marine environment have been carried out (Jernelov, 1996). Mussel is widely used as sentinel organism for the assessment of persistent organic pollutants contamination, heavy metals, organotin compounds in freshwater environments, polycyclic aromatic hydrocarbons, DDT bioaccumulation (Kramer, 2006; Binelli et al., 2006).

During a study performed at the laboratories of the Center of Water Biology, Institute of Experimental Zooprophylaxis of Abruzzo and Molise (Italy), the bivalve molluscs *Mytilus galloprovincialis* has been inserted in a device designed for the continuous monitoring of water bodies, the Mosselmonitor, and a model system, sea water contaminated by heavy metals (cadmium, lead, copper and mercury) has been used to evaluate the sensitivity of the biosensor, which resulted in a range useful to detect real pollution levels.

Pesticide residues in agriculture lands require to be continuously monitored because of their toxicity to human health and the potential hazard for the conservation of the ecological equilibrium. Pollinators, specially honeybees, are considered a reliable biological indicator (Bromenshenk et al., 1995) because they reveal the chemical impairment of the environment by the high mortality and retaining suspended particles in the air or present on the various parts of the plants. It is possible to collect the dead honeybees by means of traps and then to identify the gathered cultivation and area through pollen analysis. It is possible to univocally determine the origin of a certain pesticide by means of a net of beehives. Besides pesticides, heavy metals and radioactive elements have been monitored by analyzing the residues on honeybees body (Porrini et al., 2002).

Azinphos-methyl, [$O,O$-dimethyl $S$-[(4-oxo-1,2,3-benzotriazin-3($4H$)-yl) methyl] phosphorodithioate], is an organodithiophosphorus insecticide and acaricide introduced in 1954, and it was proved to be very toxic to honeybees. Its analysis is generally performed by liquid-solid extraction followed by gas chromatographic techniques (Mercader and Montoya, 1999).

Thiram (tetramethylthiuram disulfide) (bis (dimethylthiocarbamyl) disulfide) is a dithiocarbamate, a wide-spectrum, non-systemic, fungicide, applied on humans against scabies, as a sunscreen agent or as a bactericide in soaps (Sharma et al., 2003).

Immunochemical techniques have gained an increasing importance for screening and quantification of pesticides due to their sensitivity, speed, simplicity, and low cost (Dankwardt, 2001). The enhanced chemiluminescent (CL) reaction offers the possibility to improve the sensitivity of immunoassays compared to conventional colorimetric detection (Botchkareva et al., 2003).

We developed a sensitive indirect chemiluminescent enzyme-linked immunosorbent assay (CL-ELISA) (Figure 1) that employs luminol, horseradish peroxidase as labeling enzyme, based on monoclonal antibodies (MAbs) (Mercader and Montoya, 1999) for the detection and quantification of azinphos-methyl in honeybees (Girotti et al., 2005b), and a CL-ELISA based on polyclonal antibodies for the detection and quantification of thiram in honeybees that allow rapid and sensitive screening of large numbers of samples (Eremin et al., 2006).

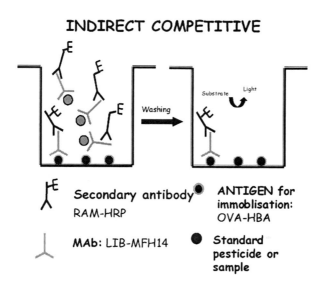

*Figure 1.* General scheme for an indirect competitive ELISA.

## 2.  Material and Methods

### 2.1.  BIOLUMINESCENT BACTERIA

Three different bioluminescent bacteria were used, two belonging to the species *Vibrio logei*: Ucibo and Russian, the first isolated from water of the Mediterranean Sea, the second one supplied by the Institute of Biophysics (Siberian Branch, Academy of Sciences), Cultures Collection IBSO, Laboratory of Bacterial Bioluminescence. Akademgorodok, Krasnoyarsk (Russia). The third was the *Vibrio fischeri* strain. Bacteria have been cultivated and employed, working at room temperature, as reported previously (Girotti et al., 2002b, 2007).

The BL analyses were made on a set of eleven samples (including that one designed as "blank") of sediments collected by the Taverna SpA (San Giorgio di Nogaro (UD), Italy) at different locations in the lagoon of Marano, Grado, and in the Aussa-Corno river, areas of intensive industrial settlement. Detailed description of all locations and the geological and physical characteristics of the samples were previously reported (Coletti et al., 2006).

The samples were divided into three groups, according to the sampling place: the Marano Channel group (samples 0078, 0080, 0108, 0114), the Aussa Corno Banduzzi group (samples 0904 and 0906) and the Foce Aussa Corno group (samples 1077, 1094, 1110 and 1119). All results were compared with the values of the "blank". This was a sample showing geological characteristics similar to the samples and harvested in a zone considered free from pollution.

Samples were extracted with water or ethanol (1 g in 5 mL) and the extracts diluted with a NaCl 2% solution to avoid osmotic problems to the marine bacteria. Luminescent measurements were done using a Wallac "Victor 1420" Multilabel Counter luminometer (Wallac, Sweden) and employing 96 wells microplates. The inhibition assay was performed by adding to each well 100 µL of the solution containing 3% of the extract and 100 µL of bioluminescent bacteria in culture broth. The emitted light was recorded, at intervals of 10 minutes, for 30–60 min in short-term assays and for 16–24 hours in long-term assays. The light emission of the bacterial suspension was tested before to start the assays in order to check if the intensity was optimal for the analysis.

## 2.2. MUSSEL MONITOR

The evaluation of mussels monitoring of sea water polluted by heavy metals was done at the Center of Water Biology, Institute of Experimental Zooprophylaxis of Abruzzo and Molise "G. Caporale", Termoli, Italy. The preparation of standard solutions and set up of the Mosselmonitor were done as previously reported (Barile et al., 2007). Briefly, the sea water introduced in the device was withdraw from the same area of the sea where the mussels naturally grow, about 500 m from the Adriatic coast in front of Termoli, Molise (Italy), and mussels have been employed as a group of 8 individuals, placed inside the Mosselmonitor (Figure 2).

The instrument Mosselmonitor has been supplied by Delta Consult (The Netherlands). The system is based on the measurement, by electromagnetic induction, of the distance between the two valves by means of a couple of coils sticked on the valves. The intensity of the measured electric tension is inversely proportional to the distance between the coils, indicating the position of the two valves.

*Figure 2.* Mussels in the Mosselmonitor device.

The sea water chemical-physical parameters (temperature, pH, oxygen and salts concentrations) have been checked each two hours by a multi parametric equipment. The whole water volume (200 L) circulated among the instrument tank and two more tanks by a pumping system, in a closed cycle. A first tank received the water going out from the Mosselmonitor® tank, from which it was pumped to a second tank where oxygen was bubbled inside. Finally, water returned to the Mosselmonitor tank.

Data produced by mussels shell movements, expressing their wellness or suffering, were collected continuously by a software, that represented them in graphs showing the medium value of the distance between the valves (percentage of valves opening) as well as the time periods of opening and closure of all mussels. Moreover, data about each mussel were available, to as certain individual anomalous behaviour.

The position of the valves was recorded each 90 seconds and the significant changes from the usual behaviour were elaborated and expressed by the instruments as "alarms" (Kramer and Foekema, 2000). The "A" type alarms express the hyperactivity of the mussels, which open and close the shell at high frequency. This behaviour reveals a condition of deep suffering of the organisms and it is very reliable. The "C" type alarm appears when a certain number of organisms remain closed for a time period longer than usual. In this study a period of 15 min has been defined as the minimum to get a "C" alarm. The "D" type of alarm is produced when the filtration activity is reduced, i.e., the average percentage of opening is reduced during a certain period of time.

Unfortunately this condition occurs not only in presence of pollutants but also when some environmental conditions changes, for example the water turbidity. For this reason this alarm is very sensitive but not reliable.

Each 24 h the molluscs were feeded with hypo dietetics rate of algae *Chaetoceros spp* collected from laboratory cultures.

The behaviour of the organisms, after they have been placed inside the Mosselmonitor, has been observed for 1 month, then the whole volume of water was changed and the toxicity test was started by adding increasing concentrations of each heavy metal. The test was divided into two phases. First, the organisms were left to stay for 36 h after the water was changed, their behaviour was observed for 12 h more and then the toxicant solution was added. After this, the alarms generated by the pollutant effects were recorded for 24 h. The concentrations were chosen in a range based on the Quality Standards defined for the year 2008 by the EU Legislation (European Community, 2000). Suitable amounts of stock solutions were added to the first tank to obtain, in the whole volume of 200 L, the following concentrations: 0.2, 10, 40, 80 and 100 ppb for cadmium; 125, 250, 500, 1000 and 2000 ppb for lead; 2.5, 5, 10, 20 and 40 ppb for copper; 0.01, 0.03, 0.5, 1 and 5 ppb for mercury. Bioluminescent measurements on the same heavy metals solutions have been performed by the Microtox kit and the luminometer Microtox MD500 (Ecotox, Italy), equipped with a cell thermostated at $15 \pm 1°C$, at the wavelength of 490 nm.

## 2.3. HONEYBEES FOR AZINPHOS-METHYL AND THIRAM IMMUNOASSAYS

Honeybees samples were collected, stored and prepared as previously described (Ghini et al., 2004). To analyze the pesticides we applied two kind of extraction: the classical liquid-liquid extraction (acetone-dichloromethane with purification by phosphate precipitation) and a new graphitized carbon extraction; the extracts were simply diluted prior to use (Ghini et al., 2004).

Standard pesticides were purchased from Dr Ehrenstorfer GmbH (Augsburg, Germany). Peroxidase-labeled goat anti-rabbit immunoglobulins were from Dako (Glostrup, Denmark). All chemicals and organic solvents used were of reagent grade or better. Black polystyrene high-binding plates were from Costar (Cambridge, USA). Before analysis, honeybee samples were lyophilized and stored at -5°C.

*Azinphos-methyl immunoassay*: The Chemiluminescent azinphos-methyl immunoassay has been previously developed (Girotti et al., 2005b). Briefly, the plates were coated overnight with 100 μl of the appropriate concentration of the heterologous OVA-HBA conjugate in coating buffer (50 mM carbonate/ bicarbonate, pH 9.6) (Girotti et al., 2005b). The plates were then washed four times with washing solution (0.15 M NaCl with 0.05% Tween 20). A volume of

50 μL/well of standard in bidistilled water or sample solution, followed by 50 μL/well of LIB-MFH14 MAb were added. The competitive reaction was allowed to take place for 1 h. Inhibition standard curve was prepared by serial dilutions from 500 to 6.40 x $10^3$ nM with a dilution factor of 5. After washing, 100 μl of a 1/2000 dilution of peroxidase-labeled rabbit anti-mouse immuno-globulins in PBS (10 mM phosphate, 137 mM NaCl, 2.7 mM KCl, pH 7.4) was added, and plates were incubated for 1 h. Plates were then washed and finally peroxidase activity was revealed by adding 100 μL/well of a freshly prepared substrate mixture (1 mM luminol, 0.5 mM p-iodophenol, 1 mM $H_2O_2$ in 0.2 M borate buffer, pH 8.5). Intensity of chemiluminescence emission (RLU) was measured by the "Victor 1420" microplates reader (Wallac, Finland), immediately after the addition of the substrate mixture.

*Thiram immunoassay*: The starting basis of the CL thiram immunoassay was the colorimetric ELISA for thiram already reported by Queffelec et al. (2001). Assays were performed in 96-well microplates as indirect competitive format (Eremin et al., 2006). Plates were coated overnight with 100 μL/well of OVA-hapten 2C (0.6 μg/mL) in carbonate-bicarbonate buffer 0.05 M pH 9.6, which corresponded to 0.06 μg/mL of OVA-hapten-2C per well and then washed three times with 0.01 M, pH 7.4 PBS supplemented with 0,05% Tween 20, as washing solution and after addition of 50 μL/well of standard or sample plus 50 μl/well of antiserum (1/30 000 in 2x fish gelatine solution in PBS) were incubated for 1.5 h at room temperature. After washing, 50 μl/well of per-oxidase-labeled goat anti-rabbit immunoglobulin, diluted 1/2 000 in 1x PBS-Gelatine, were added and incubated again for 1.5 h at room temperature. After washing, 100 μL/well of a substrate solution (1 mM luminol, 0.5 mM p-iodophenol, 1 mM $H_2O_2$ in 0.2 M borate buffer, pH 8.5) were added and intensity of chemiluminescence emission, expressed as relative light units (RLU), immediately measured in a "Victor 1420" microplate luminometer (Wallac, Finland).

RLU values from standards were mathematically fitted to a four parameter logistic equation (Botchkareva et al., 2003). The limit of detection (LOD) for ELISAs was calculated as the analyte concentration that reduced signal to 90% of the maximum. The extracts were also analyzed by gas-chromatography (Ghini et al., 2004; 2005b).

## 3. Results and Discussion

### 3.1. BIOLUMINESCENT BACTERIA

Quantitative analysis of the samples content of heavy metals were done to characterize them, and an example is showed in Figure 3.

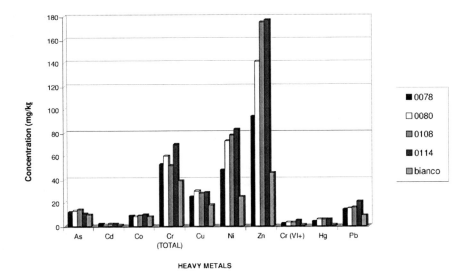

*Figure 3.* Heavy metals content in some of the analysed sludge samples.

To evaluate the limits of detection of the bioluminescent assay, that have to be lower than the Maximum Admitted Limits according to the regulatory rules, standard solutions of each heavy metal present in the sludge samples were analysed both by the short and the long term assays and the assays showed a linear behaviour in different ranges of concentration.

As reported in Table 1 the BBL assay showed good sensitivity and the limit of detection (LOD) and $EC_{50}$ values were lower than the maximum admitted levels.

TABLE 1. Limits of detection and $EC_{50}$ values of six heavy metals, obtained for the BLB assay, compared with their legal limits

| Heavy metal | LOD (ppm) | $EC_{50}$ (ppm) | Legal limits in surface water (ppm) | Legal limits in wastewater (ppm) |
|---|---|---|---|---|
| Pb (II) | 0.1 | 0.2 | <0.2 | <0.3 |
| Hg (II) | 0.005 | 0.05 | <0.005 | <0.005 |
| Cd (II) | 0.03 | 0.5 | <0.02 | <0.02 |
| Zn (II) | 1 | 3.0 | <0.5 | <1 |
| Cr (VI) | 0.2 | 5.0 | <2 | <4 |
| Cu(II) | 0.2 | 0.3 | <0.1 | <0.4 |

The extraction of heavy metals from the samples, made by water or by 2% NaCl solution gave almost the same result, in the meaning that the inhibition curves obtained by using the two extracts are very close, practically overlapped (Figure 4).

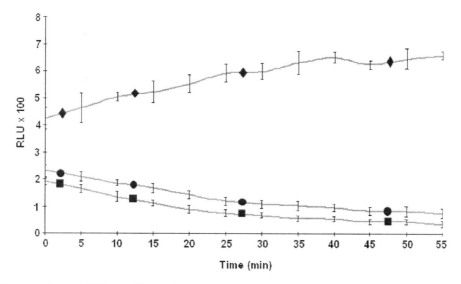

*Figure 4.* Acute inhibitory effects of the water and 2% NaCl extracts of the sample 1119 on the Ucibo light emission. ■ 1119 2%NaCl; ● 1119 water + 2%NaCl; ♦ Blank.

The HPLC analysis demonstrated that the "blank" sample was contaminated with low concentrations of heavy metals, and these low amounts did not cause a significant bacterial luminescence decrease.

The samples have been analysed by all three BLB strains for the acute and chronic toxicity (Figure 5), both at short and long term, showing different influence on the different strains.

The sample extracts showed different effects also when emitted light intensity was compared with the "blank" emission. In fact, the light emission from the different strains in contact with the same extract resulted in one case inhibited, in another one stimulated, with respect to the "blank". As shown in Figure 6 the major part of the samples inhibited the emission of *Vibrio fischeri*, while the same produced a hormesis effect (Christofi et al., 2002) on the Russian strain, probably because this strain was able to use some compounds in the sample extracts as a substrate. These differences were observed both at short and long term. The importance to test the same samples by different BLB strains is clearly confirmed by these results.

S. GIROTTI ET AL.

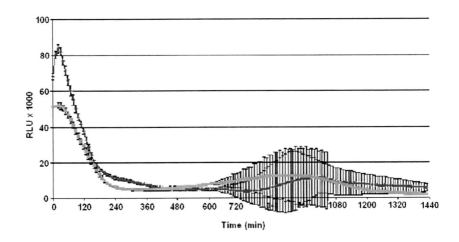

*Figure 5.* Example of chronic BLB assay, black line: blank; gray line: 0080 sample.

In the following Table 2 the respective behaviour of the three strains with respect to all samples have been summarized.

TABLE 2. Differences of the response of the various strains to the addition of the sample extracts

| Sample/ Bacteria | 78 | 80 | 108 | 114 | 904 | 906 | 1,077 | 1,094 | 1,110 | 1,119 |
|---|---|---|---|---|---|---|---|---|---|---|
| Vibrio | T | T | A | A | T | A | A | A | N | A |
| Russian | T | N | T | T | A | N | A | A | T | N |
| Ucibo | N | N | T | T | A | T | T | T | T | T |

The behaviour of the three bacteria strains, in presence of the samples where: T – typical behaviour (the bioluminescence curve in presence of the sample in below to that of the black). A – a typical behaviour (the bioluminescence curve in presence of the sample is above that one of the black). N – the bioluminescence curves are not distinguishable.

In this kind of analyses it is very important to have the possibility to measure several replicates of the same sample, to ensure reproducible results. For example, we determined that in long-term assays it is advisable to measure at least five replicates for each sample, to obtain a CV (coefficient of variation) in reasonable limits (under the 15%). By using the microplate format this number of replicates can be prepared and measured at the same time without difficulties, since small amounts of reagents are required.

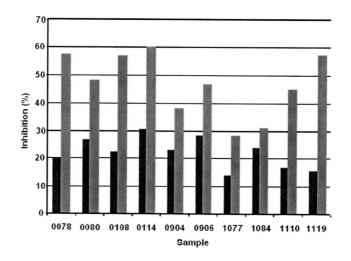

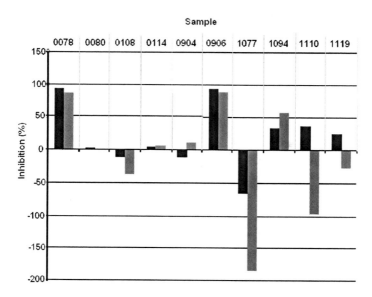

*Figure 6.* Acute (black) and chronic (gray) BLB assay, % inhibition of lagoon samples on Vibrio and Russian strain.

## 3.2. MUSSEL MONITOR

To estimate the toxicity of various concentrations of Cu, Cd, Pb, and Hg the measured parameter was the maximum length of the C alarms recorded during the period of contact with the toxicant (Figures 7–10).

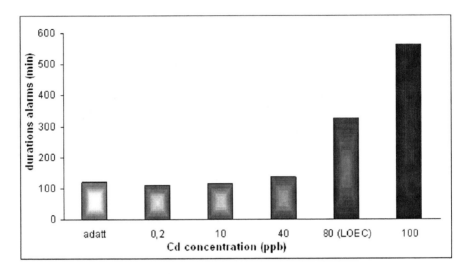

*Figure 7.* Maximum length of the C alarms corresponding to each one of the tested concentration of cadmium.

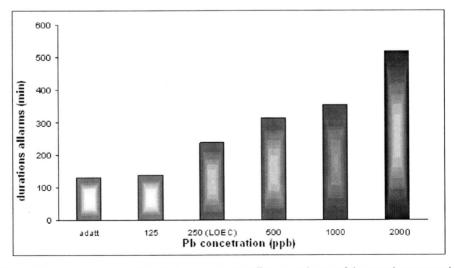

*Figure 8.* Maximum length of the C alarms corresponding to each one of the tested concentration of lead.

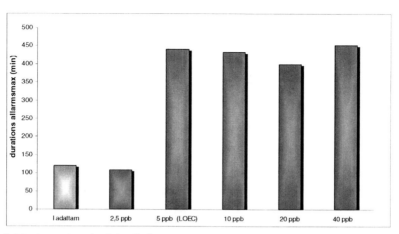

*Figure 9.* Maximum length of the C alarms corresponding to each one of the tested concentration of copper.

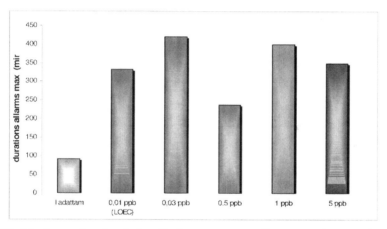

*Figure 10.* Maximum length of the C alarms corresponding to each one of the tested concentration of mercury.

It is possible obtain, for each metal, the duration of the longer alarm clearly different from the adaptation period and the corresponding concentration was defined as the Lowest Observed Effect Concentration (LOEC) (Kramer and Foekema, 2000). The LOECs obtained were: 5 ppb for copper, 0.01 ppb for mercury, 80 ppb for cadmium and 250 ppb for lead. The performances of the biosensor were compared with the response of a widely employed test for acute toxicity, the Microtox® kit, based on marine bioluminescent bacteria *Vibrio fischeri*. The luminescent test resulted less sensitive, with LOEC of 1, 0.5, 10 and 1 ppm, respectively.

Both the instrumental and the biological components showed the required characteristics to obtain a suitable biosensor. Concerning the bioluminescent assay the comparison with the mussel biosensor demonstrated that, even the first is very simple and widely employed, it is sensitive only at high concentration of the tested pollutants.

## 3.3. HONEYBEES AND PESTICIDES

### 3.3.1. *Azinphos-Methyl Immunoassay*

The ELISA was optimised and characterised by determining several parameters. Immunoreagents optimum concentrations were determined by bidimensional titration in the range 2–0.2 µg/mL for OVA-HBA and 60–7.5 ng/mL for LIB-MFH14. The ability of the chemiluminescent technique to detect lower concentrations of HRP allowed the optimal antibody concentration (30 ng/mL) to be decreased, compared to colorimetric (COL) ELISA for Azinphos-methyl, and the hapten-protein conjugate concentration was selected at 1 µg/mL for CL-ELISA) (Mercader and Montoya, 1999) (Table 3). The highest sensitivity was obtained for 60 min incubation.

TABLE 3. Comparison of the analytical performance of the CL ELISA and the colorimetric one

| ASSAY | CONJUGATE ng/mL | ANTIBODY Dilution | ng/mL | | | |
|---|---|---|---|---|---|---|
| | | | $IC_{50}$ | $IC_{20}$ | $IC_{80}$ | $IC_{90}$ |
| CL-ELISA- | 1.0 | 1:30000 | 88.5 | 100 | 71 | 39 |
| COL-ELISA by ALP* | 60 | 1:30000 | 34.0 | 90 | 11 | 5 |
| COL-ELISA by HRP* | 60 | 1:30000 | 35.0 | 110 | 11 | 5 |

*ALP = alkaline phosphatase; HRP = peroxidase enzymes.

Four organic solvents, methanol, ethanol, acetone, and dioxane, to be used for Azinphos-methyl extraction were assayed for their effects on the immunoreagents in the range 0–10%. Solvents were generally well tolerated, with only slight variations of $IC_{50}$ and RLU max. Optimized assay conditions allowed to obtain, on standard solutions in solvents, the $IC_{50}$ value of 0.40 nM, and a limit of detection of 0.01 ng/mL. A typical competitive curve for Azinphos-methyl is shown in Figure 11.

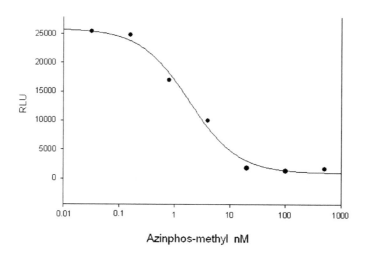

*Figure 11.* A typical competitive curve for Azinphos-methyl.

To study matrix interferences, standard curves were also prepared in extracts from non contaminated honeybees. Two different extraction procedures were tested: the liquid-liquid extraction and a graphitized carbon-based solid-phase extraction. A very important matrix effect was encountered: undiluted samples, as well as 1/5 and 1/50 dilutions greatly disturbed ELISA assay, while the 1/100 dilution allowed to obtain only slight decrease of the maximum emission, with minor increase of $IC_{50}$.

The non contaminated honeybee samples were spiked at different levels (0.1 µg/mL, 1 µg/mL and 10 µg/mL) and then extracted and analysed. The recovery proved to be higher with the graphitized carbon procedure (>60%) than with the liquid-liquid extraction (20–40%). Finally, the assay was applied to the analysis of azinphos-methyl in honeybees samples collected in Russia. 5 samples out of 19 were found positive. For two of them, the presence of azinphos-methyl was confirmed by gas chromatographic analysis, while the concentration of the other samples was smaller than the detection limit of the chromatographic technique (10 ng/mL).

### 3.3.2. *Thiram Immunoassay*

The assay was optimised and characterised by determining several parameters. Immunoreagent optimum concentrations were determined by bidimensional titrations. The most appropriate work conditions have been obtained using

0.6 μg/ml of OVA-2C (0.06 μg/well), anti-thiram serum dilution of 1/30,000 and peroxidase-labeled goat anti-rabbit immunoglobulin were diluted ½,000. Highest sensitivity, related to the anti-thiram serum incubation time was obtained for 90 min. To assess tolerance to organic solvents, acetone and methanol where added at four different final concentrations: 3%, 5%, 10% and 20%. Comparing $IC_{50}$ values and RLU max for both solvents at each concentration, acetone 3% final concentration assured the optimal conditions for CL-ELISA of thiram. The competitive assay for thiram showed an $IC_{50}$ of 120 ng/ml and a LOD of 20 ng/ml.

Also in this case the honeybee extracts were diluted 100 fold to eliminate the matrix effect. The non contaminated honeybee extracts have been spiked with 0.5 ppm and 1 ppm and the recovery was higher for liquid-liquid extraction, 54%, than for SPE, 31%, underlying the difficulties due to the complexity of the honeybee matrix.

Finally the CL-ELISA applied to real honeybees samples, 5 collected in Italy and 5 in Russia demonstrated that all Italian samples were negatives, and 3 of Russian samples were positive.

**Conclusion**

The presented BLB test can be used to evaluate the toxicity of heavy metals in solid samples also during the remediation procedure, because the decrease of light emission from the bacteria is directly related with the presence of metals affecting the cells viability. The bioluminescence test allows to analyse at the same time, by using the microplate luminometer, a lot of samples in a relatively fast way, at low costs and using small volumes of reagents.

The data obtained on the applicability of the Mosselmonitor – *Mytilus gallo-provincialis* biosensor as an Early Warning Alarm System represent interesting, positive indications for the monitoring of sea environment. Nevertheless, it must be remembered that the mussel monitoring procedure must be carried out by personnel with a deep knowledge of the ecological characteristics of the employed organisms, since this is necessary to perform a correct interpretation of the data supplied by the instrument. It must be underlined that neither the LOEC determined by the biosensor, nor the $EC_{50}$ values obtained with the bioluminescent bacteria are in agreement with the Quality Standards levels that must be reached according to the national and international legislation for the year 2008 (European Community, 2000). Those levels could be probably obtained by employing a set of organisms at different trophic levels (Dalzell et al., 2002), and among them the mussels can be surely included.

The chemiluminescent immunoassay developed for the detection and quantification of Azinphos-methyl allowed performances comparable to those of the colorimetric assay, but utilising a smaller amount of monoclonal antibodies.

The assay was applied to the analysis of honeybee samples, proving to be suitable for the use in environmental monitoring programs.

The competitive CL-ELISA developed for the detection of thiram in honeybees showed great specificity and a quite good sensitivity. The analysis of thiram in honeybees samples posed several problems due both to its instability and to the complexity of the matrix. The important matrix effect produced by the extracts can be overcome by a 100 fold dilution, that allowed to obtain reproducible results, although this resulted in a lighter decrease of the sensibility. It is important to underline that the developed chemiluminescent assay, using specific antibodies, is able to detect the thiram as itself and not as one of its degradation products, and that it is a semi quantitative method that allows a rapid and easy screening of honeybee samples.

## Acknowledgements

This work was supported by grant from the University of Bologna (Fundamental Oriented Research). P. Caputo is gratefully acknowledged.

## References

A. Alfani, G. Bartoli, F.A. Rutigliano, G. Misto, A.V. De Santo. Trace metal biomonitoring in the soil and the leaves of Quercus ilex in the urban area of Naples. Biol. Trace Elem. Res., 51:117–131 (1996).

N.B. Barile, E. Nerone, G. Mascilongo, L. Bolelli, S. Girotti. "Risultati preliminari sull'utilizzo di sensori analitici biologici nel monitoraggio delle acque marine costiere", Atti X Congresso Nazionale di Chimica dell'Ambiente e dei Beni Colturali "Conoscenza e Creatività", Acaya, Vernole (Lecce), 11–15, Giugno MI18, p. 222 (2007).

F. Batzias, C.G. Siontorou. A novel system for environmental monitoring through a cooperative/synergistic scheme between bioindicators and biosensors. J. Environ. Manag., 82:221–239 (2007).

A. Binelli, F. Ricciardi, C. Riva, A. Provini. Integrated use of biomarkers and bioaccumulation data in Zebra mussel (Dreissena polymorpha) for site-specific quality assessment. Biomarkers, 11:428–448 (2006).

A.E. Botchkareva, S.A. Eremin, A. Montoya, J.J. Manclus, B. Mickova, P. Rauch, F. Fini, S. Girotti, J. Immunol. Meth., 283:45–57 (2003).

British Standard – BS EN ISO 11348-3:. Water quality determination of the inhibitory effect of water samples on light emission of Vibrio fischeri (Luminescent bacteria test), part 3. BSI, Index House, Ascot, SL5 7EU, UK (1999).

F.M. Butterworth, L.D. Corkum, J. Guzmán-Rincón. Biomonitors and biomarkers as indicators of environmental change, Environmental Science Research, Plenum Press, London, Vol. 50 (1995).

D. Ceburnis, D. Valiulis. Investigation of absolute metal uptake efficiency from precipitation in moss. Sci. Total Environ., 226:247–253 (1999).

G. Coletti, F. Gubiani, M. Piccolo, M.D. Luque de Castro, R. Japón Luján, E. Ferri, B. Garcia Morante, L. Bolelli, S. Girotti. "Bioremediation and toxicity bioluminescent assay of contaminated soils". Luminescence, 21, 324–325 (2006).

N. Christofi, C. Hoffmann, L. Tosh, Hormesis responses of free and immobilized light-emitting bacteria. Ecotoxicol. Environ. Saf., 52:227–231 (2002).

A. Dankwardt, in Encyclopedia of Analytical Chemistry, R.A. Meyers (ed.), Wiley, Chichester, England, pp. 1–25 (2001). Y.S. El-Alawi, B.J. McConkey, G.D. Dixon, B.M. Greenberg. Measurement of short- and long-term toxicity of polycyclic aromatic hydrocarbons using luminescent bacteria. Ecotoxicol. Environ. Saf., 51:12–21 (2002).

S.A. Eremin, P. Nodet, E. Maiolini, S. Ghini, E. Ferri, F. Fini, S. Girotti. "Development of a Chemiluminescent ELISA for quantification of Thiram in Honeybees". Luminescence, 21:366–367 (2006).

European Community, 2000. Directive 2000/60/EC of the European Parliament and of the Council of 23 October 2000 establishing a framework for Community action in the field of water policy, Off. J. Eur. Comm., English edition, L 327, 43, 22 December (2000).

A. Gerhardt. Biomonitoring for the 21st century. In: "Biomonitoring of Polluted Water. Reviews on Actual Topics. Environmental Research Forum". A. Gerhardt (ed.), Trans Tech, Uetikon-Zuerich, Switzerland, Vol. 9, pp 1–13 (1999).

S. Ghini, M. Fernández, Y. Picó, R. Marín, F. Fini, J. Mañes, S. Girotti. Occurrence and distribution of pesticides in the province of Bologna, Italy, using honeybees as bioindicators. Arch. Environ. Contam. Toxicol., 47:479–88 (2004).

S. Girotti, E.N. Ferri, L. Bolelli, G. Sermasi, F. Fini. Application of Bioluminescence in Analytical Chemistry. In *Chemiluminescence in Analytical Chemistry*, Garcia-Campaña A.M. & Baeyens W.R.G. (eds.), Marcel Dekker, New York, 247–284 (2001).

S. Girotti, L. Bolelli, F. Fini, S. Ghini, C. Porrini, A.G. Sabatini, M. Musiani, G. Gentilomi, G. Andreani, E. Carpené, G. Isani. "Bioluminescent bacteria and honeybees: bioindicators in environmental monitoring". Luminescence, 17:273–274 (2002a).

S. Girotti, L. Bolelli, A. Roda, G. Gentilomi, M. Musiani. Improved detection of toxic chemicals using bioluminescent bacteria. Anal. Chim. Acta; 471:113–120 (2002b).

S. Girotti, L. Bolelli, F. Fini, D. Rinaldi, S. Ghini, E. Chirila. "Bioluminescent bacteria and honeybees: bioindicators in environmental monitoring." Environ. Engin. Manag. J., 4(2):143–148 (2005a).

S. Girotti, F. Fini, S. Ghini, S. Totti, E. Ferri, J.V. Mercader, A. Montoya, S.A. Eremin. "Development of a Chemiluminescent Immunoassay for the Detection of Azinphos-methyl in Honeybees." Analele Univeritatii Ovidius, seria: Chimie, 16:167–169 (2005b).

S. Girotti, M.D. Luque de Castro, M. Piccolo, F. Gubiani. Italian Patent UD 2005 A 000116, 11 July (2005c).

S. Girotti, S. Ghini, L. Bolelli, E. Ferri, E. Maiolini, R. Ecchia, N. Precchia, E. Preti, D. Bulgarelli, M. Sforza, "A didactic project for applied chemistry divulgation: honeybees and bioluminescent bacteria as bioindicators in environmental control." Scientific and Technical Bulletin of University of "A. Vlaicu – Arad", Series: Chemistry, Food Science & Engineering (ISSN 1582-1021), Vol. 11 (XII), 140–144, (2006a).

S. Girotti, L. Bolelli, F. Fini, M. Monari, G. Andreani, G. Isani, E. Carpené. "Trace metals in the archid clam Scapharca inaequivalvis: effects of molluscan extract on bioluminescent bacteria." Chemosphere, 65(4):627–633 (2006b).

S. Girotti, E. Maiolini, L. Bolelli, S. Ghini, E.N. Ferri. "Bioremediation of contaminated soils by hydrocarbons degrading bacteria and decontamination control". This book, 2007.

A. Henderson. Literature on air pollution and lichens XLIII. Lichenologist, 28:279–285 (1996).

M. Iriti, L. Belli, C. Nali, G. Lorenzini, G. Gerosa, F. Faoro. Ozone sensitivity of current tomato (Lycopersicon pimpinellifolium), a potential bioindicator species Environ. Pollut., 141:275–282 (2006).

K.J.M. Kramer, E.M. Foekema. The Musselmonitor® as biological early warning system-the first decade. In: Butterworth, F.M., Gunatilaka, G. & Gonsebatt, M.E. (eds.). Biomonitor and Biomarkers as Indicators of Environmental Change, Vol. II, Kluwer Academic/Plenum, pp. 59–87 (2000).

K.J.M. Kramer. Quality of data in environmental analysis. National Institute of Marine Geology and Geo-ecology, 11/2005, 15–19 (2006), www.geoecomar.ro.

J.V. Mercader, A. Montoya, J. Agric. Food Chem., 47:1276–1284 (1999).

P.L. Nimis, M. Castello, M. Perotti. Lichens as biomonitors of sulfur-dioxide pollution in La-Spezia (Northern Italy). Lichenologist, 22:333–344 (1990).

S. Parvez, C. Venkataraman, S. Mukherji. A review on advantages of implementing luminescence inhibition test (*Vibrio fischeri*) for acute toxicity prediction of chemicals. Environ. Internat., 32:265–268 (2006).

C. Porrini, S. Ghini, S. Girotti, A.G. Sabatini, E. Gattavecchia, G. Celli. Use of honeybees as bioindicators of environmental pollution in Italy. In Honey bees: The environmental impact of chemicals, J. Devillers and M.H. Pham-Delègue Routledge-Taylors & Francis Group (eds). London, 186–247 (2002).

A.L. Queffelec, F. Boisdé, J.P. Larue, J.P. Haelters, B. Corbel, D. Thouvenot, P. Nodet. Development of an immunoassay (ELISA) for the quantification of Thiram in lettuce. J. Agric. Food Chem., 49:1675–1680 (2001).

K.V. Sharma, J.S. Aulakh, A.K. Malik. Thiram: degradation, applications and analytical methods. J. Environ. Monit., 5:717–723 (2003).

D. Trott, J.J.C. Dawson, K.S. Killham, R.U. Miah, M.J. Wilson, G.I. Paton. Comparative evaluation of a bioluminescent bacterial assay in terrestrial ecotoxicity testing. J. Environ. Monit., 9:44–50 (2007).

B. Wolterbeek. Biomonitoring of trace element air pollution: principles, possibilities and perspectives. Environ. Pollut., 120:11–21 (2002).

# ACCUMULATION AND TRANSFORMATION PROCESSES OF POLLUTANTS IN SOIL

GHEORGHE COMAN[1], CAMELIA DRAGHICI[2]
AND CARMEN VRABIE[3]
[1] *Preclinical Medicine Department, Faculty of Medicine,*
*Transilvania University of Brasov,*
*36 Nicolae Balcescu str., 500019-Brasov, Romania*
[2] *Chemistry Department, Faculty of Materials Science*
*and Engineering, Transilvania University of Brasov,*
*50 Iuliu Maniu Str., 500091-Brasov, Romania*
[3] *Wastewater Treatment Laboratory, Compania Apa Brasov*
*13 Vlad Tepes Str., 500092-Brasov, Romania*

**Abstract.** The starting point of this paper is the need of knowledge on the influence of different (wastes) deposits on the soil quality. A general presentation of the accumulation and (bio)transformation of pollutants in soils, with emphasis on heavy metals behaviour will be given. A case study on heavy metals presence in a wastewater sludge deposit and their migration in the surrounding soil will be presented.

**Keywords:** soils samples, heavy metals, biotransformation, bioaccumulation, migration

## 1. Introduction

Humans have always been exposed to hazardous substances going back to prehistoric times when they inhaled noxious volcanic gases or carbon monoxide from inadequately vented fires in the caves.

Large scale production of variety of chemicals and energy and other development activities like agriculture, transport, urbanization and health care during the past decades, have lead to the release of huge quantities of wastes into the environment. Substantial amounts of these wastes are potentially pollutants to

---

*To whom correspondence should be addressed: c.draghici@unitbv.ro

L. Simeonov and V. Sargsyan (eds.),
*Soil Chemical Pollution, Risk Assessment, Remediation and Security.*
© Springer Science+Business Media B.V. 2008

the environment and are dangerous to the living organisms. These wastes may cause contamination of air, surface water, ground water, sediments, soils and biota. Some specific characteristics of wastes approach related to their impact on human body are given below (A-C):

A. The degree of hazard determined by wastes is dependent on several factors:

- Composition, reactivity, physical form and quantities
- Biochemical and ecological effects (bioavailability, bioaccumulation, ecotoxicity, toxicity)
- Mobility (transport in various environmental media or biological liquids)
- Persistence (detoxification potential, transformation and biotransformation)
- Health effects
- Local conditions (temperature, humidity, light, soil or water types)

B. The factors influencing human response to toxic chemicals present in environment are:

- Dosage (large dose may mean immediate effect)
- Body weight (is inversely proportional to toxic responses/effects)
- Age (children are more susceptible to toxic chemicals)
- Psychological status (stress increases vulnerability to toxic chemicals)
- Immunological status (health status influences the metabolism)
- Climate conditions (temperature, humidity, pressure)

For example, in Table 1 is shown the lead induced diseases to human health with the differences in ages (Hoornweg, 2000; Misra and Pandey, 2005).

TABLE 1. Lead effects on human health

| Pb concentration (mg/dL) | Children | Adults |
|---|---|---|
| 10 | Physical state decrease | No effect |
| 10–20 | Erithrocitoporphyrin | Erithrocitoporphyrin |
| 20–30 | Vitamin D decrease | Blood presure increase |
| 30–40 | *Hemoglobin decrease* | Nephropaty; Fertility decrease |
| 40–50 | Anemy | *Hemoglobin decrease* |
| 50–100 | Nephropaty | Anemy |
| 100–150 | Death | Encephalopaty |

C. The hazardous waste may be characterized based on the risk they pose:

- High risk wastes – determined by chlorinated solvents, cyanides, pesticides, dioxins, polychlorobiphenyls, polycyclic aromatic hydrocarbons
- Intermediate risk wastes – determined by heavy metals
- Low risk wastes – determined by putrescible wastes (Cui and Forssberg, 2003)

Taking into account the role of different wastes deposits on the soil quality, this paper gives a general presentation of the accumulation and transformation of pollutants in soils, with emphasis on heavy metals behavior, as an introduction to a case study on the heavy metals presence in a wastewater sludge deposit correlated with their content in the surrounded groundwater and leaches.

## 2.   Specific (Bio)processes in Soils

Bioprocesses is a generic term for different transformation occurring in soils, therefore some specific definitions are required to delimit them.

a. **Bioaccumulation** includes all processes responsible for the uptake of pollutants by living cell and includes enzymatic degradation, biosorption mechanisms, as well as intracellular accumulation and bioprecipitation.
b. **Biosorption** is the process by which pollutants (especially metals) are removed by complexation to either living or dead biomass through functional sites (carboxyl, imidazole, thiol, sulfate, carbonyl, amide, and hydroxyl).
c. **Biodeterioration** is the breakdown of economically useful products, but often the term has been used to refer to the degradation of normally resistant substances (plastics, cosmetics, wood products, metals).
d. **Biotransformation** is a step in biochemical pathway which usually leads to the conversion of a pollutant molecule into a less toxic product, but it is also possible that the resulted substances are more toxic.

The most essential requirement for biotransformation occurring in soils is the presence of specific microorganisms that posses and express those genes required to produce enzymes that are able to react with the xenobiotics. These microorganisms transform pollutants using three major processes:

- **Catabolism** (degrade xenobiotic molecules to acquire energy, carbon or nitrogen)
- **Co-metabolism** (the metabolic products are available for other processes too)
- **Extracellular enzymatic activity** (secretion of extracellular enzymes accompanied by microbial digestion of complex compounds such as amides, or phosphates); some extracellular enzymes persist in the environment after the microorganisms are dead (Blythe and Ellen, 2000)

e. **Biodegradation** is the term used for all biological breakdowns of chemical compounds and complete biodegradation leads to mineralization. Mineralization is the conversion of an organic compound to inorganic ones, such as $CO_2$, $Cl^-$, $NH_4^{4+}$, and this process is considered to be the endpoint of the biotransformation process.

f. **Bioremediation** is the removal, transformation or detoxification of any pollutant from the environment to a less toxic form, by any natural process.

Both bioaccumulation and biosorption are used as strategies for remediation of contaminated soil (with metals and radionuclide) based on the ability of bacteria and plants to concentrate metals within the cells at concentrations of 1000 times higher then the ambient concentrations. In order to use the microorganisms for boiremediation knowledge about the geochemistry of soils are imposed.

In case of soil contamination, decision making on application of remediation alternatives is an important step after a comprehensive analysis and assessment of contaminants. The costs of remediation are high and the impact on the environment or human health and agricultural productivity are multiple (Scholz and Schnabel, 2006).

g. Alternatively, any attempt to manipulate contaminated environmental areas by the addition of stimulants or additives (biostimulation) or the addition of specific microorganisms or biochemicals (bioaugmentation) can be classified as **accelerated bioremediation** (Gadd, 2000; Lederberg, 2000; Allsopp et al., 2004, Coman et al., 2006; Whiteley and Lee, 2006).

## 3.  Heavy Metals Accumulation in Soils

Climate and precipitation, particulate matter from the atmosphere that a deposited on soils, the groundwater quality, the soil nature and the plant maturity degree, all these influence the concentration of heavy metals in plants. Moreover the soil nature can be affected by the following factors:

- The application of fertilizers
- Sewage sludge
- Irrigation with wastewater (Muchuweti et al., 2006)

It is already known that heavy metals use in different economical activities and their removal in the environment, increased the contamination of the drinking water and strongly affect the human health, but little is known that they also affect the biological functioning of soils. For example, large concentrations of heavy metals in soils inhibit microbial activity, altering the enzymes conformation, blocking essential functional groups, as well as blocking other metal ions exchange process. In conclusion, heavy metals (As, Pb, Cu, Zn, Cd, Sn, Ni, Cr, Hg, Mn, Se) affect soil respiration, decrease the soil biomass, and modify the N-mineralization and nitrification. The differences in tolerance of the microorganisms to heavy metals influence the reduction degree of $NO_3^-$ to $N_2$. For example, a selective inhibition of $NO_2$-reductase could induce the accumulation of $NO_2^-$ in soils, forming soluble toxic compounds. On the other hand a selective inhibition of $N_2O$-reductase would accumulate nitrous oxide gas ($N_2O$)

and release it from soil. Nitrous oxide plays a central role in ozone decomposition in stratosphere and exerts a significant greenhouse effect, with a high global warning potential (320 higher than the effect of $CO_2$).

$$NO_3^- \rightarrow NO_2^- \xrightarrow{NO_2^- - reductase} N_2O \xrightarrow{N_2O - reductase} N_2$$

Unfortunately, metals and radionuclides cannot be biodegraded and microorganisms can interact with these contaminants and transform them from one chemical form to another by changing their oxidation state through the electrons exchange processes, reduction-oxidation (Vasquez-Marrieta et al., 2006).

The specific properties of heavy metals as super-toxicants in big cities have to be mentioned as follows (Davydowa, 2005):

- Transition of their oxidation states: Pb(II) to Pb(IV); V(IV) to V(V); Cr(III) to Cr(VI)
- Their ability to coordinate with toxic organic ligands
- Catalytic activity in polymerization, cyclysation, hydrogenation
- Extremely long times of degradation (thousand years for Pb ions)

From the great variety of heavy metals accumulation and transformation in soils as well as soil remediation processes, the behavior of chromium is given as an example.

Chromium is released into the environment by a large number of industrial operations such as electroplating, chromate manufacturing and wood preservation. Hexavalent chromium is more stable in basic medium as anionic form ($CrO_4^{2-}, Cr_2O_4^{2-}$), it is toxic, carcinogenic and mutagenic to animals as well as humans. In contrast, trivalent chromium is more stable in acidic medium as cationic form ($Cr^{3+}$), it is relatively less mobile and therefore less toxic.

The hexavalent chromium biotransformation is based on microorganisms' capacity to reduce the toxic Cr (VI) to the less soluble trivalent form, as a normal function of their metabolism. Long-term studies on this subject will be necessary to determine how reducing conditions could be maintained to prevent reoxidation of trivalent chromium to hexavalent.

The conventional treatment methodology for soil systems contaminated with Cr (VI) starts with excavation, and continues with the addition of chemical reducing agent, followed by one of the processes:

- Precipitation and sedimentation
- Ion exchange
- Adsorption

These physico-chemical methods can be selected based on different criteria as costs, efficiency, energy and chemical consumption are (Rama and Ligy, 2005).

Soil contamination with heavy metals represents one of the most pressing dangers to the human health and phytoremediation is potentially used to resolve metal contaminated sites.

One of the techniques of phytoremediation of metal-contaminated soil is phytoextraction, meaning the accumulation of metals into the plant shoots. Another technique of phytoremediation is phytostabilization, consisting in the plants capacity to minimize metal mobility in contaminated soil.

Plant metal uptake is influenced by different factors:

- Soil pH
- Soil redox potential
- Soil cation exchange capacity
- Organic substances
- Plant species and age

Heavy metals can cause severe phytotoxicity and it is possible to identify metal-tolerant plant species from natural vegetation. Plants that accumulate over 1g/kg biomass of ionic Cu, Cr, Co, Ni or Pb or over 10 g/kg biomass of ionic Mn or Zn are defined as hiperaccumulators. For example, plants known as hyperaccumulators of metals into their biomass are trees, vegetable crops, grass and weeds (Simeonov et al., 1999; Yoon et al., 2006; Simeonova and Simeonov, 2006).

## 4.  Heavy Metals Occurrence in a Sludge Deposit – Case Study

The case study is part of an environmental impact study required for the approval of closing an old sludge landfill (for conformity). The environmental impact study was done by an authorized expert from SC SOLMED SRL and the beneficiary of the study is Compania Apa RA, Brasov (Water Company), which is providing services of water treatment, potable water distribution, collection of the domestic and industrial wastewaters, wastewater treatment and release in the emissary (SC SOLMED SRL, 2006).

Our study aim is to demonstrate the heavy metals presence (in terms of occurrence and accumulation) in the sludge deposit, as well as their migration in the surrounding soils.

The experimental approach took into account the site characterization, the analytical procedures (sampling process, measurements, data processing) and the results discussions and interpretation.

### 4.1.  SITE CHARACTERIZATION

The environment impact study was done in an old sludge landfill situated in the South-South-East of the Barsa River depression, in the Bod-Feldioara-Stupini zone, Brasov County, region with alluvial soil and clay. The sludge from the Wastewater Treatment Plant of the Compania Apa RA, was deposited in a landfill located close to Brasov, opened in 1978, deposited until 2000, and for

2007 is the deadline to close it. The landfill capacity is 91,000 m³, consists in three compartments, of which one is ecological, filed at the total capacity.

## 4.2.  ANALYTICAL PROCEDURE

The soils sampling sites were selected at 30 m distance from the sludge deposit to the Barsa River side (FH1) and at the sludge landfill vicinity (FH2), at different depths, depending on the soil structure, as presented in Table 2. The sampling and measurements procedure, based on Atomic Absorption Spectrometry (AAS), were according to the Romanian regulation and performed in an accredited laboratory.

TABLE 2. Sampling sites and soil structure characterisation

| Site symbol | Depth (m) | Characterisation |
|---|---|---|
| FH1-0 | 0.5–1.3 | Brown dusty sand |
| FH1-1 | 1.8–2.4 | Dusty sand with brown gravel |
| FH1-3 | 3.9–4.1 | Sandy layer with gravel, yellow sludge |
| FH2-1 | 1.9–2.2 | Heterogeneous matter with gravel and organic matter |
| FH2-2 | 2.2–2.9 | Fine sand, dusty, grey with wastes |
| FH2-3 | 2.9–3.0 | Fine sand, with clay layer and organic matter |
| FH2-4 | 6.5–6.8 | Gravel and stone area with organic matter infiltration |
| FH2-5 | 6.8–7.1 | Sandy dust with gravel |
| FH2-6 | 7.5–9.0 | Gravel with sand and stones, grey silt layer with specific smell |

## 4.3.  RESULTS AND DISCUSSION

The data processing took into consideration the heavy metals occurrence in the two chosen sampling sites (FH1 and FH2), as given in Table 3.

TABLE 3. Heavy metals occurrence in the studied sites

| Sampling site symbol | Heavy metal (mg/kg) | | | | | |
|---|---|---|---|---|---|---|
| | Pb | Cd | Zn | Ni | Cr | Mn |
| FH2-1 | 535.6 | <LOD | 576.6 | <LOD | <LOD | <LOD |
| FH2-2 | 81.6 | <LOD | 117.0 | <LOD | <LOD | 769.2 |
| FH2-3 | 36.7 | <LOD | 122.4 | <LOD | <LOD | 532.4 |
| FH2-4 | 23.9 | <LOD | 126.6 | <LOD | <LOD | <LOD |
| FH2-5 | 21.2 | <LOD | 129.1 | <LOD | <LOD | <LOD |
| FH2-6 | 39.8 | <LOD | 97.7 | 177.8 | <LOD | 658.4 |
| FH1-0 | 72.2 | <LOD | 136.0 | <LOD | <LOD | <LOD |
| FH1-1 | 31.2 | <LOD | 98.8 | <LOD | <LOD | <LOD |
| FH1-3 | 16.0 | <LOD | 122.8 | <LOD | <LOD | <LOD |
| Alert limit | 250 | 5 | 700 | 200 | 300 | 2,000 |
| Intervention limit | 1,000 | 10 | 1,500 | 500 | 600 | 4,000 |

The results presented in Table 3 show the occurrence and accumulation of Cd, Cr, Mn, Ni, Pb and Zn in the two studied sites, as follows:

- Pb is the only heavy metal which concentration exceeded the alert limit at the sludge deposit vicinity.
- Cd and Cr are under the limit of determination (LOD) at both sites.
- Ni is present only at FH2-6 site (7.5–9.0 depth); this demonstrate an accumulation process of Ni in the soil.
- Mn is also accumulated at the investigated levels FH2-2, FH2-3 and FH2-6.

Heavy metals migration to the river side was also registered:

- Pb concentration is higher at the sludge deposit vicinity than at the river side.
- Ni and Mn are not present at the river side.
- Zn is found at almost the same concentration, at both sampling sites FH2-3 and FH1-3.

**Conclusions**

This paper presents some theoretical aspects on the accumulation and (bio)processes of pollutants in soils, with emphasis on heavy metals behaviour.

Cd, Cr, Mn, Ni, Pb and Zn presence in a sludge deposit as well as their migration to the deposit surroundings constituted a case study, as part of an environmental impact study, aiming to assess the conformity to close an old sludge deposit.

The results showed that Cd, Cr are at concentration lower than the limit of determination, Ni and Mn are present in the sludge deposit in accepted concentration limits, and Pb concentration exceeded the alert limit at the sludge deposit vicinity. In the surrounding sites was observed that heavy metals migrated, or a possible historical pollution occurred, as well as Ni and Mn accumulation in depth soil.

According to the presented results, also based on measurements done in the groundwater and river up-stream and down-stream, the beneficiary will take the adequate measure to close the old sludge deposit using the most suitable technology.

# References

Allsopp D., Seal K., Gaylarde C., 2004, Introduction to Biodegradation, Cambridge University Press, 1–256

Blythe L.H., Ellen L., 2000, Biotransformation of Pesticides in Saturated Zone Materials, Hydrogeology Journal, 8, 89–103

Coman Gh., Draghici C., Chirila E., Sica M., 2006, Pollutants Effects on Human Body-Toxicological Approach, in Simeonov L., Chirila E., Chemical as Intentional and Accidental Global Environmental Threats, Springer, 255–266

Cui J., Forssberg, E., 2003, Mechanical Recycling of Waste Electric and Electronic Equipment, Journal of Hazardous Materials B99, 243–263

Davydowa S., 2005, Heavy Metals as Toxicants in Big Cities, Microchemical Journal, 79, 133–136

Gadd G.M., 2000, Bioremedial Potential of Microbial Mechanisms of Metal Mobilization and Immobilization, Curr. Opin. Biotechnol., 11, 271–279

Hoornweg D., 2000, What a Waste Solid Waste Management in Asia. UNEP Ind. Environ, 65–70

Lederberg J., 2000, The Encyclopedya of Microbiology, Academic, San Diego, CA, 607–617

Misra V., Pandey, S.D., 2005, Hazardous Waste, Impact on Health and Environment for Development of Better Waste Management Strategies in Future in India, Environment International, 31, 417–431

Muchuweti M., Birkett J., Chinyanga E., Zwauya R., Scrimshaw M., Lester J., 2006, Heavy Metal Content of Vegetables Irrigated with Mixture of Wastewater and Sewage Sludge in Zimbabwe, Agriculture, Ecosystems, and Environment, 112, 41–48

Rama K., Ligy K., 2005, Bioremediation of Cr(VI) in Contaminated Soils, Philip Journal of Hazardous Materials, B121, 109–117

Scholz R., Schnabel U., 2006, Decision Making under Uncertainty in Case of Soil Remediation, J. of Environ. Manag., 80, 132–147

Simeonov L.I, Simeonova B.G., Nikolova J., 1999, Phytoremediation of Industrially Polluted with Heavy Metals Lands in Bulgaria (First trials), Proc. of 1999 Contaminated Site Remediation Conf.: Challenges Posed by Urban and Industrial Contaminants, Freemantle, Australia, 551–557

Simeonova B., Simeonov L., 2006, An application of a phytoremediation technology in Bulgaria. The Kremikovtzi Steel Works experiment Remediation Journal, Spring edition, Wiley Periodicals, New York, pp.113–123

Vasquez-Marrieta M., Mondragon C., Trujillo N., Herrera-Arreola G., Govaerts B., van Cleemput O., Dendooven L., 2006, Nitrous Oxide Production of Heavy Metal Contaminated Soil, Soil Biology and Biotechnology, 38, 931–940

Whiteley C., Lee D.J., 2006, Enzyme Technology and Biological Remediation, Enzyme and Microbial Technology, 38, 291–316

Yoon Y., Cao, X., Zhou, Q., Ma, L., 2006, Accumulation of Pb, Cu, and Zn in Native Plants Growing on a Contaminated Florida Site, Science of the Total Environment, 1–9

SC SOLMED SRL, 2006, Bilant de mediu nivel I, Bilant de mediu nivel II

# LASER MASS ANALYSER IN THE CONFIGURATION OF A MULTI-PURPOSE TRANSPORTABLE MASS SPECTROMETRIC SYSTEM FOR EXPRESS ENVIRONMENTAL ANALYSIS: APPLICABILITY AND ANALYTICAL LIMITATIONS

BIANA SIMEONOVA[1], LUBOMIR SIMEONOV[2*]
AND GEORGE MANAGADZE[3]
[1]*Solar-Terrestrial Influences Laboratory*
[2]*Institute of Electronics*
*Bulgarian Academy of Science, Sofia, Bulgaria*
[3]*Space Research Institute, Russian Academy of Sciences*
*Moskow, Russia*

**Abstract.** The paper is related generally to the problematic of environmental analysis. Presented are laboratory results, obtained in the process of adaptation for earth application of a miniature laser time-of-flight analyzer, originally created for the purposes of space research on board of a lander. The earth-based version of the instrument is intended to serve as a core analyzer of a universal and multi-purpose transportable mass spectrometric complex for express environmental analysis. The general conclusions about the environmental applicability of the instrument, which analytical performance is qualitative and semi-quantitative, is that the laser spectrometer is acceptable to perform an express mass analysis of environmental samples.

**Keywords:** space research instrumentation, earth-based applicability, laser mass-spectrometry, express environmental analysis, elemental and isotopic analysis

## 1. Introduction

The successful transfer of technologies and the adaptation of high-tech instrumentation, originally developed for the purposes of space research towards specific earth-based applications justifies to a great extent the huge costs of space

---

[*] To whom correspondence should be addressed: simeonov2006@abv.bg

exploration. A special place in the great variety of space instrumentation have the mass spectrometry analysers of different type and application. Generally the purpose of these instruments is to provide adequate information about the distribution of elements and isotopes in different space media, like the solar wind and interstellar environment, or the chemical composition of the surface layers of small space bodies without atmosphere, like asteroids or planetary satellites.

One of the possible earth-based implementation of instruments, initially developed for the purposes of space research, is the development of a universal mass spectrometric system for environmental applications, which is a combination of several time-of-flight analysers. The earth-based application of the instruments is a result of a technological transfer both for the devices and for the methodologies of measurement, with the introduction of options in concern to the specific needs to analyse environmental samples of air, soil, water and plant origin. Hence the classification the spheres of environmental application of the instruments should be made in connection to the aggregate condition of the samples for analysis. This classification approach is adequate and leads to a precision and differentiated evaluation of the analytical capabilities of the instruments to solve various problems.

The greatest part of the shown in the present paper results are related to the analytical evaluation of one of the main instruments of the mass spectrometric complex, namely the instrument with laser ion source. They are obtained in the process of creation of its laboratory and technological models for space research and in the process of optimization of its analytical characteristics and physical parameters. Part of the mass spectra are obtained when analyzing objects in experimental setups, dedicated for earth-based applications. The main rule in the process of adaptation of space research instrumentation in this context, is to preserve the unique characteristics, in comparison to the conventional apparatuses, whose high analytical capabilities are obtained usually at the cost of increased dimensions and mass. According to this preliminary understanding, the specific features of the mass spectrometers are the following: smaller as possible mass and dimensions; high mobility; high degree of autonomy; preserved working capabilities at different ambient conditions; minimal consumption of energy and consumables; easy maintenance and technical support; easy interchange of main functional blocks; absolutely automated cycles of analysis, registration and workout of the mass spectra. The combination of these characteristics and the high analytical capabilities of the mass spectrometers with different ion sources makes them applicable in earth-based applications in numerous cases without any competition.

The greatest chances for direct use of the analyzers are in the sphere of environmental research. The scientific tasks in connection for such implementation concern the idea to create a multi-functional and mobile mass spectrometry complex. The purpose of such mobile analysis laboratory is to solve problems of ecology and control of environmental pollution, studying the physics and chemistry of the atmosphere and the hydrosphere. The task of the complex is to perform qualitative and quantitative express analysis of environmental samples from sites and objects, polluted industrially or in cases of environmental disasters, including terrorist and criminal action. The combination of different by their functional characteristics and method of analysis instruments covers all the variants of aggregate state of the samples during the analysis of the elemental and isotopic composition of hard and dust samples, of gases and aerosols, atmosphere air, of water samples and of other liquids. The mass-spectrometry analysis, spectra manipulation and estimation of the final results has to be performed in near-to-real time. Such a mode of operation defines the need of implementation of a telemetry system for data transfer to the regional offices of people protection and earlier announcement in emergency cases of critical anthropogenic environmental pollution. The system for data analysis allows an express estimation of the initial source of pollution, including transboundary cases.

## 2.  Instrumental

The configuration of the mass spectrometry complex can be flexibly changed according to the parameters of the analysis of the environmental object of investigation. In this contest a basis system will consists of several universal analyzers with addition of instruments for accomplishment of specific tasks, as well as for express inter-calibration of the system as a whole. The choice of the analytical instruments is based on the capabilities of multi-functionality, which must enhance the solution of a maximum number of analytical problems with a minimal number of instruments and technologies. The core analyzer of the complex and which is most precisely developed at the moment is the universal laser mass spectrometer LASMA (Laser Mass Analyzer) (Schmitt, 1994; Simeonov et al., 2005). The analyzer is capable to perform an express analysis of environmental samples of solid and dust, water and liquid, and bio and plant origin.

The principle of laser time-of-flight mass spectrometer is illustrated on Figure 1 while on Figure 2 is shown the functional scheme-view of the laser mass spectrometer without the vacuum system, the power supply and the steering electronics units.

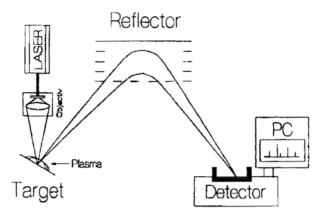

*Figure 1.* Principle scheme of laser time-of-flight mass spectrometer.

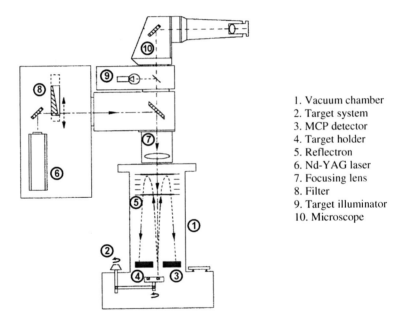

1. Vacuum chamber
2. Target system
3. MCP detector
4. Target holder
5. Reflectron
6. Nd-YAG laser
7. Focusing lens
8. Filter
9. Target illuminator
10. Microscope

*Figure 2.* Functional scheme-view of the laser mass spectrometer.

   The instrument consists of the following main functional units: a laser source with a beam-focusing system, which enables to concentrate energy into spot with diameter of 30÷50 μm on the surface of the sample in vacuum chamber. Ions, emitted as a result of an infrared irradiation influence, are reflected in the field of electrostatic reflector and directed to the detector. The detector is a special development of a standard chevron assembly of two micro-channel plates. The micro-channel plates have a central hole with a diameter big enough to let the laser beam down to the target and the flow of ejected ions up towards the time-of-flight compartment of the analyzer. After their reflection the ions of one and a same mass form expanding ring packets on their flight-back to the ring detector. With that form the mass packets arrive at the detector. The high-speed analog-to-digital converter loads a spectrum to a computer for further signal processing. The instrument is equipped with microscope and miniature TV monitor that allows observing intentional movements of laser influence area during the investigation of surface heterogeneity. The samples are introduced in the vacuum chamber through a lock – chamber, which makes it possible to replace a holder with samples on the target platform wheel within one minute without deterioration of the vacuum in the main chamber. Each sample holder contains up to ten samples situated on a metal, Teflon or a glass volumes depending on the nature of the sample material and its consistence. The possibility to mount the whole system of the instrument with the pump unit, the power supply and the personal computer on a rack and operate it on a medium-size pick-up van allows the process of sample-taking, target preparation and actual analysis to be accomplished in-field.

## 3. Estimation of the Analytical Performance of the Laser Time-of-Flight Mass Spectrometer – Experimental Examples

The analytical capabilities of the laser mass spectrometer were evaluated in laboratory experiments with laser targets with different elemental and isotopic composition, including certified samples, especially produced for testing of analytical instrumentation in material science, as well as with targets, artificially enriched with certified isotopic reference constituents to evaluate the quantification level of measurement. The mass spectra is obtained from the time-of-flight spectra with the help of a specially-written computer program, which performs a 5-fold identification scan of the recorded time-of-flight spectra in accordance to the table of natural elemental and isotopic abundances, stored in the computer. On Figure 3 is shown a magnified part of a mass spectrum in the region of Cd, which illustrates the work of the mass-identification computer program. The Cd element is identified by the relative ratios of its characteristic isotopes. On the isotopes are marked the measured by the instrument values, on which the computer program performs the identification procedure. After the mass identification process is completed, the program draws the fitting of the mass peaks and saves a mass file.

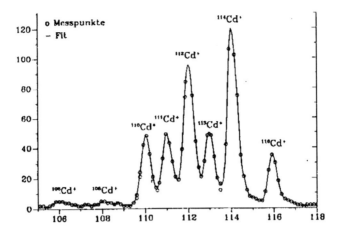

*Figure 3.* A part of a mass spectrum in the region of cadmium as illustration of the mass-identification computer program of the laser mass analyzer (Simeonov et al., 1996).

The calibration of the laser mass analyzer was accomplished according to the general calibration procedures applied to commercially manufactured mass spectrometer instruments. The industrially produced laser mass spectrometers undergo this obligatory procedure to identify and record the individual performance parameters of each particular instrument, introduced by its main building units or elements. This way of calibration is of much greater importance to be performed with every new instrumental configuration, design or concept, because besides of the preliminary evaluation of the analytical characteristics, the check with standardized targets is in fact an evaluation of the new ideas, introduced in the instrument (Laser mass analysis, 1993).

The possible sources of parameter deviations in the present laser time-of-flight mass analyzer are the solid-state laser or the micro-channel particle detector. The pulsed Nd-YAG lasers normally display an output power density with instability, which reaches up to ±12% (Moenke-Blankenburg, 1989) and affects the intensity of recorded signal. On Figures 4, 5 and 6 are presented mass spectra from a check accomplished with a target, provided by the NBS (National Bureau of Standards – USA), called NIST SRM 610-612 (Standard Reference Material) (Certificate of analysis, 1970). According to the certificate, the standard was produced and certified the development of trace analytical methods and is one of a series of four. The nominal trace element concentration is 500 ppm for each of 62 elements that have been added to the glass support matrix.

On Figure 4 is presented a consecutive series of laser shots/mass spectra, obtained at different laser beam power density. The mass spectra outlook displays a strong dependence on the laser beam power density. However the

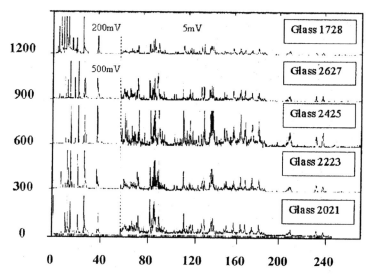

*Figure 4.* A consecutive series of laser shots/mass spectra, obtained at different laser beam power density.

intensity of recorded mass peaks is not a straight function of the laser beam power density, because the laser ablation of a solid target includes processes as evaporation and ionization, whose probabilities are strongly mass dependent.

The conclusion, which could be drawn is that the new instrument needs preliminary tuning of the laser beam power density before the actual analysis of samples. The tuning could be done with the attenuation filter 8 (Figure 2). The procedure is necessary in order to avoid unwanted high beam power, which would sum to the inevitable output power instability of ±12% of the solid-state laser and would result in a possible detector and ion counting system saturation.

When tested with the standard sample of SRM 610, which contains approximately 45 ppmA of Pb, Th and U and by averaging over 20 spectra, it was found that the spread of Th/U and Th/Pb ratios were 13% and 26% respectively. The sampling area was 30–40 μm without the possibility of the sample's non-homogeneity being taken in consideration. It is necessary to note that the registration time is 50μs and the repetition rate is one laser shot in 2 to 3 seconds. The data-handling system has the potential to accumulate an average of 100–1,000 laser shots/mass spectra. In this case the detection limit is increased from 10 to 30 and achieved 10E-8 to 3.10E-9. On Figure 5 is presented a part of the mass spectrum from the SRM 610 test in the mass region from 220 to 275 a.m.u. with Th and U.

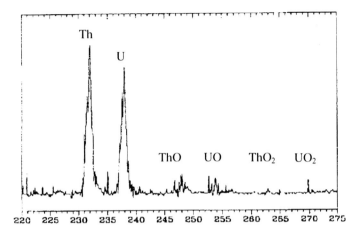

*Figure 5.* A part of the test spectrum with the standard SRM 610 in the mass region from 220 to 275 a.m.u. with Th and U.

The presence of the heavy elements Th, Pb and U was used as a check of the reproducibility of measurement. On Figure 6 are plotted the averaged Th/U and Th/Pb ratios.

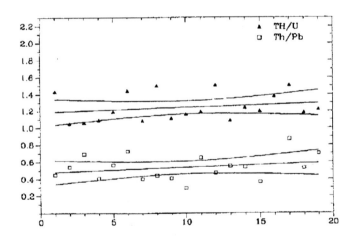

*Figure 6.* Calculated ratios of Th/u and Th/Pb from consecutive 20 laser shots/mass spectra taken with the laser time-of-flight mass analyzer in reproducibility tests with target SRM 610.

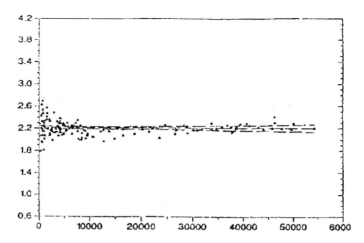

*Figure 7.* Calculated ratios of 63Cu/65Cu from 165 laser shots/mass spectra in reproducibility tests with Cu target.

On Figure 7 are presented the calculated ratios of $_{63}Cu$ to $_{65}Cu$ from from 165 consecutive laser shots/mass spectra, as the values of the copper isotopes are the correspondent areas beneath the mass peaks from the mass spectra. The averaged value of the copper isotopes ratio is calculated to be 2.24 ± 6% compared to the natural abundances ratio of 2.24. As a comparison, the calculations with the SRM 610 reproducibility test showed a ratio of the copper isotopes equal to 2.24 ± 1% on averaging from 50 laser shots/mass spectra. The difference in ratios measured with SRM 610 and Cu is due to the different homogeneity of the two laser targets.

To estimate the quantification ability of the instrument mass spectra, Woll et al. accomplished series of experiments with metal standard certified laser targets, normally used as references in material analysis. The measured and estimated by the instrument mass spectra elemental values were correlated by averaged individual Relative Sensitivity Factors (RSF), defined from the standards data and plotted as a function of the certified concentrations (Figure 8). Observed are relative deviations between the corrected concentration values from the instrument signals and the certified concentrations in the different samples.

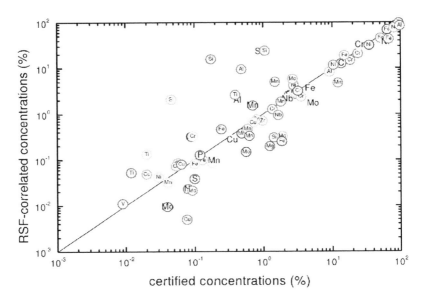

*Figure 8.* Correlation plot of measured and evaluated relative elemental concentrations for multielement metal standards. The average RSF (Relative Sensitivity Factors) are determined from the standards (Woll et al., 1999).

## Conclusions

The above presented results evaluate the small laser time-of-flight mass analyzer as instrument with a semi-quantitative ability, which could be used in compositional bulk analysis. The instrument can perform analysis in a mass range up to 250 a.m.u. with a sensitivity, sufficient enough for the purposes of the express investigation of samples of environmental origin. The instrument is small, versatile and easy operable in a transportable analysis complex.

## References

Moenke-Blankenburg, L., 1989. Laser microanalysis, Chemical analysis, V. 105, Wiley.

Schmitt, C., 1994. Handbook LASMA, ISOTOP Laseranalytische Verfaehren, Kaiserslautern.

Simeonov, L.I., K. Sheuermann, C. Schmitt, 1996. Schnelle semiquantitative Multielement-analyse wassriger Lösungen mit dem Laser-Massenaanalysator LASMA, Terratech: Zeitschrift für Altlasten und Bodenschutz, 6, 29–31.

Simeonov, L., G. Managadze, Ch. Schmitt, 2005, Miniature laser mass spectrometer for express elemental and isotopic analysis, Comptes rendus de l'Académie bulgare des Sciences, 58, 8, 903–910.

Vertes, A., R. Giebels, F. Adams (Eds.), 1993. Laser ionization mass analysis, Wiley, New York.

Woll, D.M., M. Wahl, H. Oechsner, 1999. Operation and application of a laser mass analyser (LASMA) for multielement analysis, Fresenius J. Anal. Chem., 70–75.

# BIOREMEDIATION OF CONTAMINATED SOILS BY HYDROCARBONS DEGRADING BACTERIA AND DECONTAMINATION CONTROL

STEFANO GIROTTI[1*], ELISABETTA MAIOLINI[1], LUCA BOLELLI[1], ELIDA FERRI[1], ANNA POMPEI[2], DIEGO MATTEUZZI[2], SVETLANA MEDVEDEVA[3] AND PAOLO FONTI[4]

[1] *Department of Metallurgic Science, Electrochemistry and Chemical Techniques, University of Bologna, Via S. Donato 15, 40127 Bologna, Italy*
[2] *Department of Pharmaceutical Science, University of Bologna, Via Belmeloro 8, 40126 Bologna, Italy*
[3] *Institute of Biophysics, Siberian Branch of Russian Academy of Sciences, Academgorodok, 660036 Krasnoyarsk, Russia*
[4] *Gruppo CSA SpA, Istituto di Ricerca, Via al Torrente, 22, 47900 Rimini, Italy*

**Abstract.** In this study three strains of *Vibrio* marine bioluminescent bacteria, have been employed to measure the biotoxicity of hydrocarbon contaminated soils from oil terminals. The evaluation of inhibitory effects of samples on emitted light was performed on soil before and during a four months bioremediation treatment by hydrocarbons degrading bacteria. The measurements were carried out at room temperature, using the microplate format, in medium containing 3% NaCl, and done both for short time of contact (60 minutes) and for longer time intervals (till 24 hours). The results were expressed as percentage of inhibition with respect to the blank emission. The yield of extraction of different solvents (acetone, dioxan, ethanol) was also evaluated by the bioluminescent test. After a short time of treatment (one month) the toxicity of the contaminated samples increased, since the bioremediation made available long chain hydrocarbons before in a not-soluble form. After longer period of bioremediation the toxicity decreased. The values

---

[*] To whom correspondence should be addressed: stefano.girotti@unibo.it

L. Simeonov and V. Sargsyan (eds.),
*Soil Chemical Pollution, Risk Assessment, Remediation and Security.*
© Springer Science+Business Media B.V. 2008

of the of the percentage of inhibition obtained by the bioluminescent bacteria test (BLB) test were in good agreement with the hydrocarbons content, determined by gas chromatography. The BLB test could represent a useful biomonitoring tool to evaluate the changes occurring during remediation processes, even in case of on-field conditions. Moreover, the bioluminescence test allows to analyse at the same time, by using the microplate luminometer, a lot of samples in a relatively fast way, at low costs and using small volumes of reagents.

**Keywords:** bioluminescent bacteria, bioremediation, gas chromatography

## 1. Introduction

Men's activities produce intense and multiple pollution that influence or affect the environment at different levels. The importance of these effects became clear in the last decades. The legislation concerning the protection of the environment increased the number of noxious compounds which presence and amount must be assessed, and lowered the maximum limits admitted for their residues. Analytical determinations have to become more and more sensitive, specific and rapid in order to allow the effective realization of the necessary monitoring plans (European Community, 2000; Katsivel et al., 2005; Wethasinghe et al., 2006; Girotti et al., 2005).

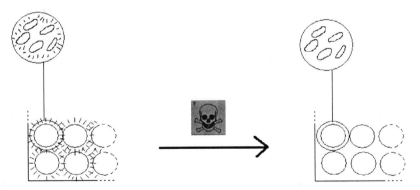

*Figure 1.* Bioluminescent bacteria emit light when surrounded by an optimal environment. If noxious substances are present, the luminescence is reduced or suppressed.

Luminescent assays, and among them light-emitting bacteria, can be a suitable tool for environmental analysis. In vivo luminescence is a sensitive indicator of xenobiotic toxicity to micro-organisms because it is directly coupled to respiration via the electron transport chain, and thus reflects the metabolic status of the cell. The higher the degree of toxicity, the less the amount of light emitted by the bacteria (Figure 1). Thus, the presence of toxic substances can be rapidly evaluated and several assays and dedicated instruments are now available (Kuznetsov et al., 1998; Girotti et al., 2002; Farré et al., 2001; Dalzell et al., 2002).

Hydrocarbons (HCs) are highly toxic to all compartments of the environment (Del Panno et al., 2005; Plaza et al., 2005; Katsivel et al., 2005) and the soil in the areas of oil terminals is heavily contaminated by these compounds. Remediation treatments in these areas are imposed also to avoid the spreading of these pollutants in the environment around and into groundwaters (Wethasinghe et al., 2006).

We applied a bioluminescent bacteria (BLB) test to the evaluation of the toxicity of specimens of soils withdrawn at an oil refinery site, treated at laboratory level by a bioremediation procedure carried out by a suitable mixture of hydrocarbons degrading bacteria, previously isolated from the soil of the same area (Gogoi et al., 2003; Girotti et al., 2001; Marin et al., 2006; Coletti et al., 2006).

## 2. Materials and Methods

### 2.1. LUMINESCENT BACTERIA

The revealing organisms were three different bioluminescent bacteria strains, two belonging to the species *Vibrio*, named *Ucibo* and *Russian*, and the third was the *Vibrio fischeri (NRRL B-111777)*, from the collection of Aerobe Bacteria of the Pasteur Institute, Paris, France, and named simply *Vibrio*. The *Ucibo* bacteria, have been harvested in the Mediterranean Sea and cultivated at the laboratories of the Department of Metallurgic Science, Electrochemistry and Chemical Techniques, Bologna (SMETEC). The *Russian* bacteria (Photobacterium phosphoreum 1883 IBSO) have been supplied by the Institute of Biophysics (Siberian Branch, Academy of Sciences), Cultures Collection IBSO, Laboratory of Bacterial Bioluminescence. Akademgorodok, Krasnoyarsk, Russia.

Bacteria were cultivated both on liquid and solid medium and lyophilised as previously reported (Girotti et al., 2001, 2006; Bolelli et al., 2006).

## 2.2.   BIOLUMINESCENCE MEASUREMENT

The BLB tests were performed using two similar microplate luminometers, the: "Victor 1420" Multilabel Counter (Wallac, Sweden) and the 1253 Luminoskan Ascent (Labsystems, Helsinki, Finland). The tests were done by using 96-wells black microplates.

To perform the bioluminescent assay each aliquot of the freeze-dried bacteria was reconstituted up to 200 µL with distilled water in the short-term assay, or of liquid broth in the long-term assay. The lyophilised bacteria can be used for the analysis or for cultivation on solid or liquid media. Lyophilised bacteria were used because they produce a constant blank signal, which is useful during the comparison of the various results (Girotti et al., 2002).

The procedure of the bioluminescent toxicity test is based on an ISO Standard one (BS EN ISO 11348-3, 1999), only slightly changed. Acute and chronic toxicity tests were performed (El-Alawi et al., 2002; Girotti et al., 2006; Bolelli et al., 2006).

Short-term assay (acute toxicity test, AT). Each well of the microplate contained: 180 µL of aqueous sample + 20 µL of bacteria suspension in distilled water. Long-term assay (chronic toxicity test, CT). Each well was filled with: 100 µL of aqueous sample + 100 µL of inoculated broth. For this assay, 10 mL of broth were inoculated with 100 µL of lyophilised bacteria. The lectures, in RLU, were repeated each 10 minutes, for 60 minutes in short term assays, from 10 till 24 hours in the long term ones.

Before to start any assay, light emission intensity of the bacterial suspension was evaluated, to determine if it was optimal for the analysis.

The results obtained from both acute and chronic toxicity tests were calculated as the percentage of inhibition of the blank emission, a relative way that was considered the more correct one, mainly during bioremediation treatment. In case of the bioremediated samples the blank was represented by the corresponding, not treated, samples. In toxicity tests by not contaminated soil samples:

$$\frac{B_e - S_e}{B_e} \times 100$$

where $B_e$ is the emission of the not treated or not contaminated sample and $S_e$ that one of the treated or contaminated sample.

## 2.3. SAMPLES CHARACTERISTICS

Concerning the geological characteristics of the samples, they are mainly silty sand, gravel and calcarenite. Samples were collected from 8 geognostic holes drilled until the depth of 24–30 m. The macroscopic structure of the samples showed differences among them, with not similar size of their particles. These differences influenced negatively, in part, the analysis, mainly because produced different affinity to the hydrocarbons. All samples had a hydrocarbons smell. The total hydrocarbon content (THC) was determined, by gas chromatography, as the sum of the content of hydrocarbons with a chain length lower or higher of 12 (C > 12 and C < 12) and the aromatic ones, evaluated separately (Table 1). As the blank was used a not contaminated soil showing geological characteristics and composition similar to majority of the samples.

TABLE 1. Characteristics of the analysed sample

| Depth (m) | THC (ppm) | Sample |
|-----------|-----------|--------|
| 5 | 1,592 | S6.1 |
| 7 | 1,319 | S6.2 |
| 9 | 889 | S6.3 |
| 2 | 8,700 | S7.1 |
| 7 | 15,025 | S7.2 |
| 9 | 17,033 | S7.3 |
| 4 | 831 | S8.1 |
| 8 | 434 | S8.2 |
| 4 | 5,189 | S9 |
| 5 | 855 | S10 |
| 6.5 | 329 | S11 |
| 7 | 702 | S12 |
| 6.5 | 5,327 | S13 |
| 5 | 2,355 | S14 |

## 2.4.  EXTRACTION PROCEDURE

The extraction procedure employed on the soil samples is reported in Diagram 1.

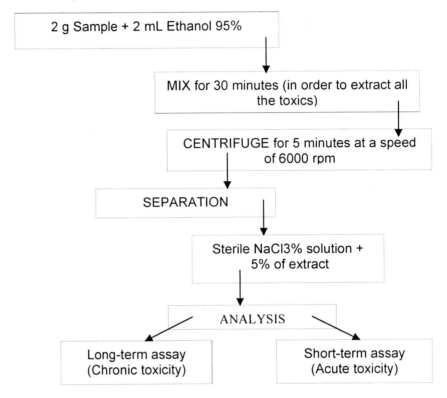

*Diagram 1.* Standard extraction procedure.

To improve the extraction yield on these samples were tested also solvents other than ethanol, such as acetone and dioxane.

## 2.5.  GAS CHROMATOGRAPHY ANALYSES

Gas Chromatography determinations were made according the US Environmental protection Agency (EPA) methods: EPA 3510C 1996 and EPA 8270D 1998 (http://www.epa.gov/).

## 2.6. BIOREMEDIATION PROTOCOL

The bioremediation experience has been started, on a group of five samples (S6.1, S7.1, S7.2, S8.1 and S8.2) which content of THC (Table 1) was in a wide range of concentration, by employing a mixture of hydrocarbons degrading bacteria (MBDH), isolated from analogous soil samples and then prepared as enriched cultures in our laboratory.

For each sample were prepared two bioreactors, containing the same culture medium. To the first (TR) were added the sample solution and the hydrocarbon degrading bacteria, to the second one (NOTR) only the sample solution. Both were incubated under similar oxygen supply. The (TR) bioreactor was incubated at room temperature and in microaerofilic conditions, to reproduce the "in situ" real conditions. The (NOTR) bioreactor was maintained at 4°C.

After 1, 2, 3 and 4 months of treatment the samples were again analysed by bioluminescent bacteria to evaluate their bio toxicity and by gas-chromatography to assess the exact content and kind of hydrocarbons.

## 3. Results and Discussion

The short-term bioluminescence inhibition assay is a widely used, highly standardized biotest. It is used for the determination of the acute toxicity of pure substances or mixtures of chemicals in different environmental matrices. The advantages of this test are the: ease of use, speed, low costs. The disadvantage is the lack of information about long term effects.

The main advantage of the chronic bioluminescence inhibition assay is the more realistic assessment of the overall toxic effect of many compounds, especially those acting through a slow mechanism of action. The main disadvantage is the longer time required to perform the assay.

## 3.1. SAMPLES EXTRACTION

The effects on light emission of various solvents and of solutions containing different percentage of them were tested on the three bacterial strains. The 3% acetone in NaCl 3% contained a measurable amount of extracted, non polar, compounds producing an inhibition of about 5% on *Ucibo* bacteria emission. On the other hand, the 3% dioxane in NaCl 3% was the most suitable solution for *Vibrio* and *Russian* strains, both in case of acute and chronic toxicity tests. In Figure 2 are reported, as an example, the results obtained measuring the light emission of the *Vibrio* strain in presence of the different solvents.

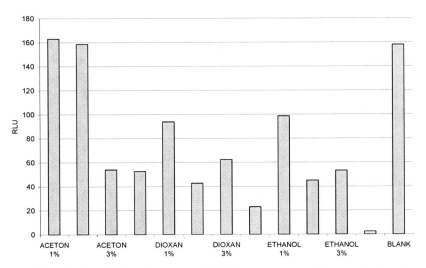

*Figure 2*. Light emission of *Vibrio* in contact with different amounts of the tested solvents.

Acetone and dioxane inhibited less the bioluminescent bacteria, but the extracts showed a higher imprecision with respect to ethanolic extracts. The repeatability (coefficient of variation) for acetone and dioxane extracts was 25% and 30%, respectively, while for ethanol extracts was only 15%. Ethanol as extracting solvent was always used after these tests.

To ascertain that no saturation problems existed during extraction, different volumes of ethanol were used to extract the HCs. Moreover, ultrasounds treatment was applied, but no improvement of extraction yield was observed, and then in the following analysis a volume of 2 mL of ethanol per 2 g of sample, without sonication, was used.

## 3.2. ACUTE TOXICITY AND CHRONIC TOXICITY TESTS

When no blank real sample was available, it was substituted by a NaCl 3% solution mixed with the appropriate amount of pure solvent.

This kind of "blank" was employed, for comparison purposes, in all experiments and its emission values reported in all the figures. Sometime it was possible to observe clearly the enhancing effect provoked by the various compounds extracted from a complex matrix like soil samples (see Figure 3). Due to this effect, various samples with low level of contaminants showed higher light emission than the blank and no inhibition was reported. This phenomenon is due to the components of the samples that act as substrates for the bacterial metabolism, and for this reason it is important to have a suitable

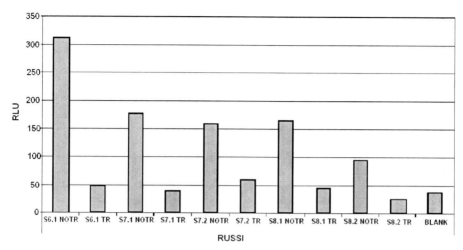

*Figure 3.* CT light emissions for Russian strain with 5% ethanol in solution.

blank sample collected at the same site, that reasonably should contain the same stimulating substances but not the toxic compounds.

Using as blank a soil sample which was not contaminated, but had the same characteristics, the hormesis phenomenon is not present and all the samples analysed by BLB resulted toxic, inhibiting the light intensity with respect to those of this real blank.

The behaviour of the three strains was similar, but it was necessary to optimise the test conditions specifically for each strain: for example the bacteria need different period of time to reach the maximum of emission, and then it is not correct to compare the data obtained, at the same analysis time, from different BLB strains.

The differences in sensitivity of the three strains represent an advantage, since the use of different strains, and not only one like in the commercial kits, allows revealing more carefully the presence of the various toxic components, to which the different strains are more or less sensitive. The biotoxicity of the hydrocarbons, extracted from the soil samples (3% ethanol in 2% NaCl), is shown in Table 2. The percentage of inhibition is usually in good agreement with the total hydrocarbons content (THC) determined by gas chromatography.

TABLE 2. Percent of inhibition induced by soil samples on BLB emission

| Sample (ppm THC) | Acute toxicity (% inhibition) | | | Chronic toxicity (% inhibition) | | |
|---|---|---|---|---|---|---|
| | Ucibo | Vibrio | Russian | Ucibo | Vibrio | Russian |
| S6.1 (1,592) | 74 | 52 | 61 | 60 | 65 | 63 |
| S6.2 (1,319) | 74 | 46 | 67 | 63 | 60 | 68 |
| S6.3 (889) | 69 | 34 | 62 | 40 | 56 | 59 |
| S7.1 (8,700) | 72 | 41 | 64 | 83 | 83 | 82 |
| S7.2 (15,025) | 92 | 88 | 93 | 84 | 78 | 70 |
| S7.3 (17,033) | 69 | 43 | 60 | 87 | 97 | 83 |
| S8.1 (831) | 35 | 22 | 40 | 5 | 44 | 39 |
| S8.2 (434) | 35 | 33 | 32 | 9 | 28 | 21 |
| S9 (5,189) | 99 | 88 | 99 | 99 | 93 | 74 |
| S10 (855) | 32 | 26 | 40 | 7 | 32 | 28 |
| S11 (329) | 15 | 20 | 18 | 5 | 10 | 8 |
| S12 (702) | 52 | 29 | 57 | 4 | 34 | 7 |
| S13 (5,327) | 98 | 87 | 99 | 98 | 78 | 98 |
| S14 (2,355) | 58 | 37 | 74 | 17 | 19 | 43 |

## 3.3. BIOREMEDIATION BY HYDROCARBONS DEGRADING BACTERIA

The bioremediation process has been followed by determining changes in hydrocarbons content and composition by gas chromatography. The content, before and after 4 months of treatment, is reported in Table 3.

TABLE 3. THC content determined by gas chromatography before and after 4 months bioremediation treatment by MBDH

| Sample | Initial content of THC (ppm) | THC after 4 months of treatment (ppm) |
|---|---|---|
| S6.1 | 1,592 | 421 |
| S7.1 | 8,700 | 879 |
| S7.2 | 15,025 | 7,087 |
| S8.1 | 831 | 52 |
| S8.2 | 434 | 16 |

During the bioremediation it was possible to observe that the aspect of treated and not treated samples was clearly different. The treated samples were more opaque, as a result of the fragmentation and emulsifying activity of the MBDH (Gogoi et al., 2003, Del Panno et al., 2005; Marin et al., 2006).

The bio toxicity analysis performed during the 1–4 months treatment period initially showed increasing values, which decreased in the last, fourth month (Figure 4).

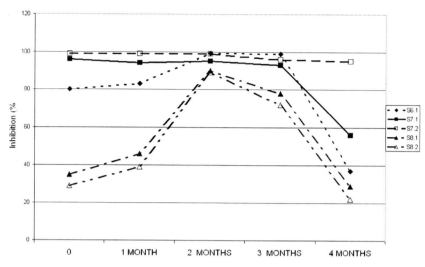

*Figure 4.* Biotoxicity values during bioremediation treatment determined by *Ucibo* bacteria.

All samples, with the only exception of S7.2, after 4 months of treatment presented an hydrocarbons content below the allowed limits, as reported in Table 3, demonstrating a good degrading action of MBDH.

The percentage of inhibition recorded from both acute and chronic toxicity tests, referred to each one of the three strains, are reported in the following Tables 4–6. The percentage of inhibition is expressed with respect to the not treated samples, that was considered the more correct way. It must be take into account that the composition of treated and not treated samples was different also because the presence, in treated samples, of the hydrocarbons degrading bacteria, which metabolism certainly produces compounds not present in the non treated samples.

TABLE 4. Inhibition of *Vibrio* strain emission by samples treated for 4 months with the MBDH

| Sample | Bioluminescent bacterial inhibition | | | |
|---|---|---|---|---|
| | Acute toxicity (%) | | Chronic toxicity (%) | |
| | Before | After | Before | After |
| S6.1 | 76 | 16 | 61 | 36 |
| S7.1 | 77 | 25 | 84 | 48 |
| S7.2 | 89 | 78 | 98 | 88 |
| S8.1 | 51 | 19 | 67 | 39 |
| S8.2 | 22 | 15 | 69 | 24 |

TABLE 5. Inhibition of *Russian* strain emission by samples treated for 4 months with the MBDH

| Sample | Bioluminescent bacterial inhibition | | | |
|---|---|---|---|---|
| | Acute toxicity (%) | | Chronic toxicity (%) | |
| | Before | After | Before | After |
| S6.1 | 76 | 39 | 98 | 45 |
| S7.1 | 83 | 32 | 97 | 39 |
| S7.2 | 96 | 92 | 96 | 97 |
| S8.1 | 75 | 281 | 78 | 28 |
| S8.2 | 49 | 15 | 99 | 19 |

TABLE 6. Inhibition of *Ucibo* strain emission by samples treated for 4 months with the MBDH

| Sample | Bioluminescent bacterial inhibition | | | |
|---|---|---|---|---|
| | Acute toxicity (%) | | Chronic toxicity (%) | |
| | Before | After | Before | After |
| S6.1 | 86 | 21 | 83 | 37 |
| S7.1 | 77 | 33 | 94 | 56 |
| S7.2 | 91 | 90 | 99 | 95 |
| S8.1 | 50 | 22 | 46 | 29 |
| S8.2 | 40 | 12 | 39 | 23 |

From data showed in the tables above it is clear that after this treatment the acute and chronic bio toxicity decreased for all samples except for S7.2. In general, the CT was slightly higher than the AT, even the THC after the treatment was low. It should demonstrate that an increase in the content of short chain hydrocarbons ($C < 12$), highly toxic for the bioluminescent bacteria, was produced by the degrading activity of the MBDH on the long chain hydrocarbons ($C > 12$). The short chain hydrocarbons could permeate better the bacterial wall, producing damages which are more evident in the long period.

## 3.4. GC ANALYSES

The results of the GC analysis (Table 3) confirmed that the bioremediation treatment produced important changes in the composition of the hydrocarbons. The measurable HCs resulted definitely higher in treated samples, since the degrading action of the MBDH transformed the practically solid contaminants in more soluble hydrocarbons, that were extracted and measured. This higher availability of "soluble and extractable" hydrocarbons is the reason for the increased toxicity of the treated samples.

Observing the shape of the chromatographic spectra, reported in Figure 5 for a treated sample, and in Figure 6 for a not treated one, it is possible to realize that important changes occurred in the composition of treated samples. The spectra in Figure 5 shows numerous peaks corresponding to hydrocarbons, distributed in a wide interval of retention times, while in the same area of spectra in Figure 6 those peaks are totally absent or definitely lower. The MBDH was clearly working in a good way at the optimal conditions created in our laboratory, included the employment of a specific cultural medium developed by us specifically for this application.

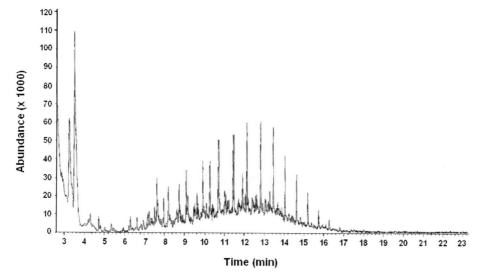

*Figure 5*. Chromatogram of the S7.1 sample treated by MBDH for 40 days.

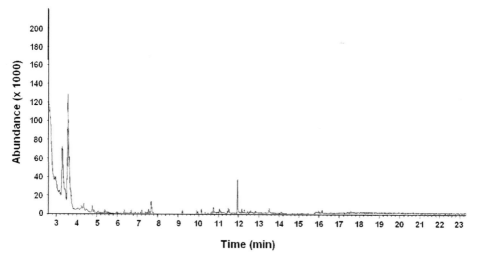

*Figure 6*.  Chromatogram of the S7.1 sample not treated by MBDH for 40 days.

## Conclusions

The bioluminescent test here presented can be used to evaluate the toxicity of several pollutants in complex matrices, because the decrease of bacterial light emission is directly related to the effects of compounds affecting the cells viability.

The bioluminescence test allows to analyse at the same time, by using the microplate luminometer, a lot of samples in a relatively fast way, at low costs and using small volumes of reagents.

The BLB acute and chronic toxicity test resulted useful to measure and control the contaminated samples, since it was able to distinguish the bio toxicity of the contaminated samples from that of samples considered non contaminated according to the regulatory rules.

Concerning the sample that resulted slightly toxic according to the BLB test, but classified as non toxic according to the chemical data, this could be due to the toxic effects of compounds different from hydrocarbons, even present at very low concentration in the sample. This shows once again the need for bio toxicity analyses, like those by BLB, as an important support to the chemical analyses, even they cannot be considered an alternative, since bio and chemical assays offer different information.

The MBDH developed and applied by our laboratory seems to be effective in the decontamination of hydrocarbons polluted soils, and it can be applied to the bioremediation of the soil samples.

## Acknowledgements

This work was supported by grant from the University of Bologna (Fundamental Oriented Research). P. Caputo is gratefully acknowledged.

## References

Bolelli L., Bobrovova Z., Ferri E., Fini F., Menotta S., Scandurra S., Fedrizzi G., Girotti S., "Bioluminescent bacteria assay of veterinary drugs in excreta of food-producing animals". Journal of Pharmaceutical and Biomedical Analysis, 42, 88–93, 2006.

British Standard – BS EN ISO 11348-3: 1999. Water quality determination of the inhibitory effect of water samples on light emission of Vibrio fischeri (Luminescent bacteria test), part 3. BSI, Index House, Ascot, SL5 7EU, UK.

Coletti G., Gubiani F., Piccolo M., Luque de Castro M.D., Japón Luján R., Ferri E., Garcia Morante B., Bolelli L., Girotti S. "Bioremediation and toxicity bioluminescent assay of contaminated soils". Luminescence, 21, 324–325, 2006.

Dalzell D.J., Alte S., Aspichueta E., de la Sota A., Etxebarria J., Gutierrez M., Hoffmann C.C., Sales D., Obst U., Christofi N., A comparison of five rapid direct toxicity assessment methods to determine toxicity of pollutants to activated sludge, Chemosphere, 47, 535–545, 2002.

Del Panno M.T., Morelli I.S., Engelen B., Berthe Corti L., Effect of petrochemical sludge concentration on microbial communities during soil bioremediation. FEMS Microbiology Ecology, 53(2), 305–316, 2005.

El-Alawi Y.S., McConkey B.J., George Dixon D., Greenberg B.M., Measurement of short- and long-term toxicity of polycyclic aromatic hydrocarbons using luminescent bacteria. Ecotoxicological and Environmental and Safety, 51(1), 12–21, 2002.

EPA 3510C Separatory funnel liquid-liquid extraction 1996 and EPA 8270D Semivolatile Organic Compounds by Gas Chromatography/Mass Spectrometry (GC/MS) 1998, http://www.epa.gov/.

European Community, Directive 2000/60/EC of the European Parliament and of the Council of 23 October 2000 establishing a framework for Community action in the field of water policy, Off. J. Eur. Comm. L, 327, 43, 22 December 2000 English edition.

Farré M.L., Ferrer I., Ginebreda A., Figueras M., Olivella L., Tirapu L., Vilanova M., Barcelo D., Determination of drugs in surface water and wastewater samples by liquid chromatography-mass spectrometry: methods and preliminary results including toxicity studies with Vibrio fischeri. Journal of Chromatography A, 938(1–2), 187–197, 2001.

Girotti S., Ferri, E.N., Bolelli, L., Sermasi, G., Fini, F., Applications of bioluminescence in analytical chemistry. In: "Chemiluminescence in analytical chemistry", A.M. Garcia-Campaña, W.R.G. Baeyens, E.D.S. New York, Marcel Dekker, 247–284, 2001.

Girotti S., Bolelli L., Roda A., Gentilomi G., Musiani M., "Improved detection of toxic chemicals using bioluminescent bacteria". Analytica Chimica Acta, 471, 113–120, 2002.

Girotti S., Bolelli L., Fini F., Rinaldi D., Ghini S., Chirila E., "Bioluminescent bacteria and honeybees: bioindicators in environmental monitoring". Environmental Engineering and Management Journal, 4 (2), 35–40, 2005.

Girotti S., Bolelli L., Fini F., Monari M., Andreani G., Isani G., Carpené E., "Trace metals in the archid clam Scapharca inaequivalvis: effects of molluscan extract on bioluminescent bacteria". Chemosphere, 65 (4), 627–633, 2006.

Gogoi B.K., Dutta N.N., Goswami P., Krishna Mohan T.R., A case study of bioremediation of petroleum-hydrocarbon contaminated soil at a crude spill site. Advances in Environmental Research, 7, 167–782, 2003.

Katsivel E., Moore E.R.B., Moroukli D., Strompl C., Pieper D., Kalogerakis N., Bacterial community dynamics during in-situ bioremediation of petroleum waste sludge in landfarming sites. Biodegradation, 16(2), 169–180, 2005.

Kuznetsov A.M., Rodicheva E.K., Medvedeva S.E., Biotesting of sewage and river water by lyophilized luminous bacteria biotest. Field Analytical Chemical Technologies, 2(5), 267–275, 1998.

Marin J.A., Moreno J.L., Hernandez T., Garcia C., Bioremediation by composting of heavy oil rafinery sludge in semiarid conditions. Biodegradation, 17(3), 251–261, 2006.

Plaza G., Nalecz-Jawecki G., Ulfig K., Brigom R.L., The application of bioassays as indicators of petroleum-contaminated soil remediation. Chemosphere, 59(2), 289–296, 2005.

Wethasinghe C., Yuen S.T.S., Kaluarachchi J.J., Hughes R., Uncertainty in biokinetic parameters on bioremediation. Health risks and economic implications. Environmental International, 32(3), 312–323, 2006.

# APPLICATION OF CAPILLARY ELECTROPHORESIS FOR SOIL MONITORING

CAMELIA DRĂGHICI[1*], GHEORGHE COMAN[2]
AND ELISABETA CHIRILA[3]
[1] *Preclinical Medicine Department, Faculty of Medicine,
Transilvania University of Brasov,
36 Nicolae Balcescu Str., 500019-Brasov, Romania*
[2] *Chemistry Department, Faculty of Materials Science and
Engineering, Transilvania University of Brasov,
50 Iuliu Maniu Str., 500091-Brasov, Romania*
[3] *Chemistry Department, Faculty of Physics, Chemistry and Oil
Technology, Ovidius University of Constanta,
124 Mamaia Bdv, 900527-Constanta, Romania*

**Abstract.** Soils are complex systems, in direct contact with plants and animals, surface waters and ground waters, where very complicated biotransformation processes of the pollutants occur. Soils characterization as one of the most challenging subject for researchers, no only the environmentalists, due to the complexity of this media, combined with the fact that land is used with so many and different purposes. Soil chemical composition, inorganic, organic and biochemical compounds, can nowadays be determined by use of a large variety of analytical methods. This paper presents capillary electrophoresis as an alternative and competitive method to the spectrometric and chromatographic ones, for soils composition characterization.

**Keywords:** capillary electrophoresis, soil samples, inorganic and organic pollutants

## 1. Introduction

Soil solution or soil extract analysis evidenced its composition, inorganic ions, low-molecular-weight aliphatic carboxylic acids (LACA) and their related

---

* To whom correspondence should be addressed: c.draghici@unitbv.ro

L. Simeonov and V. Sargsyan (eds.),
*Soil Chemical Pollution, Risk Assessment, Remediation and Security.*
© Springer Science+Business Media B.V. 2008

anions, humic and fulvic acids (HA, FA), pesticides/herbicides and their metabolites, as well as uncharged compounds, depending on soil acidity, organic matter content and land use (Westergraad et al., 1998, 1999; Schmitt-Kopplin et al., 1998; Aga et al., 1999; Groom et al., 2000; Hilmi and Luong, 2000; Strobel et al., 2001).

Soil composition studies start with method optimization and validation using standard solutions (Aga et al., 1999; Berrada et al., 2001; Lanyai and Dinya, 2002), and continues with complex analysis, including forest and soil monitoring (Strobel et al., 2001). A great variety of analytical methods and techniques is used and sometimes soils composition studies require combined and coupled techniques for a complete characterisation (Schmitt-Kopplin et al., 1998; Strobel et al., 2001).

Spectrometric methods offered good selectivity and sensitivity for both inorganic and organic compounds analysis and are used independently, as detection devices for chromatographic and electrophoretic equipment, and more recently in hyphenated-coupled systems (Schmitt-Kopplin et al., 1998; Aga et al., 1999). Flame atomic absorption spectrometry (F-AAS) or graphite furnace atomic absorption spectrometry (GF-AAS) and inductively coupled plasma atomic emission spectrometry (ICP-AES) were used to detect inorganic cations and sulfur from different soil samples (Westergraad et al., 1998, 1999; Dieffenbach and Matzner, 2000; Strobel et al., 2001; Ahumada et al., 2001; Li et al., 2001; Birghila and Popescu, 2002). Inductively coupled plasma-mass spectrometry (ICP-MS) have also been used, with excellent results for oxoanionic and organo-metallic species of As, Se, Sb, and Te in soil matrixes (Casiot et al., 1998).

Gas chromatography (GC) was widely used to analyse different pesticides accumulated in soils, using either flame ionisation detector (GC-FID) (Schmitt-Kopplin et al., 1998; Lanyai and Dinya, 2002) or coupled with mass spectrometry (GC-MS) (Schmitt-Kopplin et al., 1998; Berrada et al., 2001; Romero et al., 2001).

Cations and metal complexes accumulated in soils were determined by ion chromatography (IC) either with conductivity detector (IC-CD) or coupled with mass spectrometry (IC-MS), which increases the separation sensitivity (Collins et al., 2001). High performance liquid chromatography (HPLC) was applied to inorganic and organic compounds determination, using diode array detector (HPLC-DAD) (Romero et al., 2001), UV-Vis-multi-wavelength detector (HPLC-UV-Vis-MWD) and coupled with mass spectrometry (HPLC-MS) (Groom et al., 2000) or with inductively coupled plasma-mass spectrometry (HPLC-ICP-MS) (Bissen and Frimmel, 2000).

This paper is completing our studies focused on capillary electrophoresis (CE) principles and practical applications in environmental analysis (Draghici, 2002, 2003) as well as particular approaches for atmospheric (Draghici et al.,

2003a), soil (Draghici et al., 2003b) and biological samples (Draghici et al., 2004). Our study aims to present CE as a powerful analytical technique with a broad environmental potential for soil sample analysis and characterisation.

## 2.  Principles of Capillary Electrophoresis

Capillary electrophoresis is known as a chromatographic related group of techniques. The main separation mechanism is based on the analyte migration in a fused silica capillary, electrically driven by a background electrolyte with buffering properties, when a high voltage is applied at the capillary ends. CE separation principle differs to the chromatography one, which is based on the repartition of the analyte between two non-miscible phases, but the graphic presentation of the analytical result (electropherograms) and, thus, the evaluation criteria (selectivity, sensitivity, efficiency, and resolution) are similar to chromatography.

Modern CE set-up consists in a fused silica capillary, introduced in two electrolyte vials connected via two electrodes (anode and cathode) to the power supplier. Analytes are injected in the capillary and migrate at the cathodic end of the capillary, where a detector is placed and gives a specific signal to the computer.

The electroosmotic flow (EOF), generated by the electrolyte, carriers the analytes to the detector window, and the elution order of the analytes is cations, neutral compounds and anions. EOF can be suppressed (SF-CZE) by use of amines, or can be reversed (RF-CZE) adding in the electrolyte cationic surfactants in a concentration close to their double-molecular layer. In this case the elution order is also reversed, anions, neutrals and cations. Ionic or non-ionic surfactants, added in the electrolyte at a concentration higher than the critical micellar concentration (CMC), will form micellar aggregates that act as pseudo-stationary phases in micellar electrokinetic chromatography (MEKC) techniques, so that non-charged compounds can also be separated. Several reviews on CE techniques principles and its applications to inorganic, organic and biochemical compounds determination are published in the specialized literature (Janini et al., 1997; Janos, 1999; Majidi, 2000; Issaq, 2000; Brocke et al., 2001). The authors are emphasising the advantages and disadvantages of using CE for complex matrixes separation.

## 3.  Applications of Capillary Electrophoresis for Soil Analysis

Soils analysis is carried out either from soil solution or from soil extracts, depending on the water solubility of the analyte. One of the most important stages of the soil analysis is sampling (including sample pretreatment), being known that sampling brings the highest contribution to the analysis uncertainty.

It was reported that no standardized method for the extraction of metalloids from soils is available (Casiot et al., 2002). On the opposite, the International Humic Substances Society (IHSS) edited a guideline also containing sampling of the humic substances (Schmitt-Kopplin et al., 1998). Therefore analysts are usually developing sampling and sample pretreatment methods, based on different complexation, ion exchange or extraction processes, adapted to the soil composition and the compounds of interest (Chirila et al., 2006). Some examples of sample pretreatment/preparation methods used for capillary electrophoresis analysis are given in Table 1.

TABLE 1. Sample pretreatment/preparation methods used for CE analysis

| Analyte | Sample preparation/pretreatment methods | Ref. |
|---------|------------------------------------------|------|
| Cr(III), Cr(VI) | complexation, extraction, centrifugation and filtration | Chen et al., 2001 |
| As, Se, Sb, Te (organic or inorganic compounds) | extraction, centrifugation and filtration | Casiot et al., 2002 |
| organic anions | centrifugation and filtration | Westergraad et al., 1998, 1999 |
| HA, DHA | ion exchange and centrifugation (guidelines of the International Humic Substances Society-IHSS) | Schmitt-Kopplin et al., 1998 |
| explosives (TNT and their metabolites) | extraction (in ACN), centrifugation and filtration | Groom et al., 2000 |
| LACA | ion exchange, centrifugation and filtration | Strobel et al., 2001 |

There are several reviews available, on capillary electrophoresis application to environmental samples, mainly addressed to groups of compounds of interest, as mutagenic heterocyclic amines (Kataoka, 1997), inorganic ions (Fukushi et al., 1999), surfactants (Heinig and Vogt, 1999), organic pollutants (Sovocool et al., 1999, Martinez et al., 2000), haloacetic acids (Urbansky 2000), arsenic species (Jain and Ali 2000), algal toxins (Pierce and Kirkpatrik 2001), inositol phosphate (Turner et al., 2002), pesticides and their degradation compounds (Andreu and Pico, 2004). CE has been reported as a successful method for environmental sample analysis, with different separation modes CZE, MECK, CEC, even in non-aqueous electrolytes (NACE) (Heinig and Vogt, 1999).

The electrophoretic systems used for soil compounds separations show the adaptability of the electrolytes and detection conditions to the large group of analytes from soil samples (see Table 2).

The reviewed studies show that inorganic buffered electrolytes (chromate, phosphate, borate), as well as organic ones (acetate, citrate, ascorbic acid, pyromellitic acid, trimellitic acid, *p*-hydroxy-benzoic acid) were used to determined different compounds from soil samples, covering both acidic and basic pH, between 4.5–11.2.

Organic additives were sometimes introduced in buffers, with different purposed:

- Complexation reagents – 18-crown-6-ether
- Chiral selectors – γ-CD
- Chromophores for indirect UV detection – chromate, 4-methylamino-phenole sulfate, trimellitic acid, p-hydroxy-benzoic acid
- Amines for EOF suppressors – DETA and TEMED
- Cationic surfactants foe the EOF reversal – TTAB and TTAOH
- Anionic surfactant for micellar systems – SDS

The detection modes connected to the CE system are also different, depending on the analytes detectable properties. Usually UV absorption is used, unless the analytes do not have their own chromophores, when indirect UV detection is used, by introducing chromophoric substances in the electrolyte. In order to increase the detection sensitivity, hyphenation techniques are used, like coupling the reversed flow capillary zone electrophoresis with inductively coupled plasma-mass spectrometry (RF-CZE-ICP-MS).

Inorganic cations as well as organic substances (HA, FA, herbicides) from different sampling sites were determined by capillary zone electrophoresis (Blattner et al., 2000; Schmitt-Kopplin et al., 1998; Aga et al., 1999), and γ-CD was used to separate racemic mixtures of some chiral herbicides (Aga et al., 1999).

In order to separate anions and organo-metallic species, reversed flow capillary zone electrophoresis was applied. Cationic surfactants were added in the electrolyte, like an organic ammonium hydroxide (TTAOH) (Casiot et al., 2002) or an organic ammonium salts, bromide form, (TTAB) (Westergraad et al., 1998, 1999; Strobel et al., 2001; Sovocool et al., 1999; Heining and Vogt, 1999; Turner et al., 2002; Chen et al., 2001).

Introducing different amines in the electrolyte, in order to reduce the electroosmotic flow (EOF), the suppressed flow capillary zone electrophoresis technique was applied. For example, TEMED (Blattner et al., 2000; Owens et al., 2000) and DETA (Westergraad et al., 1998, 1999; Strobel et al., 2001) were used to separate inorganic anions and LACA too, respectively by SF-CZE.

TABLE 2. Soil sample characterisation by capillary electrophoresis techniques (Adapted from Draghici et al. 2003b)

| Sample | Analytes | CE mod | CE buffer (BGE) and conditions | Remarks | Ref. |
|--------|----------|--------|-------------------------------|---------|------|
| humic soil [1] | HA, FA | CZE | 50 mM acetate, pH 5.3, 20 kV | HA composition | Schmitt-Kopplin et al., 1998 |
| Standard herbicides | ESA and OXA | *Chiral-CZE* | 30 mM citrate, *γ-CD* 2%, pH 5.5, 30 kV, 210 nm; | CE advantages | Aga et al., 1999 |
| limestone with aeolic loess soil (luvisoil) | inorganic ions | CZE | cat: 5 mM Metol, 1 mM ascorbic acid, 2 mM 18-crown-6-ether; | | Blattner et al., 2000 |
| | | SF-CZE | an: 3 mM PMA, *TEMED*, pH 8.0, 30 kV, 225 nm | | |
| moist soil material [2] | inorganic ions | CZE, SF-CZE | see Blattner et al., 2000 | | Andreu and Pico, 2004 |
| arable and forest soil [3] | inorganic acids, LACA | *SF-CZE* | inorganic and di-, tri-carboxylic acids: TD buffer: 3 mM TMA, 0.01–0.03% *DETA* (v/v), pH 5.8, -30 kV; | LOD: 0.26-1.77 µM | Westergraad et al., 1998 |
| | | *RF-CZE* | monocarboxylic acids: TTT buffer: 8 mM TRIS, 2 mM TMA, 0.3 mM *TTAB*, pH 7.6, 30 kV; 254 nm; | | |
| podzolized soil [4] | inorganic acids, LACA | *SF-CZE, RF-CZE* | see Westergraad et al., 1998 | LOD: 0.28-1.77 µM LOQ: 0.93-5.91 µM | Westergraad Strobel et al., 1999 |
| alluvial soil | organic acid anions | *RF-CZE* | 10 mM *p*-hydroxy-benzoic acid, 0.5 mM *TTAB*, pH 4.5, -20 kV, 254 nm; | organic anions influence on Cd/Cu soil solubility | Sovocool et al., 1999 |
| soil extracts | As, Sb, Se, Te species | *RF-CZE*-ICP-MS | 5 mM chromate/phosphate, 0.5 *TTAB*, pH 11.2, -20 kV | $DL_{hydr}$: 6–58 µg/L $DL_{ek}$: 0.08–633 µg/L; | Heining and Vogt, 1999 |
| industrial soil, leaches | As, Se, Sb, Te species | *RF-CZE* | chromate, 0.1–2 mM *TTAOH*, pH 8–11.5, -10 kV, 254 nm | $DL_{hydr}$: 52–3,900 µg/L $DL_{ek}$: 13-509 µg/L; | Owens et al., 2000 |
| forest floor soil [5] | inorganic acids, LACA | *SF-CZE, RF-CZE* | see Westergraad et al., 1998 | tree and soil type influence on DOC reactivity | Strobel et al., 2001 |

| Standards | metallo NTA, EDTA | *RF-CZE* | 50 mM phosphate, 0.5 mM *TTAB*, pH 6.86, -20 kV, 185/254 nm; | DL: 2–510 µM | Casiot et al., 2002 |
|---|---|---|---|---|---|
| anaerobic sludge [6] | TNT metabolites | *MEKC* | 2.5 mM borate, 12.5 mM boric acid, pH 8.5, 50 mM *SDS*, 30 kV, 225 nm; | MEKC comparison with LC/MS | Groom et al., 2000 |
| contaminated soil [7] | explosives | *MEKC*-chips | 15 mM borate, pH 8.7, 25 mM *SDS*, 0.9-2 kV | DL: 150–200 µg/L N: 1,500–2,100 | Hilmi and Luong, 2000 |

[1]Forschungsverbund Agrarokosystem Munchen, Scheyern-South Germany; [2]Fichtelgebirge, NE Bavaria, Germany; [3]Christianssæde, southern Danemark; [4]Ravnsholt forest, north-eastern Zealand, Danemark; [5]four sampling sites: Christianssæde, Løvenholm, Stenholtsvang and Tisted Nørskov; Denmark; [6]wastewater anaerobic sludge nearby a food manufacturer, Champlain Industries, Cornwall, Canada; [7]local contaminated soil, Montreal, Canada;

*Abbreviations: γ-CD-*γ-cyclodextrin; ***DETA***-diethylenetriamine; **DHA**-dissolved humic acid; **DL**-detection limit; **DL$_{ek}$**-detection limit by electrokinetic injection; **DL$_{hidr}$**-detection limit by hydrostatic injection; **DOC**-dissolved organic carbon; **EDTA**-ethilenediaminotetraacetic acid; **ESA**-ethanesulfonic acid; **FA**-fulvic acid; **HA**-humic acid, **LACA**-low-molecular-weight aliphatic carboxylic acids; **LOD**-limit of detection; **LOQ**-limit of quantification; **Metol**-4-methylamino-phenole sulfate; **N**-theoretical plates number; **NTA**-nitrilotriacetic acid; **OXA**-oxanilic acid; **PMA**-pyromellitic acid; ***SDS***-sodium dodecyl sulfate; ***TEMED***-N,N,N',N'-tetramethyl-ethylenediamine; **TNT**-2,4,6-trinitrotoluene; **TMA**-1,2,4-benzenetricarboxylic acid (trimellitic acid); **TRIS**-tris-(hydroxymethyl)-aminomethane; ***TTAB***-tetradecylmethylammonium bromide; ***TTAOH***-tetradecylmethyl-ammonium hydroxide.

Due to the fact that MEKC is suitable for uncharged compounds separations, SDS was added in the BGE and used for some explosives and their metabolite analysis (Groom et al., 2000; Hilmi and Luong, 2000; Halasz et al., 2002).

Flow Injection-capillary electrophoresis (FI-CE) was also used for on-site simultaneous determination of anions and cations from soil matrixes (Kuban et al., 2004).

## Conclusions

After a short presentation of some spectrometric and chromatographic techniques for soil complex matrixes, CE principles and their applications in environmental analysis, and a group of reviewed papers presenting CE for soil samples characterization, one can conclude that CE is an advantageous separation technique, because it offers selective systems to resolve complex samples.

CE is suitable for inorganic, organic and biological matrixes, charged and uncharged compounds, small and large molecules, with simple equipment and no need for capillary changes, only using adequate background electrolytes, with different composition, ionic strength and pH. The sensitivity requirements, for trace and ultra-trace analysis, are solved by mean of different detection systems, CE being suitable for hyphenation techniques. One of the most challenging subject that is now preoccupying the chromatographers' community, are miniaturized systems for separation and detection purposes.

CZE-chips are already an affordable systems and devices, applicable for environmental analysis too, by using MEKC chips with microelectrochemical detection (MEKC-chips-MED) for explosives detection in contaminated soils. The future for chromatographic and electrophoretic separation technique research seems to be focused on CEC development and miniaturized systems.

## References

Aga, D.S., Heberle, S., Rentsch, D., Hany, R., Muller, S.R., 1999, Environ. Sci. Technol., 33, 3462–3468.

Ahumada, I., Mendoza, J., Escudero, P., Ascar, L., 2001, Commun. Soil Sci. Plant Anal., 32(5–6), 771–785.

Andreu, V., Pico, Y., 2004, Trends in Anal. Chem., 23(10–11), 772–789.

Berrada, H., Font, G., Molto, J.C., 2001, Chromatographia, 54, 253–262.

Birghila, S., Popescu, V., 2002, Environmental Engineering Management Journal, 1, 355–359.

Bissen, M., Frimmel, F.H., 2000, Fresenuis J. Anal. Chem., 367, 51–55.

Blattner, M.S., Augustin, S., Schack-Kirchner, H., Hildebrand, E.E., 2000, J. Plant Nutr. Soil Sci., 163, 583–587.

Brocke, von A., Nicholson, G., Bayer, E., 2001, Electrophoresis, 22, 1251–1266.

Casiot, C., Alonso, M.C.B., Boisson, J., Donard, O.F.X., Potin-Gautier, M., 1998, Analyst, 123, 2887–2893.

Casiot, C., Donard, O.F.X., Potin-Gautier, M., 2002, Spectrochimica Acta, Part B, 57, 173–187.

Chen, Z., Naidu, R., Subramanian, A., 2001, J. Chromatogr. A, 927, 219–227.

Chirila, E., Draghici, C., Dobrinas, S., 2006, "Sampling and sample pretreatment in environmental analysis" in L. Simeonov, E. Chirila, (Eds) "Chemicals as Intentional and Accidental Global Environmental Threats", Springer, 7–28.

Collins, R.N., Onisko, B.C., McLaughlin, M.J., Merrington, G., 2001, Environ. Sci. Technol., 35, 2589–2593.

Dieffenbach, A., Matzner, E., 2000, Plant Soil, 222, 149–161.

Drăghici, C., 2003, Capillary electrophoresis, in Pollution and Environmental Monitoring, Colbeck, I., Drăghici, C., Perniu D. (Eds), Transilvania University of Brasov, Brasov, Romania.

Drăghici, C., 2002, Capillary Electrophoresis. Principles and Environmental Analysis Application (Electroforeza capilara. Principii si aplicatii in analize de mediu, in Romanian language), Transilvania University of Brasov, Brasov, Romania.

Drăghici, C., Coman, Gh., Şica, M., Perniu, D., Badea, M., 2003a, Environmental Engineering and Management Journal, 4, 1–12.

Drăghici, C., Coman, Gh., Şica, M., Perniu., D., Ţică, R., Badea, M., 2003b, International Conference on Materials Science and Engineering, BRAMAT 2003, 494–501.

Drăghici, C., Şica, M., Chirila, E., 2004, Bulletin of the Transilvania University of Brasov, 10(45) Series B2, 201–208.

Fukushi, K., Takeda, S., Chayama, K., Wakida, S.-I., 1999, J. Chromatogr. A, 834, 349–362.

Groom, C.A., Beaudet, S., Halasz, A., Paquet, L., Hawari, J., 2000, Environ. Sci. Technol., 34, 2330–2336.

Halasz, A., Groom, C., Zhou, E., Paquet, L., Beulieu, C., Deschamps, S., Corriveu, A., Thiboutot, S., Ampleman, G., Dubois, C., Hawari, J., 2002, J. Chromatogr. A, 963, 411–418.

Heining, K., Vogt, C. 1999, Electrophoresis, 20, 3311–3328.

Hilmi, A., Luong, J.H.T., 2000, Envion. Sci. Technol., 34, 3046–3050.

Issaq, H., 2000, Electrophoresis, 21, 1921–1936.

Jain, C.K., Ali, I., 2000, Wat. Res., 34, 4304–4312.

Janini, G.M., Issaq, H.J., Muschik, G.M., 1997, J. Chromatogr. A, 792, 125–141.

Janos, P., 1999, J. Chromatogr. A, 834, 3–20.

Kataoka, H., 1997, J. Chromatogr. A, 774, 121–142.

Kuban, P., Reinhardt, M., Muller, B., Hauser, P.C., 2004, J. Environ. Monit., 6, 169–174.

Lanyai, K., Dinya, Z., 2002, Chromatographia Supplement, 56, S-149–153.

Li, S., Lin, B., Zhou, W., 2001, Common. Soil Sci. Plant Anal, 32(5&6), 711–722.

Majidi, V., 2000, Microchemical Journal, 66, 3–16.

Martinez, D., Cugat, M.J., Borrull, F., Calull, M., 2000, J. Chromatogr. A, 902, 65–89.

Owens, G., Ferguson, V.K., McLaughlin, M.J., Singleton, I., Reid, R.J., Smith, F.A., 2000, Environ. Sci. Technol., 34, 885–891.

Pierce, R.H., Kirkpatrik, G.J., 2001, Environ. Toxicol. Chem., 20, 107–114.

Romero, E., Matallo, M.B., Pena, A., Sanchex-Rasero, F., Schmitt-Kopplin, Ph., Dios, G., 2001, Environmental Pollution, 111, 209–215.

Schmitt-Kopplin, P., Hertkorn, N., Schulten, H.-R., Kettrup, A., 1998, Environ. Sci. Technol., 32, 2531–2541.

Sovocool, G.W., Brumley, W.C., Donnelly, J.R., 1999, Electrophoresis, 20, 3297–3310.

Strobel, B.W., Hansen, H.C.B., Borggaard, O.K., Andersen, M.K., Raulund-Rasmunssen, K., 2001, Biogeochemistry, 56, 1–26.

Turner, B.L., Paphazy, M.J., Haygarth, P.M., McKelvie, I.D., 2002, Phil. Trans. R. Soc. Lond. B, 357, 449–469.

Turner, B.L., Paphazy, M.J., Maygarth, P.M., McKelvie, I.D., 2002, Phil. Trans. R. Soc. Lond. B, 357, 449–469.

Urbansky, E.T., 2000, J. Environ. Monit., 2, 285–291.

Westergraad Strobel B., Bernhoft, I., Borggaard, O.K., 1999, Plant Soil, 212, 115–121.

Westergraad, B., Hansen, H.C.B., Borggaard, O.K., 1998, Analyst, 123, 721–724.

# SUBJECT INDEX

Printed in the United States
115119LV00002B/72/P